2nd Mineral Processing and Technology International Conference 2018

IOP Conference Series: Materials Science and Engineering
Volume 478

Tangerang, Indonesia
1 November 2018

Editor:

Erik Prasetyo

ISBN: 978-1-5108-8215-7
ISSN: 1757-8981

Printed from e-media with permission by:

Curran Associates, Inc.
57 Morehouse Lane
Red Hook, NY 12571

Some format issues inherent in the e-media version may also appear in this print version.

Printed by Curran Associates, Inc. (2019)

For permission requests, please contact the Institute of Physics at the address below.

Institute of Physics
Dirac House, Temple Back
Bristol BS1 6BE UK

Phone: 44 1 17 929 7481
Fax: 44 1 17 920 0979

techtracking@iop.org

Additional copies of this publication are available from:

Curran Associates, Inc.
57 Morehouse Lane
Red Hook, NY 12571 USA
Phone: 845-758-0400
Fax: 845-758-2633
Email: curran@proceedings.com
Web: www.proceedings.com

Table of contents

Volume 478

The 2nd Mineral Processing and Technology International Conference

1 November 2018, Tangerang, Banten Province, Indonesia

Accepted papers received: 28 December 2018
Published online: 26 February 2019

Preface

Papers

Preface

Mineral Processing and Technology International Conference (mineprocèt) 2018 was a scientific meeting that is technically organized by Research Unit for Mineral Technology in conjunction with Indonesian Science Expo (ISE) 2018 organized by Indonesian Institute of Sciences (LIPI). mineprocèt 2018 was held on November 1, 2018 at Indonesia Convention Exhibition (ICE), BSD City, Serpong, Tangerang, Indonesia.

This symposium aimed to bring together researchers, academicians, and scholars all around the world to share information and their research results. This event was also expected to become a beneficial media in exchanging their innovative ideas and build research collaborations in the future. The symposium covered all scientific and technical activities in mineral processing and technology – from basic to the advanced applications – to pitch ideas, insight and findings for optimal and sustainable use of sub-economic mineral resources especially those related to mineral beneficiation, extractive metallurgy (bio-hydrometallurgy, pyrometallurgy, electrometallurgy), separation and purification, physical metallurgy and foundry, powder and nano technology, mineral industry application, energy and coal technology, mineral economy and policy, phytoextraction and environmental remediation, control and modelling for mineral processing.

We specially invited 4 scientists as keynote speakers from various countries to share their knowledge to all participants. The total participants attending this symposium were 51 people. This symposium has successfully attracted 46 submitted papers. Papers that have been presented in the mineprocèt 2018 were eligible for publishing in IOP Journal of Material Science and Engineering: Conference Series. The editorial board has decided to accept 41 papers from Indonesia and Philiphines. The accepted papers have passed through strict reviewing process by scientific committee of this symposium. EDAS system has been used to facilitate the submission, reviewing, and revision processes.

We would like to appreciate all of our co-organizers, supporting societies, sponsors, and Research Unit for Mineral Technology staffs who have contributed generous amount of grants and effort to support us during the symposium. We hope this symposium has successfully enabled us to share and communicate our knowledge and skills and to build collaboration among the participants. We are looking forward to welcoming you in Indonesia again. See you in the next mineprocèt!

Erik Prasetyo, Ph.D.
Editor-in-Chief
Research Unit for Mineral Technology
Indonesian Institute of Sciences

IOP Conf. Series: Materials Science and Engineering **478** (2019) 011001 doi:10.1088/1757-899X/478/1/011001

Conference Organizers

Scientific Organizing Committee

Dr.Eng. Widi Astuti
Conference Chair/Editor
Research Unit for Mineral Technology
Indonesian Institute of Sciences

Erik Prasetyo, Ph.D
Conference Co-chair/Editor-in-Chief
Research Unit for Mineral Technology
Indonesian Institute of Sciences

Fika Rofiek Mufakhir, ST, MT
Conference Co-chair/Editor
Research Unit for Mineral Technology
Indonesian Institute of Sciences

Editorial Board

F. Nurjaman, Research Unit for Mineral Technology, Indonesian Institute of Sciences
I. W. Warmada, Geological Engineering Department, University of Gadjah Mada

Secretary

Slamet Sumardi, Research Unit for Mineral Technology, Indonesian Institute of Sciences
Yoka Pramadi, Research Unit for Mineral Technology, Indonesian Institute of Sciences
Nur Firdaus, Research Unit for Mineral Technology, Indonesian Institute of Sciences
Yusup Hendronursito, Research Unit for Mineral Technology, Indonesian Institute of Sciences
David Candra Birawidha, Research Unit for Mineral Technology, Indonesian Institute of Sciences

Treasure

Ulin Herlina, Research Unit Mineral for Technology, Indonesian Institute of Sciences
Sri Okfarina, Research Unit Mineral for Technology, Indonesian Institute of Sciences

Under auspice of

Balai Penelitian Teknologi Mineral – Lembaga Ilmu Pengetahuan Indonesia
Research Unit for Mineral Technology – Indonesian Institute of Sciences

LIPI

Jl. Ir. Sutami km. 15 Tanjung Bintang
Lampung Selatan 35361, Indonesia
Ph. +62721350054 Fax +02721350056
http://bptm.lipi.go.id/

IOP Conf. Series: Materials Science and Engineering **478** (2019) 011001 doi:10.1088/1757-899X/478/1/011001

Keynote Speaker I

Assoc. Prof. Dr. Sabine Willscher
Martin Luther University Halle-Wittenberg, Germany

Subject Area

Phytoremediation; Phytoextraction; Bioremediation; Mining Remediation; Microbiotechnology; Biometallurgy

Biography

Sabine Willscher was born in Eisenach, Germany. She received her Ph.D degree in Technology University Mareburg in 1988. Her education and scientific career was started as Scientific Assistant and Lecturer in Technology University Mareburg. She was also a visiting scholar in Federal Institute Geophysical Society & Natural Resources, Hanover, Germany, 2000 and Colorado School Mines, Golden, Colorado, 2001—2002. From 2001-2016, she was a science researcher and lecturer in Technology University Dresden, Pirna, Germany. Now, she is a senior lecturer in Martin Luther University Halle-Wittenberg, Germany. She also has some other scientific and education activities such as a Member curriculum committee United Nations Environmental Program, Dresden, since 2004 and a member International Science Committee Biohydrometallurgy, Germany, since 2005. She is an author/co-author of some publications in International Journals. She is also a reviewer of some International Journal and Conference Proceedings.

IOP Conf. Series: Materials Science and Engineering **478** (2019) 011001 doi:10.1088/1757-899X/478/1/011001

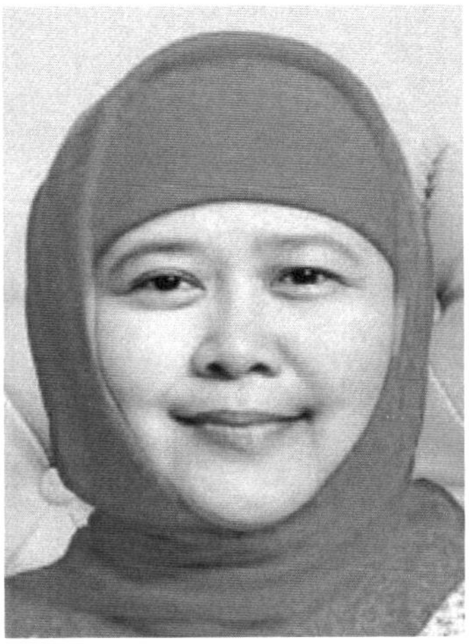

Keynote Speaker II

Assoc. Prof. Ir. Siti Khodjah Chaerun, Ph.D.
Institut Teknologi Bandung (ITB), Indonesia

Subject Area

Environmental Geomicrobiology; Mineral Biotechnology; Bioremediation; Biocorrosion

Biography

Siti Khodijah Chaerun received her Bachelor and Master of Engineering in Environmental Engineering from Institut Teknologi Bandung (ITB), Bandung, Indonesia. Her Ph.D degree in Environmental Geomicrobiology was received from Kanazawa University, Japan in 2004. After finished her doctoral studies, she obtained DAAD Fellowship to conduct the post doctoral researchs at the Leibniz Centre for Agricultural Landscape Research (ZALF), Institute of Landscape Matter Dynamics & Institute of Soil Landscape Research, Muncheberg, Germany in 2006. She was also awarded as Visiting Scientist at some research centers that are the Graduate School of Bio-applications and Systems Engineering, Tokyo University of Agriculture and Technology, Japan in 2007; the Department of Earth Sciences, Faculty of Science, Kanazawa University, Japan in 2008; and Department of Energy and Resources Engineering, College of Engineering, Peking University, Beijing, China in 2012-2014. In 2008-2009, she also received IDB Fellowship for Postdoctoral Associate at the Department of Microbiology, the University of Georgia, Athens, GA, USA. Now, she is a senior lecturer in Department of Metallurgy Engineering, Institut Teknologi Bandung (ITB), Indonesia with the special field of biominerolgy, biomining, biometallurgy, and bioremediation. She is also a head of Laboratory of Geomicrobiology, Biomining, and Biocorrosion. She is an author/co-author of more than 50 publications in International Journals. She is also a reviewer of some International Journal and Conference Proceedings.

IOP Conf. Series: Materials Science and Engineering **478** (2019) 011001 doi:10.1088/1757-899X/478/1/011001

Keynote Speaker IV

Bouman Tiroi Situmorang, S.T., M.T., IPU
Asosiasi Perusahaan Industri Pengolahan & Pemurnian Indonesia (AP3I)

Subject Area
Metallurgy; Mining

Biography
Bouman Tiroi Situmorang received his Bachelor of Engineering in Metallurgy from Institut Teknologi Bandung (ITB), Bandung, Indonesia in 1996. Next, he pursued his Master of Engineering in Industry from Institut Teknologi Sepuluh November (ITS), Surabaya, Indonesia in 2006. He also obtained the prominent professional engineer degree from Indonesian Engineer Society in 2018. He is a metallurgist in industry especially in smelting industries. Since 1997, he works in PT. Smelting Gresik that is a biggest copper smelting industry in Indonesia. Now, he is a Senior Manager of Technical External Relation in PT. Smelting Gresik. Besides his professional work, he also has many activites that are related to his professional epertise. He is member of The Minerals, Metals, and Material Society (TMS), Indonesian Engineer Society (PII), and Perhimpunan Ahli Pertambangan Indonesia (PERHAPI). Now, he is a chairman of Metallurgical Engineering Alumni Association of ITB, a co-chairman of PERHAPI East Java, a chairman of Copper Committee of Asosiasi Perusahaan Industri Pengolahan & Pemurnian Indonesia (AP3I), and also a member team of professional sertification of metallurgy in PII. He has great attention to the development of metallurgical engineering education, so that he also join in the Department of Metallurgical Engineering, ITB as a member of Advisory Board.

Keynote Speaker III

Assoc. Prof. Dr. Anwaruddin Hisyam
Universiti Malaysia Pahang, Malaysia

Subject Area
Process Integration; Rare Earth Separation Processes; Gold Mining and Processing

Biography
Anwaruddin Hisyam received his Bachelor of Engineering in Chemical Engineering from Universitas Gadjah Mada (UGM), Yogyakarta, Indonesia. He pursued his Master of Science in Process Integration from University of Manchester Institute of Science and Technology (UMIST), England. His Ph.D degree in Chemical and Natural Resources Engineering was received from Universiti Malaysia Pahang, Malaysia. Now, he is a senior lecturer in the Faculty of Chemical and Natural Resources Engineering. He is also a Principal Fellow and Reserach Fellow in Rare Earth Research Center. In addition, he is also an Editor and Reviewer some International Journals and Conference Proceedings. He has delivered some Keynote Speeches in some International Conferences such as Engineering International Conference (Universitas Negeri Semarang, Indonesia) in 2018 and International Conference Industrial, Mechanical and Chemical Engineering (Universitas Sebelas Maret, Indonesia) in 2017.

Peer review statement

All papers published in this volume of *IOP Conference Series: Materials Science and Engineering* have been peer reviewed through processes administered by the proceedings Editors. Reviews were conducted by expert referees to the professional and scientific standards expected of a proceedings journal published by IOP Publishing.

The role of sodium-based additives on reduction process of nickel lateritic ore

F Nurjaman[1], A Rahmahwati[2], M F Karimy[3], N Hastriana[1], A Shofi[1], U Herlina[1], B Suharno[2], and D Ferdian[2]

[1]Research Unit for Mineral Technology, Indonesian Institute of Sciences, South Lampung, Indonesia
[2]Dept. of Metallurgy and Materials Engineering, Faculty of Engineering, Universitas Indonesia, Depok, Indonesia
[3]Research Unit for Natural Product Technology, Indonesian Institute of Sciences, Yogyakarta, Indonesia

Email: fajar.nurjaman@lipi.go.id

Abstract. In this present work, the selective reduction of nickel laterite was conducted by using 5 to 15 wt. % of sodium carbonate, sodium chloride and sodium sulfate as additives. The 5% wt. of palm shell charcoal was used as the reductant. The selective reduction process for nickel laterite was carried out at 1150°C for 60 minutes followed by wet magnetic separation using 500 Gausses permanent magnet. The characterization of reduced ore was performed by using X-ray Diffraction, optical microscope and Scanning Electron Microscope with Energy Dispersive X-ray Spectroscopy, while the chemical composition of concentrate or ferronickel was identified by X-ray Fluorescence. The results showed that the increase of nickel grade and recovery of nickel laterite was significantly more affected by the addition of sodium sulfate rather than sodium carbonate and sodium chloride. The ferronickel particle was agglomerated and growth with the increasing of reduction temperature and dosses of additives, thus it provided favorable conditions for the separation of ferronickel from impurities in the magnetic separation process.

1. Introduction

Nickel is a commercially important metal and used mainly in the production of stainless steel. About 68% of the nickel consumed to make stainless steel and another 32% goes into alloyed steels, castings, electroplating, and rechargeable batteries. About 70% of world land-based nickel resources are contained in laterites, they currently account for only about 40% of the world nickel production [1]. Thus, nickel laterite ore processing has attracted much greater attention in recent years and economical processing of nickel laterite ores is becoming imperative due to the increasing demand for nickel which is driven by the rapid development of stainless steel industry [2].

Nickel is closely associated with a complex compound in lateritic nickel ore, such as olivine (magnesium and silicate oxide) and goethite (iron hydroxide). The complex compound of nickel ore makes their inability to improve its metal oxide concentration using physical beneficiation process, thus pyrometallurgy or hydrometallurgy is a must. Pyrometallurgy includes a processing with a blast furnace or rotary kiln-electric arc furnace smelter [3], while hydrometallurgy includes high-pressure acid leaching and Caron process [4]. Nevertheless, it is costly and produces a lot of chemical waste.

A reduction process of nickel ore was conducted at lower than smelting process temperature had been conducted. A small amount of additives were involved in this reduction process to enhance the nickel grade and recovery at low reduction temperature [5]. Therefore, a low cost and less chemical waste in nickel ore processing can be realized. Sodium sulfate is commonly used as an additive in the selective-reduction of nickel ore processing. Jiang et al. (2013) reported that the addition of 10 wt. % of sodium sulfate and 2 wt.% of coal in the reduction of nickel ore containing 1.49% Ni at 1200°C for 50 minutes resulted in 10.86% and 88.56% for nickel grade and recovery, respectively [6].

However, there is still less information for the using of others sodium-based additives in selective-reduction of nickel ore processing. In this present work, the addition of sodium carbonate and sodium chloride were used as additives in the selective-reduction process. The effect of both additives on nickel grade and recovery was investigated clearly comparing with the sodium sulfate additives. The ferronickel growth during the reduction process was also observed carefully.

2. Material and Methods

The nickel laterite ore was from South East Sulawesi, Indonesia. The X-ray fluorescence (XRF) was performed to determine the chemical compositions of the nickel ore, as showed in table 1.

Table 1. Chemical analysis of laterite ores

Element	Fe	Ni	Al	Si	Ca	Mg	Cr	Mn	Co
wt. %	50.5	1.4	4.86	6.5	0.18	1.81	2.68	0.85	0.07

The nickel laterite was ground to less than 0.147 mm and mixed together with 10% wt. of palm shell charcoal as reductant and 5-15% wt. sodium-based additives. The proximate and calories analysis of the reductant were listed in table 2. The additives were sodium carbonate, sodium chloride and sodium sulfate. The mixture was agglomerated into 10-15 mm in diameter of pellets. The pellets were drying at 100°C for 4 hours in an oven and continued with the reduction process, which took place in a muffle furnace at 1050°C for 60 minutes.

Table 2. Chemical analysis of palm kernell shell charcoal.

Proximate analysis (wt. %)				Calories (cal/g)
Moisture	Volatile matter	Ash content	Fixed carbon content	
0.43	22.57	21	77	6,244

The reduced pellets were cooling rapidly in a water and then dried at 100°C for 4 hours. The pellets were ground again into less than 74 μm before the wet magnetic separation process was performed and separated into the concentrate (magnetic) and the tailing (non-magnetic).

The iron and nickel grade of concentrate were analyzed by using X-ray Fluorescence (XRF), while the morphological of ferronickel was observed using Scanning Electron Microscopy-Energy Dispersive Spectroscopy (SEM-EDS). The X-ray Diffraction (XRD) was used to analyze the mineralogical phases in the reduced pellet.

3. Results and Discussion

3.1. Effect of additives on nickel and iron grade and recovery

Figure 1(a-d) presents the result of various types and amount of additives on iron-nickel grade and recovery in the concentrates resulting from the reduction process at 1150°C for 60 minutes followed by the magnetic separation process. The reduction of nickel limonite ores without additives only produced nickel grade and recovery of 2.5% and 29.3%, respectively. The nickel grade and recovery significantly increased by adding 15% wt. of sodium sulfate additive which was 5.3% and 83.7%, respectively. Nevertheless, the nickel grade tended to be constant, while the recovery seemed to be decreased during the addition of sodium carbonate and sodium chloride additives.

Iron grade increased towards the increasing of sodium sulfate additive, as shown in figure 1(c). The optimum iron grade was 81.9% resulted from 10% wt. sodium sulfate addition. For sodium carbonate additive, the optimum iron grade was 72.1% resulted from the addition of 5% wt. of additive. Nevertheless, the addition of sodium chloride additive decreased the iron grade. From figure 1(d), the increasing of additives from 5% wt. to 15% wt. decreased the iron recovery. The addition of 15% wt. of additives produced iron recovery of 46.2%, 45.4%, and 35.2% for sodium chloride, sodium carbonate, and sodium sulfate, respectively. The low recovery of iron indicated a low of iron metalization in the selective reduction process of nickel laterite ore, thus the concentrate contains a small amount of iron and resulted in high nickel grade.

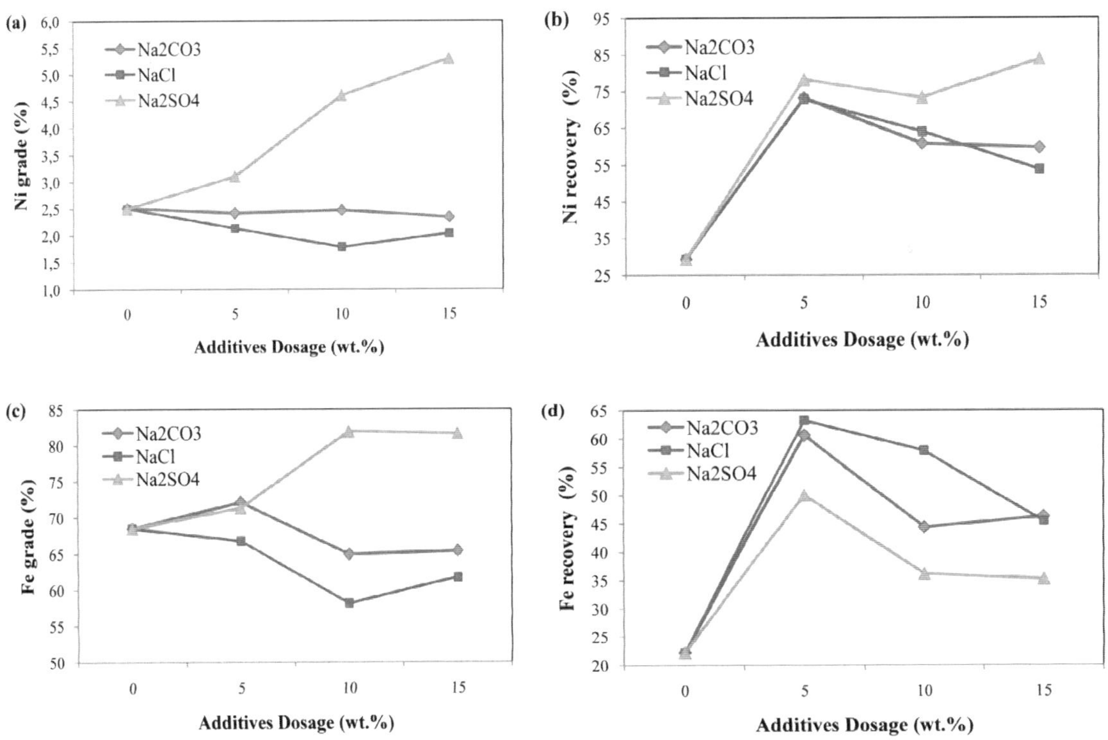

Figure 1. Effect of different additives on nickel and iron grade in concentrate

3.2. Characterization of reduced laterite by XRD and SEM-EDX

The XRD analysis was performed in nickel laterite for reduction temperature at 1150°C for 60 minutes with various additives in 5, 10, and 15% wt. as given in figure 2. From the XRD patterns of reduced nickel laterite without additives, as given in figure 2(a), showed that there is only two dominant peaks, they are fayalite ((Mg, Fe)$_2$ SiO_4) and iron nickel (FeNi). Fayalite was formed due to the lizardite dehydroxylation ($Mg_3(Si_2O_5)(OH)_4$) and silicate mineral recrystallization of nickel laterite ore [7]. Moreover, goethite (FeO.OH) and olivine ($MgNiSiO_4$) in nickel laterite were reduced to be ferronickel in the reduction process. The 2.5% of nickel grade in concentrate was resulted from the selective reduction of nickel ore without additives.

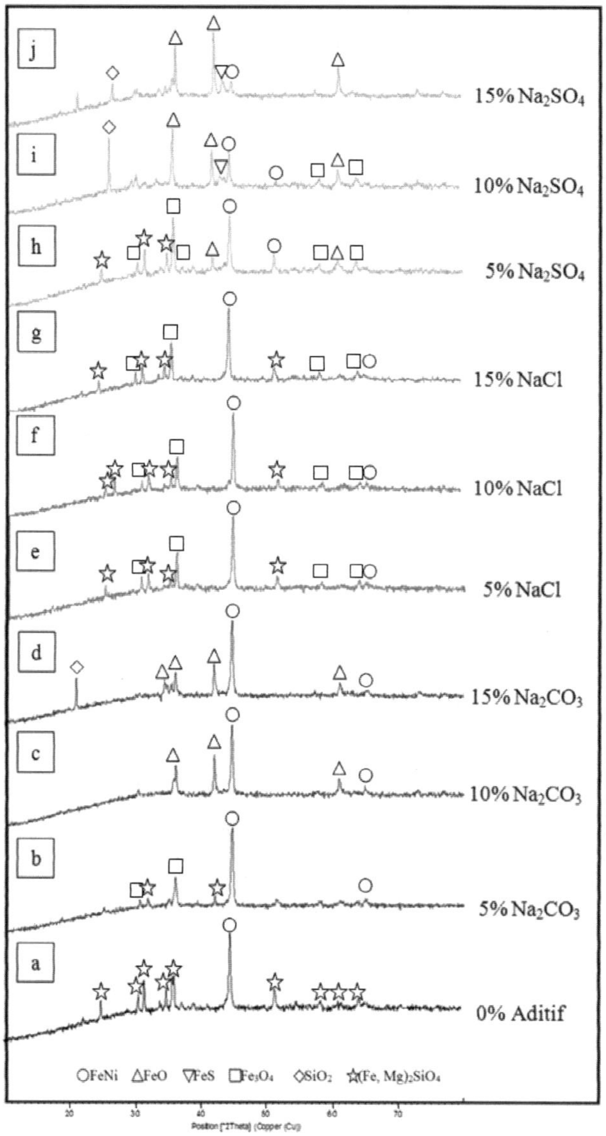

Figure 2. XRD pattern of the reduced laterite with different additives: (a) without additive, (b-d) sodium carbonate, (e-g) sodium chloride, and (h-j) sodium sulfate.

In addition of 5% wt. of sodium carbonate additive, the peak of fayalite (Fe_2SiO_4) diffraction in reduced ore decreased and replaced by the peak of magnetite (Fe_3O_4). By increasing the sodium carbonate additive, the intensities of wustite (FeO) and ferronickel (FeNi) peaks were increased, as presented in figure 2(b-d). It was explained that by adding the sodium carbonate additive, the decomposition of sodium carbonate producing CO_2 occurred and promotes the bouduard reaction producing CO gas. The CO gas could increase the reduction of oxides in laterite ores [8]. However, if the sodium carbonate additive increased up to 15% wt., the peak of quartz (SiO_2) was found in the reduction process of nickel laterite due to the reduction of fayalite into wustite and quartz at high temperature. Quartz as an impurities inhibit the reduction of nickel oxide, thus the nickel recovery at the reduction process with 15% wt. sodium carbonate at 1150°C was very low.

Figure 2(e-g) presents the XRD pattern of the reduced nickel laterite ores with the addition of sodium chloride additive. The peak diffraction of 5%, 10%, and 15% wt. sodium chloride had a tendency of having the similar peak diffraction consists of fayalite (Fe_2SiO_4), magnetite (Fe_3O_4), and ferronickel (FeNi).

The XRD patterns with 5%, 10%, and 15% wt. of sodium sulfate additive are given in figure 2(h-j). The increasing of wustite and troilite was observed in the increasing of sodium sulfate thus it was decreasing the iron recovery in concentrate. Wustite was more difficult to reduce than nickel oxides and troilite. In addition, the sulfur, resulting from decomposition of sodium sulfate, reacted with iron to form troilite which was non-magnetic thus increasing nickel concentration in concentrate while undergoing beneficiation through the magnetic separation process [6]. Furthermore, there was a decrease in peak diffraction of nickel along the increased temperature as shown in figure 2(h-j). This indicated that ferronickel in concentrates containing more nickel than iron and it also confirm that the decrease of iron recovery in the increasing dose of sodium sulfate.

Figure 3 and figure 4 shows the micrograph of reduced nickel laterite using sodium carbonate additive at temperature 1150°C for 60 minutes performed by SEM and phase identification of microstructure using EDX analysis. The white area, which is shown by point A, C, and G, indicated as ferronickel (FeNi) particles.

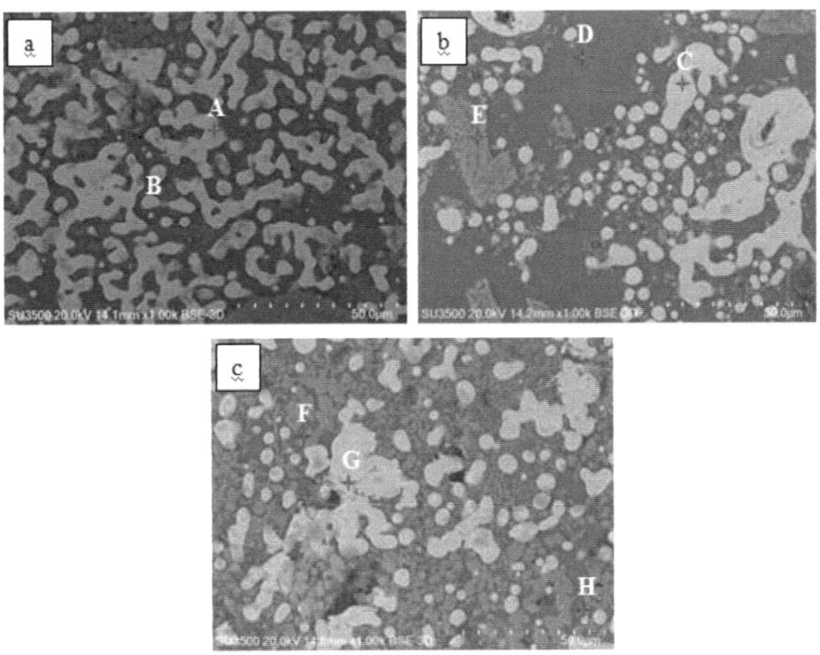

Figure 3. SEM analysis of reduced laterite in presence of sodium carbonate additive (a) 5% wt., (b) 10% wt. and (c) 15% wt. (magnification 1000BSE)

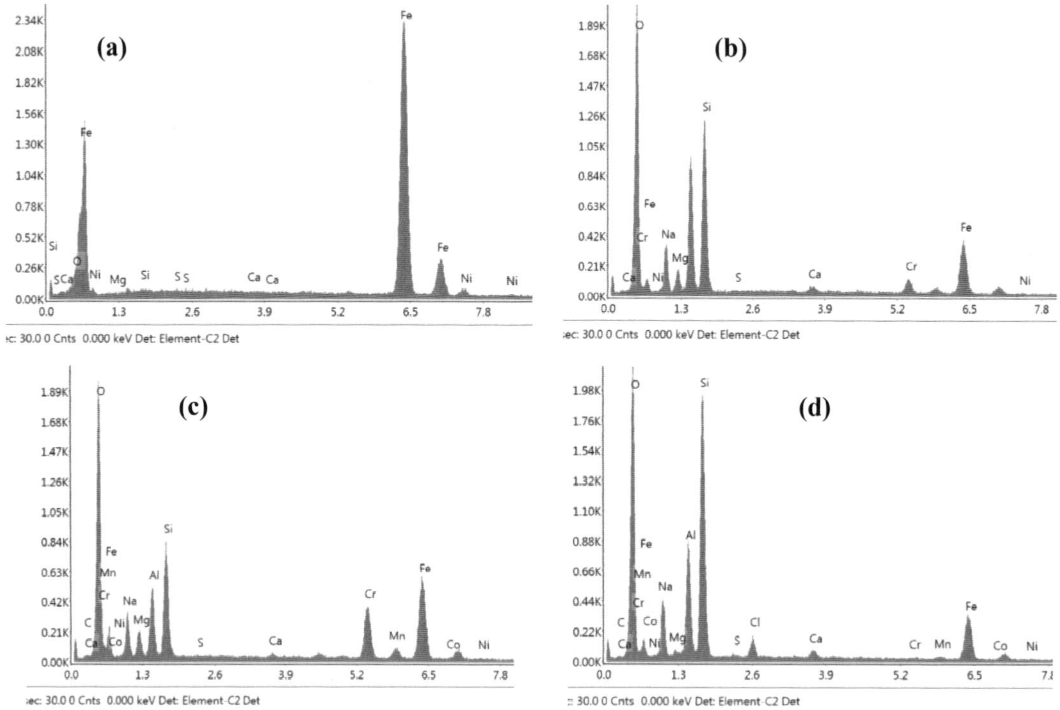

Figure 4. Spectrum of EDX analysis of reduced laterite in the presence of sodium carbonate for point: (a) A; (b) B; (c) E; (d) H

In the addition of 5% wt. sodium carbonate additive, the particle size of ferronickel was fine and distributed uniformly. The increasing of sodium carbonate addition to 10% wt. had an effect on transforming the ferronickel particle size to be larger. The average optimum size of ferronickel particles was found at 10% wt. of sodium carbonate which was 12.8 μm. Nevertheless, the increasing to 15% wt. of sodium carbonate would lead to the dispersion on ferronickel particles as shown in figure 3(c). More sodium carbonate addition will result in more CO_2 gas. As reported by Zhou et al. (2017), that the CO_2 gas resulted in the expansion of volume, so the structure of polluters and particles in the ores resulting from reduction was exposed to the tension due to the expansion of volume and particle shapes started to decay and be dispersed [8].

From micrograph analysis (figure 3), the ferronickel particles were surrounded by impurities, which is shown by point B, E, D, F and H. From the XRD analysis, the impurities are wustite (point E) and fayalite (point B). At point H, it was found that the impurities contained high of silicon and oxygen. It was confirmed from the XRD analysis with 15% wt. of sodium carbonate additive which has found the peak diffraction of quartz.

Figure 5(a-c) shows the micrograph in reduced nickel laterite ores with sodium chloride additive. The points A, C, and E show ferronickel as a white area which is dominated by Fe and Ni element (figure 6(a)). Meanwhile, the points B, D, and F are impurities minerals surrounding the ferronickel particles. The impurities consisted of Mg, Si, and O which indicated fayalite as the gangue mineral. This could confirm the result analysis of XRD pattern with the addition of additive sodium chloride, as shown in figure 2(e-f). The optimum size of ferronickel particles was 8.0 μm with the addition of 10% sodium chloride. The increasing of sodium chloride up to 15% wt. decreased the ferronickel particles size. This was occurred because there was an indication of pyrohydrolysis process towards sodium chloride to be HCl as an agent of chlorination that had been achieved. Thus, adding further additives up to 15% did not significantly affect the formation of ferronickel particles.

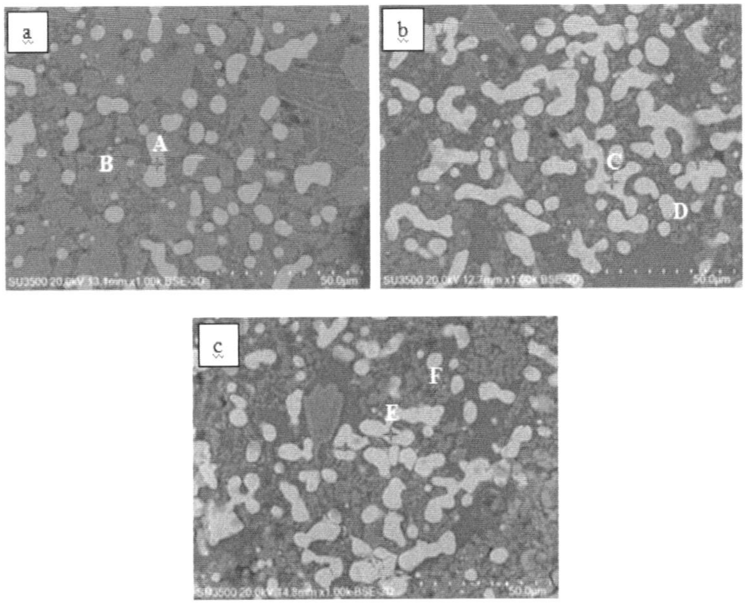

Figure 5. SEM analysis of reduced laterite in the presence of sodium chloride additive (a) 5% wt., (b) 10% wt. and (c) 15% wt. (magnification 1000BSE)

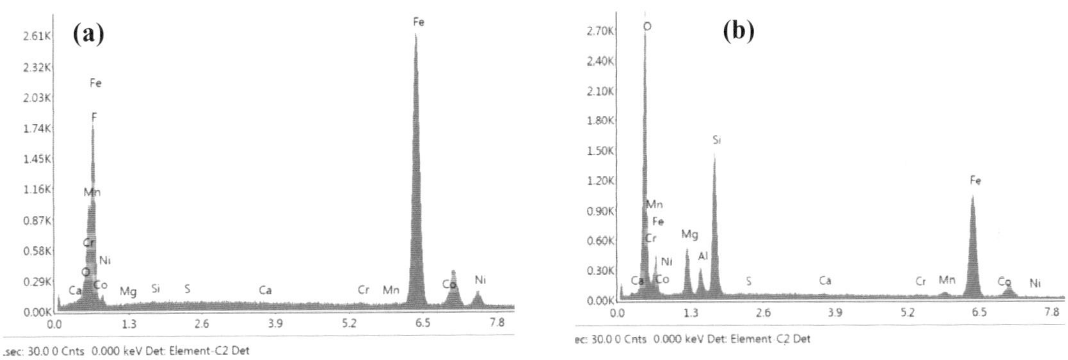

Figure 6. Spectrum of EDX analysis of reduced laterite in the presence of sodium chloride for point: (a) A; (b) B

Figure 7 shows the micrograph of reduced nickel laterite with the addition of sodium sulfate additive. The points B, D, and G presented the ferronickel particles enfolded by its impurities. There was found fayalite, wustite, and quartz as impurities, as shown in XRD pattern in figure 2(h-j), which is shown in point A, E and H. The addition of 10% until 15% wt. of sodium sulfate additive resulted the ferronickel particles which surrounded by troilite (FeS), which is shown by point C. The interaction of troilite and metallic iron was formed during the process of reduction to promote ferronickel particle growth due to the low Fe-FeS eutectic melting point (985°C). The liquidus state will promote the rate of mass transfer of metal ions during the process of reduction. Thus, ferronickel particles growth and agglomerates surrounded the troilite [5]. The increasing addition of sodium sulfate additive increases the reduction reaction of sodium sulfate with these following reactions in equation (1-2) [9].

$$1/4\ Na_2SO_4 + CO \leftrightarrow 1/4\ Na_2S + CO_2 \quad (1)$$

$$Na_2S + FeO + 2SiO_2 \leftrightarrow Na_2Si_2O_5 + FeS \quad (2)$$

Sodium sulfate was reduced by CO gas produced Na_2S which would react with wustite into troilite. This FeS would enfold the ferronickel particles as resulting from SEM-EDS analysis in figure 7.

Figure 9 showed the ferronickel size after selective reduction using various additives at a reduction temperature 1150°C for 60 minutes. The ferronickel particle size was observed to be larger from 6.3 µm to 43.7 µm with the increasing sodium sulfate from 5% to 15%. While, the optimum of ferronickel particle size was around 12,8 µm and 8 µm with the addition of 10% sodium carbonate and sodium chloride additives, respectively. The further increasing amount of more than 10% additives sodium carbonate and chloride reduced the ferronickel particle size. The sodium sulfate additives resulted the optimum size of ferronickel due to the formation of FeS at the low melting point which promotes the agglomeration of ferronickel particles during the reduction process.

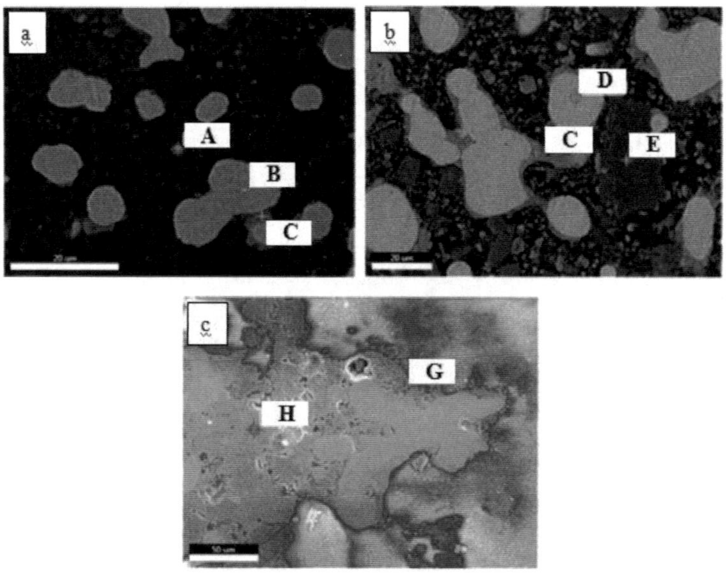

Figure 7. SEM analysis of reduced laterite in the presence of sodium sulfate additive (a) 5% wt., (b) 10% wt., and (c) 15% wt. (magnification 1000BSE)

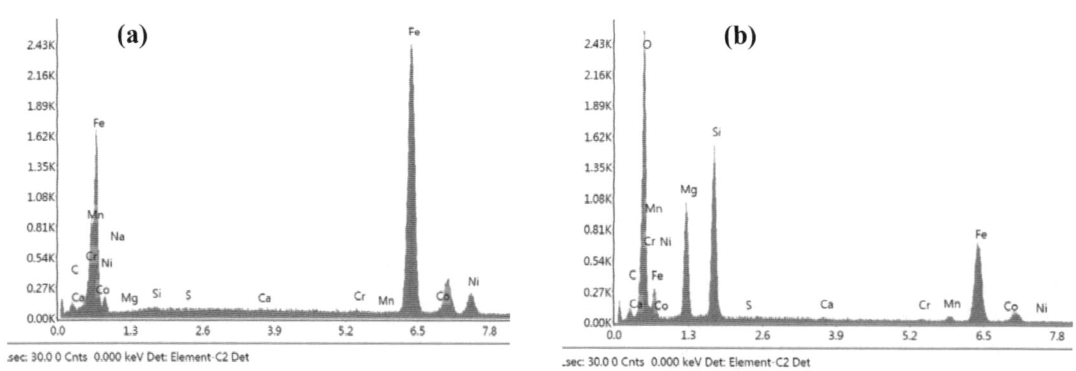

IOP Conf. Series: Materials Science and Engineering **478** (2019) 012001 doi:10.1088/1757-899X/478/1/012001

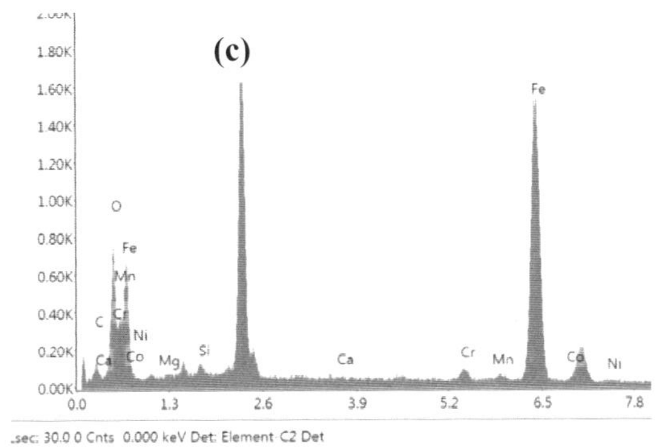

Figure 8. Spectrum of EDX analysis of reduced laterite in the presence of sodium sulfate for point: (a) D, (b) E, (c) C

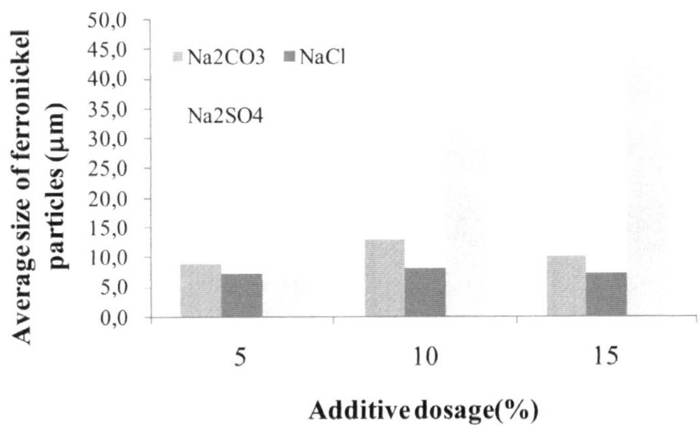

Figure 9. Ferronickel particle size of reduced laterite with different additives (reducing temperature 1150°C for 60 minutes)

4. Conclusion

The addition of additives was potentially increased the grade and recovery of nickel in concentrate compared with the nickel laterite ore resulting from the selective reduction process without additives. Adding additive sodium sulfate could significantly increase nickel and iron grade and recovery compared to the sodium carbonate and sodium chloride additives. At reduction temperature of 1150°C for 60 minutes, the nickel laterite ore with addition 15% wt. of sodium sulfate resulted in the optimum nickel grade and recovery, which was 5.3% and 83.7%, respectively. The addition of 15% wt. sodium sulfate also produced the optimum ferronickel particle size, which was 43.7 μm. Sodium sulfate promotes the FeS formation in the reduction of nickel laterite ore, thus increase the recovery of nickel and ferronickel particle size.

Acknowledgments

The authors would like thanks to the Ministry of Research, Technology and Higer Education for funding this research through Indonesia's National Innovation System (INSINas) Programme with Grant Number: 06/INS-1/PPK/E4/2018 and The Research Unit for Natural Product Technology for SEM-EDX utilization in this experiment.

References

[1] Dalvi A, Bacon W G and Osborne R C 2004 *PDAC International Convention* (Toronto) p. 1
[2] Pickles C A, Forster J and Elliot R 2014 *Miner. Eng.* **65** 33
[3] Rao M, Li G, Jiang T and Luo J 2013 *JOM* **65** 1573
[4] Mudd G 2009 *Proc. Int. Conf. on 48th Conference of Metallurgists* (Montral: Metsoc) p.1
[5] Li G, Shi T, Rao M, Jiang T and Zhang Y 2012 *Miner. Eng.* **32** 19
[6] Jiang M, Sun T, Liu Z, Kou J, Liu N, and Zhang S 2013 *IJMP* **123** 32
[7] Li B, Wang H and Wei Y 2011 *Miner. Eng.* **24** 1556
[8] Zhou S, Dong J, Lu C, Li B, Li F, Zhang B, Wang H, Wei Y 2017 *Mater. Trans* **58** 790
[9] Rao M, Li G, Zhang X, Luo J, Peng Z and Jiang T 2016 *Sep. Sci. Technol.* **51** 1727

IOP Conf. Series: Materials Science and Engineering **478** (2019) 012002 doi:10.1088/1757-899X/478/1/012002

Effect of the first thermal stage temperature and retention time in two-stage thermal upgrading of low-grade nickel lateritic ore

A Shofi[1], C Ramadini[1], R Samuel[1], F Nurjaman[2], B Suharno[1], and R Subagja[3]

[1]Metallurgical and Materials Engineering Department, Faculty of Engineering Universitas Indonesia, Kampus Baru UI Depok, 16424 – Indonesia
[2]Research Unit for Mineral Technology, Indonesian Institute of Sciences, Lampung,35361, Indonesia
[3]Research Center for Metallurgy and Material, Indonesian Institute of Sciences, South Tangerang, Banten, Indonesia

E-mail: achmad.shofi@lipi.go.id

Abstract. Selective reduction process and magnetic separation of nickel lateritic ore wereconducted through the two-stage thermal upgrading processwith the addition of sodium sulphate and sodium chloride as the additives agent. Effect of the first thermal stage temperature has been investigated at various of temperature of 500, 600 and700°Cfor 60 minutes and continued with the reduction process at temperature of 1050°C for 30 minutes. The mineralogical composition of the reduced samples was performed by XRD analysis. The optimum temperature was 500°C, which produced a concentrate withnickel grade of 2.98% and iron recovery of 40.17%. Effect of the first thermal stageretention time also has been investigated at temperature of 500°C for 30, 60, and 90 minutes, and continued with the reduction process at temperature of 1150°C for 60 minutes. The optimum temperature was 90 minutes, which produced a concentrate with 3.69% nickel grade, and 90.61%nickel recovery.

1. Introduction

Nickel is being a precious metal in infrastructure and technology field, especially inmaking industries of stainless steel, Ni-based alloy, alloy steels,electroplating, batteries, etc [1]. Nickel was used either in the form of metallic nickel or form of NPI/ferronickel with various of iron content [2].More than 65% of the nickel consumption is used in stainless steel-making industry[11]. The extraction process of nickel lateritic process required high intensive energy and usually nickel is melted to produce low-grade ferronickel with the high amount of slag. Furthermore, this pyrometallurgy process merely is used for low-grade nickel lateritic ore containing nickel grade above 1.5%. Nevertheless, the grade of nickel lateritic ore around the world including Indonesia is about 1.45%, making this conventional process is ineffective[3]. Recently many researchers are focussing on the method for upgrading the effectiveness and efficiency of pyrometallurgy process to generate ferronickel by using selective reduction process followed by a magnetic separation process [4,5]

Thermal upgrading is a method to promote the value of nickel in the ore by applying physical beneficiation. This technique refers to the reduction process of nickel contained in limonite ore to reduce the metallic and oxide producing ferronickel that can be separated from the gangue/non-metallic oxideby using magnetic separation process[6].

In this present work, the nickel oxide extractionof low-grade nickel ore from Desa Morombo, Konawe Utara District, Sulawesi Tenggara Province through selective reduction process followed by magnetic separation with the addition of the combination of sodium sulphate (Na_2SO_4) and sodium chloride (NaCl) as additive and palm kernel shell as reductant was investigated clearly.

2. Experimental

The raw materials used in this research are low-grade nickel limonitic ore, palm kernel shell as reductant, sodium sulphate and sodium chloride as the additives. The XRF analysis of nickel lateritic ore and proximate analysis of the reductantarelisted in Table 1 and Table 2, respectively.

Table 1. Main chemical composition of the laterite ore samples (mass fraction, %)

Element	Ni	Fe	Si	Mg	Al	Ca	Cr	Mn	Co
MassFraction (%)	1.4	50.5	16.5	1.81	4.86	0.18	2.68	0,85	0.07

Table 2. Proximate analysis of palm kernel shell (wt.%)

Reductor Type	**Volatile Matter**	**Ash**	**Fixed Carbon**	**Moisture**
Palm ShellCharcoal	22.57	21	77	0.43

The nickel ore was dried in an oven at temperatureof 105°C for 24hours to reducethe water content untill less than 5 wt.%. Thedried limonitic ore and the reductant were crushed and shieved to gain the particle less than 147 μm. Afterwards, 50 g of the ores, 5.5g of reductants, and 5g of each additives were mixed and pelletized. The pellets, which were 10-15 mm in size, were dried in an oven at the temperature of 100°C for 2 hours.

The two-stage thermal upgrading of selective reduction process wasperformed in this research.The first stage of reduction was done in various of temperature, which were 500°C, 600°C, and 700°C for 60 minutes and the second stage was 1050°C for 30 minutes. To investigate the effect of the first thermal stageretention time, another samples were added with the first stage of reduction temperature 500°C for 30 minutes, 60 minutes, and 90 minutes and the second stage was 1150°C for 60 minutes. After the selective reduction was done, the pellets were cooled rapidly in water. Then the pellets were dried and crushed to achieve particle size of 74 μm.The phase transformation was determined by XRD (*X-ray Diffraction*) analysis to observe the transformation of mineral composition and phase change after the selective reduction process. Afterwards, the wet magnetic separation process was conductedby dissolving the reduced ore into water and separating the concentrate and tailing product using 500 gauss magnet. The ratio of the reduced ore and water in this magnetic separation process was 1:10

3. Results and discussions

3.1. Effect of the first thermal stage temperature

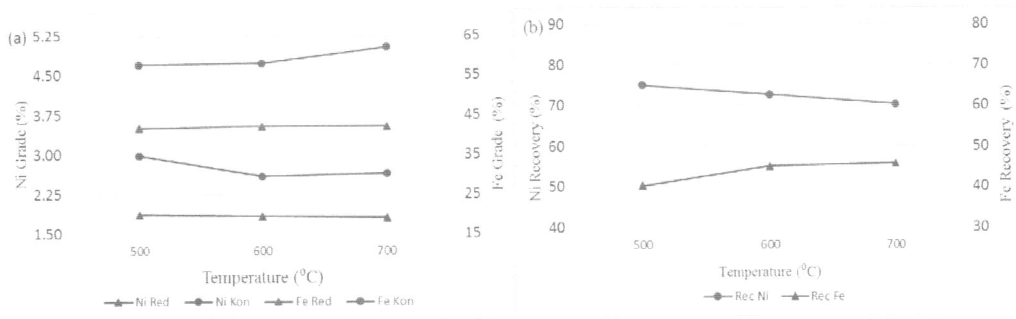

Figure 1. Effect ofthe first thermal stage temperature on: (a) grade of nickel and iron, (b) recovery of nickel and iron in concentrate at reduction temperature of 1050°C for 30 minutes.

From Figure 1, it shows that the highest nickel grade and recoveryproduced at the first stage temperature of 500^{0}C with 2.98% and 73.20%, respectively. The decreasing nickel grade followed by increasing iron grade in the temperature interval of 500^{0}C - 700^{0}C was due to the selective reduction mechanism of trevorite. At the temperature of 500^{0}C, trevorite had been reduced into nickel oxide and partially metallic nickel, but the iron oxide had not been reduced yet. It was because iron oxide would be reduced into metallic iron at 557°C. Hence, the nickel grade was relatively high with less iron grade contained. Higher temperature would result in higher metallic iron being produced from iron oxide and causing the increase in iron recovery. The higher the first thermal stage temperature resulted in decreasing of nickel recovery to 70.25% and increasing of iron recovery.

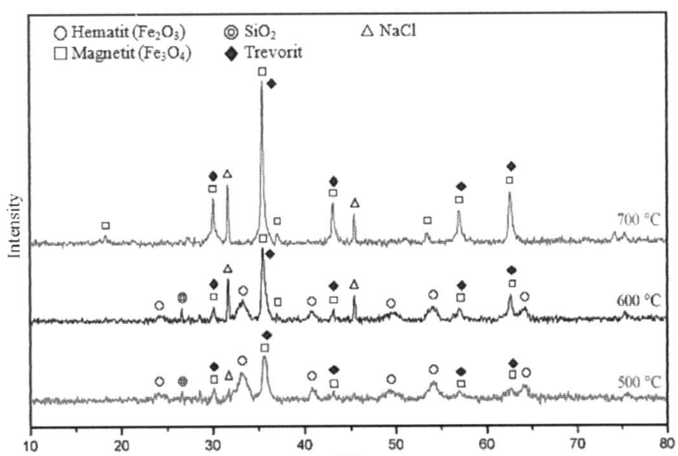

Figure 2. XRD patterns of reduced nickel ore with the various of the first thermal stage temperature

Figure 2 shows that the observed phase at the temperature of 500°C wastrevorite, magnetite, silicate, and natrium chloride.Goethite could not be observed due to its free-crystal water released and transformed to hematite phase. The peak of lizardite, which less amount in ore, also could not be observed due to the hydroxylated process, even though it was not completely transformed yet. The transformation of magnetitewas intensified at the temperature of 600°C due to the more hematite phase being reduced. Hematite phase could not be observed at the temperature of 700°C. It shows that hematite had been completely transformed into magnetite. Silicate also could not be observed due to its reaction with sodium sulphate as follows:

$$Na_2SO_4 + SiO_2 + CO \rightarrow Na_2SiO_3 + SO_2 + CO_2 \quad T \geq 670\ °C \quad (1)$$

3.2. Effect of the first stage retention time

The aim of the holding process at the first stage is to let the selective reduction process effectively applied to nickel which were contained in goethite phase in the ore[8]. Dehydroxylation mechanism plays a crucial role in achieving optimum nickel recovery. In the limonitic ores, nickel is subtituted into the crystal structure of goethite[9]. Goethite is a crystal containing hexagonal closed-packed oxygen with iron occupying the octahedral interstices[10].

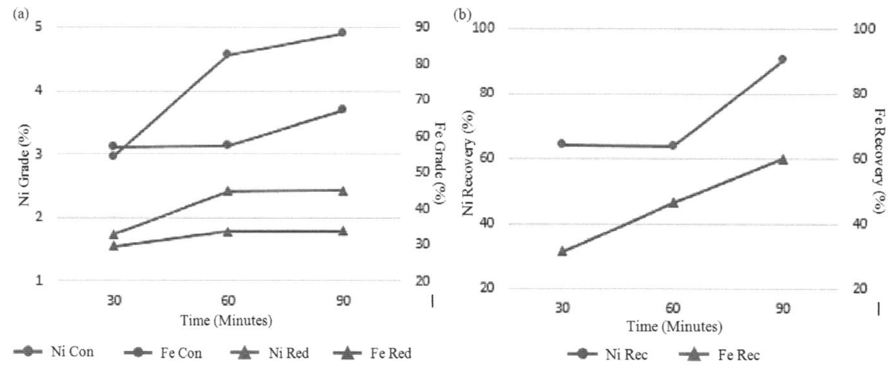

Figure 3. Effect ofthe first thermal stage retention time on: (a) grade of nickel and iron, (b) recovery of nickel and iron in concentrate at the first thermal stage temperature of 500°Cand reduction temperature of 1150°C for 60 minutes

Longer retention time of the first thermal stage process will increase the reduction of hematite to trevorite and finally becoming nickel oxide. Figures 3 shows that longer retention time of the first thermal stage will increase the nickel grade in the concentrate. This was due to reduction reaction of nickel oxidehaving lower free energy value than the reduction reaction of iron oxide, thus it makes the reduction of nickel oxide will takes place before the reduction of iron oxide. Longer retention time will facilitates more complete reduction of nickel oxide but not allowing iron oxide to occur. It means that there will be more nickel reduced from goethite phasethus increasing nickel grade.And it also reduce the amount of iron oxide to be reduced and thus lowering the iron recovery. Another effect of the longer retention time is the increasing carbon monoxide as the residue from the reduction process of hematite. It will promote the transformation of some phases, such as sodium silicate and nickel sulfide, which was being prepared for the reaction at higher temperature. Nickel sulfide will be reacted with wustite to form nickel oxide and FeS. Nickel oxide is more capable to be reduced at higher temperature ($\geq$ 700°C) due to more formation of gaseous CO. The longer retention time also affected to the addition of sodium chloride which will promote the reaction process between nickel oxide and gaseous hydrochloric acid to form nickel chloride. Nickel chloride, afterwards, will be segregated and forming metallic nickel. Figures 3 shows that longer retention time of the first thermal stage will increase the nickel grade inthe concentrate. It was due to the more nickel that had been completely reduced from goethite phase. The optimum time was 90 minutes for the first thermal stage with nickel grade of 3.69% and nickel recovery of 90.61%.

4. Conclusions

The first thermal stage at the two-stage thermal upgrading process in selective reduction of nickel ore wasfocusedin the reduction process of nickel form goethite phase and hinderthe reduction process of iron. It affectsthe decreasing iron recovery and the increasingnickel grade. The longer retention time at the first thermal stage promoted the increasing grade and recovery of nickel due to the more reducction process of nickel occurred.

Acknowledgement
The authors thank to DRPM UI for their financial support as PITTA grant funds that has been offered to perform this work.

References

[1] Mudd G M 2010 *Ore Geol. Rev* **38(1-2)** 9-26
[2] Kim J, Dodbiba G, Tanno H, Okaya K, Matsuo S, and Fujita T 2010 *Minerals Engineering* **23(4)** 282–288
[3] Norgate T and Jahanshahi S 2011 *Minerals Engineering* **24(7)** 698–707
[4] Zhu D Q 2012 *Int. J. Miner. Process.* **106-109** 1-7
[5] Li Q 2010 *Int. Miner. Process. Congr.* September 1549–1556
[6] Elliott R, Rodrigues F, Pickles C A, and Peacey J 2015 *Can. Metall. Q.***54(4)** 395-405
[7] Kobayashi Y, Todoroki H, and Tsuji H 2011 *Iron steel Int. J.* **51(1)** 35-40
[8] O'Connor F, Cheung W H, and Valix M 2006 *Int. J. Miner. Process.* **80(2–4)** 88–99
[9] Elliott R, Pickles C A, and Forster J 2016 *J. Miner. Mater. Charact. Eng.* **04(06)** 320-346
[10] Harris C T, Peacey J G, and Pickles C A 2011 *Miner. Eng.* **24(7)** 651–660
[11] Rhamdhani M A, Chen J, Hidayat T, Jak E, and Hayes P 2009 *Proc. of 5th European Metallurgical Conference* **EMC2009(4)** 1401-17

The aspects of technology and economy in processing aluminum scrap into fence ornament products at small and middle sized and enterprises in Lampung Province

K Isnugroho, D C Birawidha, and Y Hendronursito

Research Unit for Mineral Technology – Indonesian Institute of Sciences
Jl. Ir. Sutami Km 15 Tanjung Bintang Lampung Selatan

Email: yusu016@lipi.go.id

Abstract. Increasing building in property sectors causes chain effects in many of its supporting fields. One of them is the fence construction. Fence nowadays is not only functions as separator between a main building and other open spaces but it also has prestige values for the owner of the building. Fence ornaments continue to develop over time from its form, raw material, to its making process. Fence makers start to gradually abandon cast iron for fence ornament raw material. The aluminum becomes an alternative raw material in the fence making process. A research on the technological and economy aspects of aluminum scrap processing into fence ornament was conducted. From the technological aspect, small and middle sized industries were suggested to use batch system (50 kg/batch) with solid-fuel-based crucible furnace. The furnace dimension was 390 mm diameter and 440 mm height. Furnace isolation used fire brick type SK 32 with 240 mm thickness and 700 mm of furnace height. The molding process used sand molds method and used local raw material. Ratio of scrap raw material to ornament product was 1.1-1.3, while ratio of solid fuel use was 1.5-2.1. It took 6 hour /batch for smelting process with temperature of 600-710^{0}C. From economy aspect, it took investment cost of Rp. 210,000,000 to start aluminumscrap smelting business into fence ornament product. The financial analysis obtained net present value (NPV) of Rp. 176,699,790; net B/C (benefit/cost ratio) of 1.84; and internal rate of return (IRR) of 31.12%.

1. Introduction

The positive growth in property sectors indirectly cause chain effects to other supporting industries, and one of them is the constructions of fences with ornaments (for houses, apartments, and office buildings). The fence function now is not only a partition between main building and other open spaces (including roads), but it is also a symbol of prestige for the building owner. The fence ornaments continue to develop along the developing era from their forms, materials, and their making processes. Iron is often used for the fence ornament raw material. It is made by metal casting to produce casted iron. The fence ornament making with iron raw material needs bigger energy with high temperature process of ± 1400°C, while the product characteristic tends to be fragile with less resistance to corrosion. Energy source availability in form of coke for iron smelting is getting more difficult to obtain with more expensive price and these make casted iron ornament is started to abandon by small and middle sized industries of metal casting. The metal smelting and casting processes are processes to change the metal from solid phase in liquid phase by using smelting furnace, and the type of metal raw material and type of furnace to smelt it should be matched [1]. The factors to consider in selecting type of furnace include type of metal to cast, design of smelting temperature and casting temperature, ability or capacity of furnace to smelt, required operational cost, easiness in operating, easiness in maintenance and minimum pollution side effect for the environment [2].

Aluminum is a light metal and it has a good resistance to corrosion, so that is suitable to use as a raw material in the fence ornament making process in the small and middle sized industries, the aluminum scrap smelting is recycled into varying types of products. One of the products is the fence ornament as it is shown in Figure 1. There are many aluminum scrap recycling technologies from conventional (simple technology) by using crucible furnace to modern (advanced) technology by using reverberatory furnace, open hearth furnace, rotary furnace, and induction furnace [3]–[5]. The basic principle of aluminum scrap smelting process starts with pre-melting process by separating the raw material from impurities. The pre-melting process is conducted by such technologies using magnet, air separation, and color spectrum. The pre-melting process is then continued to purifying process. Based on the energy to use, aluminum scrap smelting can be done with electricity, gas, oil, and solid energy (coal, coke, and charcoal) [6]–[9], but aluminum scrap smelting process by using biomass energy (charcoal waste) is rarely done.

To find out the real, clear, and measured descriptions, a research on aspects of technology and economy of aluminum scrap processing into fence ornament product at small and middle sized industries in Lampung province was conducted. Lampung location was selected for this research because aluminum scrap in this province was currently only sent to metal smelting factories in Java Island. If the aluminum scrap was recycled in Lampung into fence ornament product, then this would improve additional value of the product and in the same time it would supply the fence ornament product demands in Lampung province, where these fence ornament products were usually imported from Java Island. Besides improving additional value of aluminum scrap into fence ornament product, this activity had an objective to improve the use of charcoal waste from charcoal factories in Lampung province. In Lampung province, there are exporters of wood charcoal and active charcoal, where in the product classifications, there is a charcoal product with size of < 2.5 cm and it is categorized as a byproduct or charcoal waste.

Figure 1. Various aluminum fence ornaments

2. Methodology

2.1 Aluminum scrap smelting

In this research, the aluminum scrap processing into fence ornament product was done by using crucible furnace. The furnace dimensions were 390 mm diameter and 440 mm height. The furnace isolation was from fire brick type SK 32 with 700 mm of furnace height, 240 mm of fire brick thickness, and 3 mm mild steel plate was used for outer wall of the furnace. The furnace design is shown at Figure 2.

The raw material and energy source should be considered to have optimal results. The raw material and energy source availabilities were adapted with the technology to use. The aluminum scrap to use was adjusted with conditions in Lampung province. In general, the types of aluminum scraps commercially available are shown in Table 1. The energy source came from wastes from charcoal factory byproducts, from briquette factories and active charcoal factories in Lampung Province. These materials were made into briquette to obtain a compact fuel form.

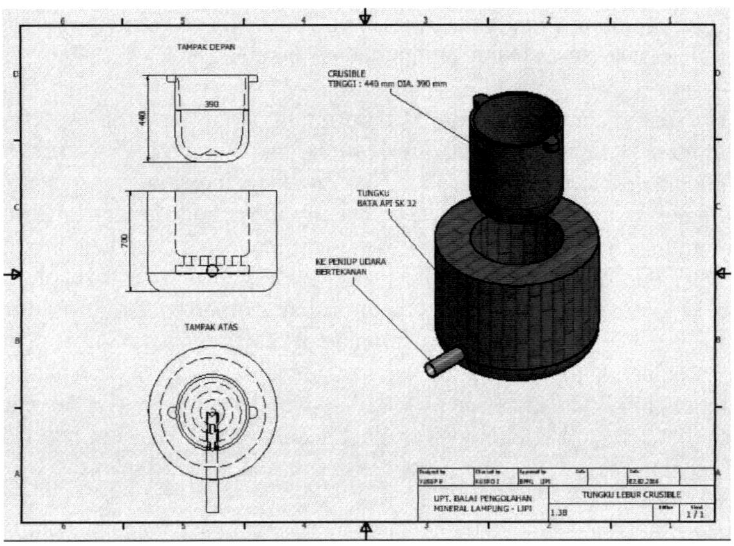

Figure 2.Technical drawing of aluminum furnace design

The Miscellaneous 1 material type contained of metal waste from machinery processes or components from machinery processes. The Miscellaneous 2 material type contained of automotive components; both two wheels and four wheels vehicles[10].

Table 1. Scrap aluminum composition from scrap seller

Types of materials	Composition						
	Cu	Mg	Si	Fe	Mn	Ni	Zn
Piston	1,46	0,71	12,34	0,24	0,46	0,006	0,08
Wire	0,12	0,01	0,78	0,47	0,06	0,01	0,02
Plat	0,05	0,007	0,13	0,37	0,015	0,004	0,005
Hanger	0,48	0,11	12,28	0,70	0,31	0,02	0,72
Miscellaneous 1	5,87	0,01	0,14	0,23	0,01	0,003	0,11
Miscellaneous 2	2,86	0,26	10,96	0,77	0,19	0,1	0,65
Propeller blade	3,66	0,16	9,28	0,94	0,23	0,16	2,07
Pan	0,17	0,08	0,22	0,50	0,09	0,004	0,01

The fence ornament making process is started by selecting raw material physically where material is separated from its impurities by macro appearance. The process continues with pattern (model) making of the fence ornament. Material of the pattern (model) comes from wood material. After the pattern (model) is done, aluminum casting process is conducted into fence ornament product by using sand casting method. Investment-casting and die-casting methods require more equipment and investment cost than sand casting method. For small and middle sized industries of fence ornaments in Lampung province, sand casting method is more suitable to apply. This is supported by the fact that silica sand availability as the raw material for sand casting can be found easily in Lampung province. Sand casting, though it is least accurate, is the most common and most widely used method of casting which involves the pouring of molten metal into a mould including typically 80% silica sand, 5–10% clay binders (mostly bentonite), 2–6% tempering water and a variety of additives. Sand casting is mostly used for high volume production where surface finish and dimensional accuracy are not critical[11]. In this research, aluminum scrap smelting was done in 650^0C – 730^0C. Temperature control was conducted with thermocouple. Pouring temperature was maintained being not too high to prevent hydrogen gas absorption to the liquid metal. The metal pouring process into sand mold was done with ladle where slag had to be removed first to prevent product defect from slag. After the liquid metal had frozen, the mould was removed to get the fence ornament product. Final stage was the finishing process including runner cutting, refining, and surface coating.

The gating system of aluminium smelting, the aplication different depends on simplicity or complexity the casting product. For simple shape and small in dimension, the riser used not too much but with complex shape and big in dimension require to much riser and ventilation hole in many place to flow out the air and the casting fluids can fulfill the hole of model.

2.2 *Study of financial aspect*

The objective of financial aspect study was to find out the funding estimation and the business cash flow to find out the feasibility and properness of the business plan. The financial aspect concerned with the fund need and its source, determination of cash flow policy, study about capital cost, analysis of sensitivity, assessment on the business plan appropriateness through varying methods (PI, NPV, IRR, PP), determining rent or buy for the fixed assets, and business priority selection. The need for investment fund was classified based on the tangible fixed assets (land, building, factory and machinery), and intangible fixed assets (patents, license, pre-costs and costs before operation). Investment funds were required for fixed assets and gross working capital (indicating all required investment for current assets)[12].

2.2.1 *Net present value (NPV) method*

The Net Present Value (NPV) is the difference between the present value of present investment and future values of net cash inflows (operational cash flow and terminal cash flow)[12]. The Net Present Value method formula is as follows:

$$NPV = \sum_{t=1}^{n} \frac{CF_t}{(1+K)^t} - I_0 \tag{1}$$

where NPV is questioned NPV value, t is year –t, n is number of economic year, CF_t is net cash inflow during year -t periode, K is discount rate, and I_0 is initial investment at year 0.

Assessment criteria:

- If NPV > 0, business investment proposal is feasible
- If NPV < 0, business investment proposal is not feasible
- If NPV = 0, the company value remains the same either the business investment proposal is feasible or not feasible

2.2.2 *Internal rate of Return (IRR) Method*

IRR method is used to find out a discount rate that makes the net present value of all cash flows to expect in the future, or cash inflows and cash outflows of initial investment[12]. The formula is as follows:

$$I_0 = \sum_{t=1}^{n} \frac{CF_t}{(1+IRR)^t} \quad (2)$$

Where I_0 is initial investment at year 0, t is year-t, n is number of economic year, CF_t is net cash inflow during year -t period, IRR is questioned discount rate value.
The IRR value can be obtained with trial and error. It is conducted by estimating present value of cash flow from an investment with a reasonable interest rate, for instant 10%, and then it is compared with the investment cost. If the investment value is smaller, the estimation can be done with higher interest rate, and it is continued the same way until the investment cost equals. Alternatively, it can also be done with smaller interest rate until the estimation produced the equal investment value with the present value. The IRR formula for interpolation is the following:

$$IRR = P_1 - \left(C_1 \; x \; \frac{P_2 - P_1}{C_2 - C_1} \right) \quad (3)$$

Where P_1 is interest rate 1, P_2 is interest rate 2, C_1 is NPV 1, C_2 is NPV 2.
The assessment criteria are that when IRR value is bigger than predetermined rate of return, then the investment is acceptable.

3. Result and discussion

3.1 Aluminum scrap smelting

The aluminum scrap smelting technologies in common are shown in Table 2. Table 2 shows that smelting technology with crucible furnace is selected for fence ornament product making for the following reasons. The crucible furnace is proven to be reliable. It has simplest construction with fixed position, where liquid metal can be picked up with ladle. This furnace is very flexible and multifunction for small and middle sized smelting compared to other types of furnaces.
The testing result by using the smelting furnace as it is shown in Figure 2 produced maximum capacity of 50 kg liquid aluminum. This maximum capacity was obtained by putting the raw material (aluminum scrap) gradually into the furnace, and it took 6 hours for the aluminum scrap to melt. The ration of material use was 1.1-1.3, while the ration of fuel use (wood charcoal waste with size of <2.5 cm) was 1.5-2.6.

The lowest raw material use ratio of 1.1 was obtained by using aluminum scrap material from pan, and the highest ratio of 1.3 was obtained by using propeller and miscellaneous 1 aluminum scrap materials. The temperature of smelting process based on the use of varying types of aluminum scrap is shown in Figure 3. The raw material from propeller and miscellaneous 1 (Table 2) required higher processing temperature compared to raw materials from pan, wire and so on. The Fe and Cu contents were relative high compared to other types of aluminum scraps, and this required higher demand of energy and processing temperature. However, in the fence ornament product making, there were no special specifications of types of aluminum scraps. The material use considered the availability of the raw materials. The small and middle sized business of metal casting paid more attention to obtain raw materials which were easy to smelt with cheapest technology of processing.

The lowest fuel use ratio of 1.5 was obtained by making fuel (wood charcoal waste) into briquette. The highest fuel use ratio of 3.0 was obtained when using wood charcoal with size of < 2.5 cm. These two types of fuel were required initial burning process (startup) with coke at ratio of 0.05 from aluminum scrap raw material. The use of startup fuel was required to maintain processing temperature of 750°C and to ease initial burning of fuel feeding the process was going on.

Table 2. Aluminum scrap smelting technologies

Type	Description
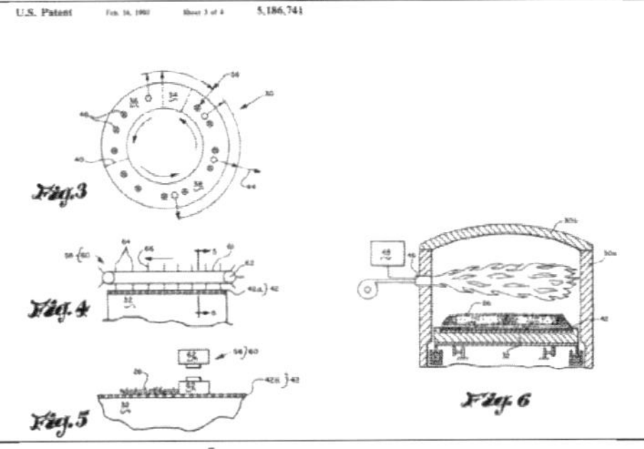	Heart Furnace Heart furnace is certainly the classical furnace in metallurgy and has been used early times by our forefathers for processing their bronze or, in later periods, their iron[6]
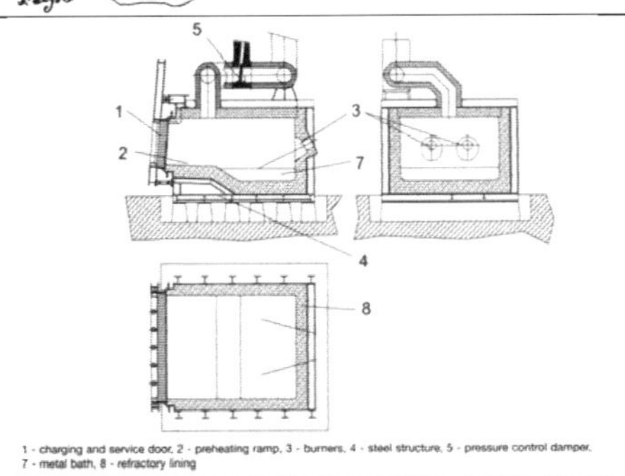	Reverberatory Furnace The basic technology for the fuel-heated furnace is the reverberatory furnace. Its name originates from the kind of heat transfer. It is assumed that the most of the energy is passed onto the metal by radiation following the principle of reverberation.In practice, this is, however, only the case when furnace is filled with liquid metal[6], [7]
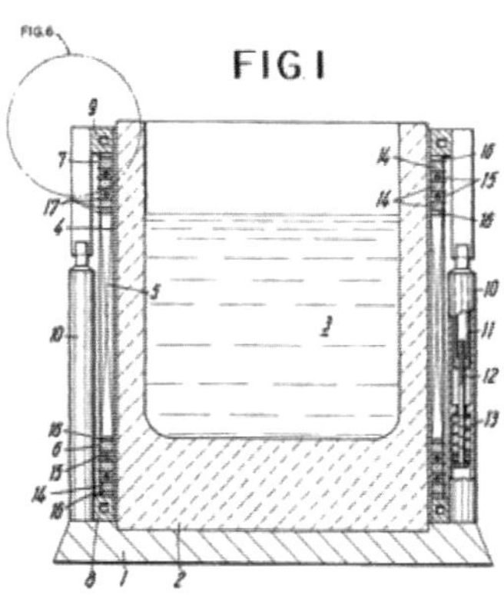	Induction Furnace An induction crucible furnace having a crucible surrounded by an induction coil member that is prestressed in axial direction by ring members disposed at opposite ends of the coil member and connected together by a plurality of tie rods extending parallel to the axis of the coil member and distributed uniformly about the periphery of the crucible one of the ring members being fixed to a base frame of the furnace, includes spacer means disposed between the ring members and the coil member and having inner and outer diameters matching those of at least one of the members[7], [8]
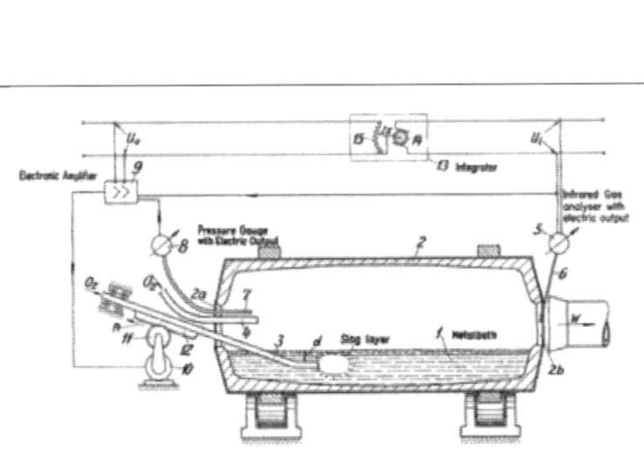	Rotary Furnace These furnaces follow a design principle that differs very much from that of the hearth furnaces. It comprises a refractory-lined steel vessel that rotates around central axis. Scrap is charged trough one of the front ends. The only burner is also arranged at one front end. Traditionally the rotary drum furnace has a fixed axis

	design. That means the centre line of the rotation remains in the horizontal position[7], [9]
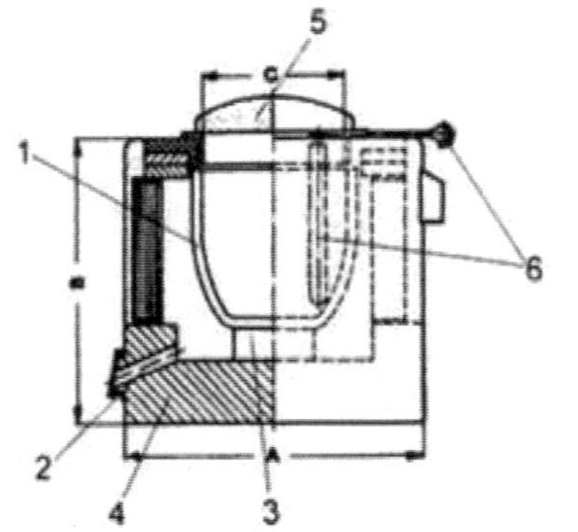 1 - crucible, 2 - burner, 3 - crucible support, 4 - furnace housing, 5 - lid	Crucible Furnace Crucible furnace major application is in small or midsize capacities, having a wide range of alloys. They are used as holding and casting furnaces in sand or die-casting plants. Heating is by means of charcoal, gas or oil burners that heat the metal in crucible from outside. The furnace can be designed as tiltable unit for easy pouring. The crucible furnace may also be heated by electrical resistors[6], [7]

The aluminum scrap smelting occurred at 650-750°C. The 5°C temperature decrease occurred at the time of feeding wood charcoal briquette process, while in feeding wood charcoal of size < 2.5 cm the temperature decreased insignificantly. This condition was because briquetting required mixture of starch powder and water as adhesive agent, so that the water content increased and this was different with wood charcoal (size <2.5) with lower water content. The briquette size was compact, dense, and bigger than wood charcoal (size < 2.5 cm), and it made more difficult for initial burning, so that temperature decrease occurred in the fuel feeding process. The experiment result showed that the aluminum scrap smelting process by using crucible furnace with wood charcoal briquette and charcoal waste (size < 2.5 cm) required higher ratio (1.55 and 2.65) compared to the same system with different types of fuels, where the fuel ratios were 0.5 for LPG, 1.3 for coal briquette, 1.17 for diesel fuel, 1.4 for lubricant waste, and 1.28 for kerosene.

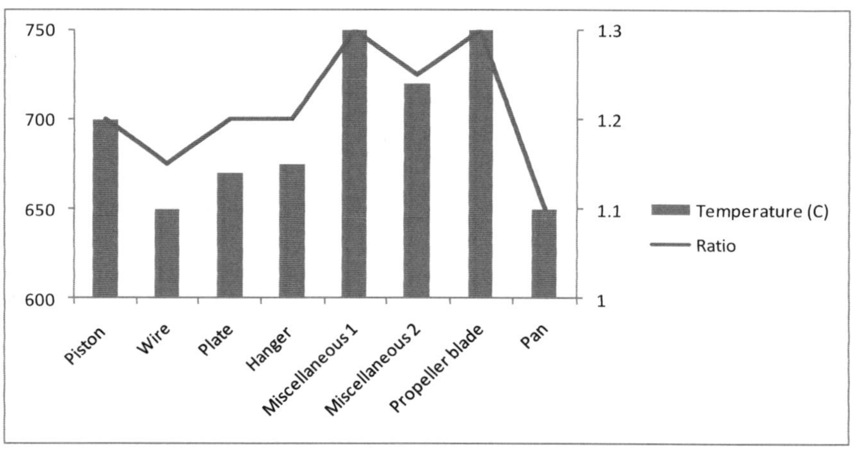

Figure 3. Fuel consumption ratio and temperature process

The lowest fuel use ratio of 1.5 was obtained by making fuel (wood charcoal waste) into briquette. The highest fuel use ratio of 3.0 was obtained when using wood charcoal with size of < 2.5 cm. These two types of fuel were required initial burning process (startup) with coke at ratio of 0.05 from aluminum scrap raw material. The use of startup fuel

3.2 Economy aspect

The review on the economy aspect of aluminum scrap smelting into fence ornament product was conducted with 50 kg liquid metal capacity production. Based on field studies in ornamental shops and welding workshops with face to face questionnare, the demand for fence ornaments reachedapproximately 5,000 kg / month and most of the ornaments were imported from Java with high tax costs and long shipping.To be seen from initial investment, the processing technology by using crucible furnace was lower than by using other types of technologies. The required investment value is IDR 210,000,000. This investment value is distributed for the following costs shown in Table 3.The estimation of annual operational cost for both fixed cost and variable cost is presented in Table 4. The estimation base uses 25 working days/ month and 12 effective months in one year.

The labors were two persons, one person was a supervisor and administration altogether, and another man was making fence ornament pattern model and product finishing. The smelting process required special labor with separate labor contract system.

Table 3. Investment cost components

Type of Cost	Total Cost	Volume
Building	IDR 144.000.000,00	120 m^2
Land	IDR 40.000.000,00	400 m^2
Equipment		
- Crucible	IDR 3.000.000,00	1 unit
- Blower	IDR 4.000.000,00	1 unit
- Furnace	IDR 10.000.000,00	1 unit
- Sand	IDR 5.000.000,00	10 m^3
Other equipment	IDR 4.000.000,00	1 package

Table 4. Operational cost components

Fixed Cost	
Electricity	IDR 24,000,000
Employee's Salary	IDR 44,200,000
Transportation	IDR 15,800,000
Land and Building Taxes	IDR 65,000
Total of Fixed Cost	IDR 84,065,000
Variable Cost	
Raw Material for Production	IDR 439,500,000
Employment wage	IDR 150,000,000
Total of Variable Cost	IDR 589,500,000
Total of Operational Costs (FC+VC)	IDR 673,565,000

The cost for finished fence ornament product was IDR 10,000/kg. The ratio of aluminum scrap raw material use was 1.3 and the fuel ratio to the finished product was 2.1. the average price of aluminum scrap in Lampung province in the collector trader was IDR 13,000 – 18,000/kg. The wood charcoal price was IDR 2,000 – 2,500/kg. The minimum selling price of fence ornament from producer level in Lampung province was 52,000/kg, it could be higher when particular model or detailed products asked by customers were able to supply by the producers. The market potential for this product in Lampung province is widely open, because current products are supplied from by producers from Java Island. The financial analysis derived net present value (NPV) of IDR 176,699,790 and internal rate of return (IRR) of 31.12%.

Comparing the fuel use ratio and processing technology by using crucible furnace, this experiment result is still economically possible, despite of higher ratio of fuel use as it is shown in Table 5.

Table 5. Cost of various fuel

Type of fuel	Ratio	Cost Unit	Total of Cost
Coal briquette	1.30	IDR 3,000	IDR 3,900
Diesel fuel	1.17	IDR 8,000	IDR 9,360
Oil Waste	1.40	IDR 3,500	IDR 4,900
Kerosene	1.28	IDR 7,000	IDR 8,960
LPG	0.50	IDR 7,500	IDR 3,750
Experiments			
- Wood charcoal briquette	1.50	IDR 2,200	IDR 3,300
- Wood charcoal of size <2.5cm	2.60	IDR 1,000	IDR 2,600

If startup burning cost with coke ratio of 0.05 is added with coke price of IDR 7,000/kg, the total of fuel cost pe4r kilogram of raw material was IDR 3,650 by using wood charcoal briquette and IDR 2,950 by using charcoal waste with size of <2.5 cm. Table 6 shows that the lowest total of fuel per kilogram of raw material is by using wood charcoal waste of size < 2.5 cm.

4. Conclusion

The aluminum scrap processing into fence ornament product can be conducted with a simple technology by using solid fuel based crucible furnace. The raw materials of aluminum scraps may come from varying types, while solid fuel can be obtained from waste of charcoal factory which is available in Lampung province. Financial analysis result shows that the fence ornament making in Lampung province in small and middle sized industries is viable and proper to do.

Acknowledgment

The authors acknowledged financial and facility supports from Research Unit of Mineral Technology, Indonesian Institute of Sciences and the authors would like to thank Slamet Sumardi as researcher from Research Unit of Mineral Technology, Indonesian Institute of Sciences for his valuable assistances.

References

[1] Ana G, Wanga Y, Lia W and Liub J 2012 *Procedia Enviromental Sci* **16** 202
[2] Hill R C 1979 *Design, Construction and Performance of Stick-wood Fire Furnace for Residental and Commercial Application*(Maine: University of Maine Orono)
[3] Verran G O and Kurzawa U 2008 *Resour. Conserv. Recycle* **52** 731
[4] Boeckenhauer K and Kaczmarczyk K 2008 *Modern Furnace for Alumunium Scrap Recycling* (Poland: Keith Boeckenhauer, SECO/WARWICK Corporation)
[5] Alexopoulos S O, Dersch J, Roeb M and Pitz-paal R 2015 *Appl. Therm. Eng.* **78** 387
[6] Kotraba N L and Holley C A 1993*Direct Reduction Process in A Rotary Hearth Furnace*(United States Patent)
[7] Schmitz C 2006 *Handbook of Aluminum Recycling* (Vulkan-Verlag GmbH)
[8] Hegewaldt F, Gillhaus H and Erdman O 1976 *Induction Crucible Furnace* (United States Patent)
[9] Graef R F and Von Bogdandy R L 1961 *Method for Refining Metal Baths in Rotary Drum Furnace* (United States Patent)
[10] Widodo R, 2012 *Foundry J* **2(1)** 1–8
[11] Boylu F 2011 *Appl. Clay Sci.*, **52** 104
[12] Husein U 2007 *Studi Kelayakan Bisnis* (Jakarta: Gramedia Pustaka Utama)

IOP Conf. Series: Materials Science and Engineering **478** (2019) 012004 doi:10.1088/1757-899X/478/1/012004

Microstructure transformation of Cr-Al-BN coating on low carbon steel prepared by ball milling method

A S Wismogroho[1], W B Widayatno[1], M I Amal[2], Y Affandi[3], G S T Sinaga[1], A Fitroturokhmah[1], and R Kusumaningrum[1]

[1]Research Centre for Physics, Indonesian Institute of Sciences, Setu 15314, Tangerang Selatan, Banten, Indonesia
[2]Research Centre for Metallurgy and Material, Indonesian Institute of Sciences, Setu, 15314
[3]Center for Technology for Mineral Resources, Agency for The Assessment and Application of Technology, Setu 15314 Tangerang Selatan, Banten, Indonesia

Email: agus.sukarto.wismogroho@lipi.go.id

Abstract: Cr-Al-BN powder was successfully applied as a coating agent on low carbon steel substrate by mechanical alloying method. The addition of BN as a doping material is considered as a feasible option because of excellent thermal and chemical stability presented by BN. In this paper, the compositions of Cr-Al-BN are varied Cr49.5-Al49.5-BN, Cr48.5-Al48.5-BN3, and Cr47.5-Al47.5-BN5. Each powder was milled by shaker mill for 2h in a stainless steel chamber with stainless steel balls to powder ratio 10:1. Subsequently, each powder compositions were mechanically alloyed onto substrate surface in air for 2h. The 2h heat treatment at temperature 800°C was conducted to each coated sample in vacuum furnace. In order to achieve the characteristics of phase composition and microstructure of the coating before and after heat treatment, XRD and Optical Microscope were performed, while the automatic microhardness tester was carried out to get the hardness of coating layer. The oxidation behaviour of coated substrates was also studied by heating treatment at 800°C for 8 cycles where each cycle is 20 h. The results show that the ball milling induces the formation of homogeneous Cr-Al-BN coating structure with a thickness of about 62.5 μm The optimum coating hardness and oxidation resistance was achieved by Cr47.5-Al47.5-BN5 coating composition. The addition of higher concentration ofBN increases the tendency onthe formation of intermediate phase of Cr_2B and Cr_2N. The detailed result of this study is shown in this paper.

1. Introducion

Low carbon steel with (0.025-0.25% C) nowadays has become a famous commodity among all industries bearing to its performance and market opportunity [1]. Low carbon has good flexibility, easy to shape and low cost. However, low carbon steel manufactures encounter limitation in term of oxidation. Hence, coating is the common way to prevent this problem [2]. Various researches and developments have been conducted in order to improve the properties of low carbon steel.

Al-Cr based-mixtures are the most common element for protective layer. Various elements have been co-doped with Al-Cr to get strong and heat resistant coating. The Cr–Al–N, Cr–Si–N, and Cr–Al–Si–N have been successfully developed and the result shows the improvement of hardness level reduction with respect to friction coefficient [3]. Furthermore, previous study of Cr–Al–N, TiC–C, and Cr–B–N coating also confirmed that the coating composition successfully improve the hardness and wear resistance of the substrate [4]. The mechanical alloying process is a method for producing composite metal powders. Mechanical alloying is an entirely solid state process, permitting dispersion of insoluble phases such as refractory oxides and addition of reactive alloying elements such as aluminum and titanium. The process was originally invented for the production of oxide dispersion strengthened (ODS) Ni-base super alloys and later extended to other ODS alloys [5-9]. Recently, many studies about coating are carried out with mechanical alloying method [10-12]. Mechanical alloying can be used to synthesize nanocrystalline-Si powder with different bandgaps, [10] and soluble catalyst systems may be rapidly produced by mechanical alloying [12].

Prior to this research, an Al-Cr base coating with addition of Tungsten (W) by mechanical alloying technique demonstrated that the mechanical alloying successfully subjected Al-Cr-W coating onto low carbon steel [13]. Presently, Boron Nitride (BN) was investigated as alternative composite in order to improve the properties of Al-Cr alloy. Heat treatment was carried out to refine the microstructure of the coating. The metallographic profiles of the coating before and after heat treatment were studied by means of XRD and Optical Microscope. Meanwhile, the hardness and oxidation resistance test were conducted to get the detail properties of the coating.

2. Experimental Details

The experiment was intended to investigate the physical properties from Cr-Al-BN varied coating composition. The substrates prepared were low carbon bar with dimension (10x10x3) mm which was polished using abrasive SiC paper up to 1500 grit. Before the coating deposition applied, the substrate was ultrasonically cleaned in methanol solution and dried by hot flowing air.

The followed step was processing the high purity Cr-Al-BN mixture. The Cr-Al-BN powder were mixed in stainless steel vial using shaker mill for 1 h with powder to steel balls ratio 1:10. The composition of Cr-Al-BN mixture was varied Cr49.5-Al49.5-BN1, Cr48.5-Al48.5-BN3, and Cr47.5-Al47.5-BN5 at%, respectively, where the total mass of each variation was 4 gr. The carbon steel specimens were then charged into the vial containing the Cr-Al-BN mixture. The specimen was mechanically alloyed under air atmosphere for 2 h. The deposition of coating powder on the steel surface is the result of collision between steel balls and powder on low carbon surface.

The annealing treatment was carried out at 800°C to improve the coating microstructure. The coated specimens were heated in vacuum atmosphere of 5.6 Pa for 2 h. Thereafter the specimens were then allowed to cool down to room temperature. The morphology of the coated sample was characterized by optical microscope (Best Scope). Analysis of phase composition of coating was conducted by X-ray diffraction (Smart Lab Rigaku) with Cu Kα radiation at 40 kV and 30 mA. The hardness of coating layer was measured using automatic microhardness tester. The cyclic oxidation test was also performed in order to measure the oxidation resistance of coating layer in high temperature condition. The measuring of dimensional and weight of the samples were prepared using digital calipers and electronic balance with a sensitivity of 1 x 10-5 g prior to oxidation cyclic process. Each sample was heated at 800°C for 20 h. The weight of each sample was measured after the oxidation test, including the weights of spalled oxide. The sample was then oxidized again for 20 h. This procedure was repeated for up to 8 times and each cycle was 20 h.

3. Results and Discussion

The cross-sectional optical images of Cr49.5-Al49.5-BN1, Cr48.5-Al48.5-BN3, and Cr47.5-Al47.5-BN5 coatings before heat treatment are shown in figures 1 respectively. The figures exhibit that the coatings were successfully deposited on the low carbon steel surfaces. Mechanical alloying was effectively used to cover all substrate although the density of the coating is subsided bearing to the increment of BN concentrate. The Cr49.5-Al49.5-BN1 composition shows denser structure followed by Cr48.5-Al48.5-BN3, while the Cr47.5-Al47.5-BN5 looks more porous.

IOP Conf. Series: Materials Science and Engineering **478** (2019) 012004 doi:10.1088/1757-899X/478/1/012004

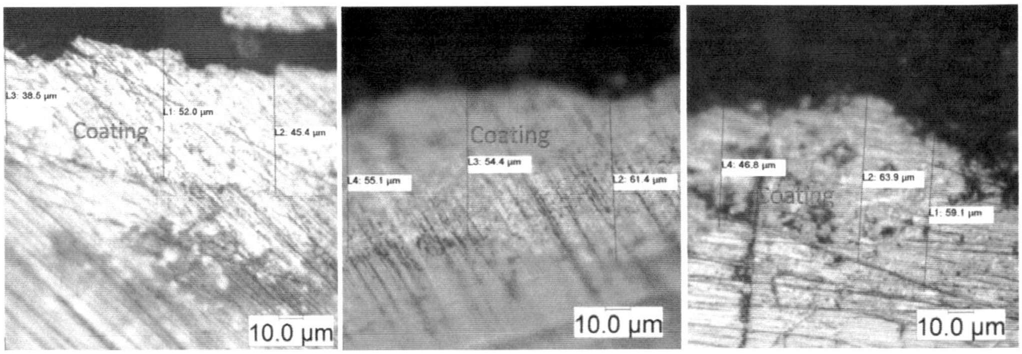

(a) (b) (c)

Figure 1. cross-sectional optical images of (a)Cr49.5-Al49.5-BN1 (b) Cr48.5-Al48.5-BN3 (c) Cr47.5-Al47.5-BN5 before heat treatment

Figure 2 shows the cross-sectional optical images of Cr49.5-Al49.5-BN1, Cr48.5-Al48.5-BN3, and Cr47.5-Al47.5-BN5coatings after heat treatment at 800°C for 2 h.

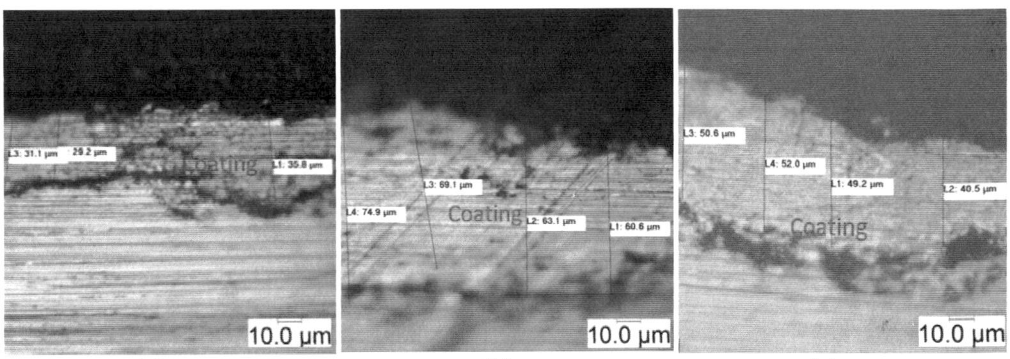

(a) (b) (c)

Figure 2. cross-sectional optical images of (a) Cr49.5-Al49.5-BN1; (b) Cr48.5-Al48.5-BN3; (c) Cr47.5-Al47.5-BN5 after heat treatment

Figure 3 presents the X-ray diffraction patterns of Cr49.5-Al49.5-BN1, Cr48.5-Al48.5-BN3, and Cr47.5-Al47.5-BN5 coating before heat treatment. The XRD analysis reflects that all elements in the coating remain in their original phase. Diffraction peak of Cr, Al and BN are detected on all sample, even though the diffraction peak of Al for Cr47.5-Al47.5-BN5 sample is relatively lower than that of the other two. There is no trace of intermediate phase, as usually seen for the mixture which experience mechanical milling. Interestingly, the first Al peak around 38° is not observed for Cr49.5-Al49.5-BN1 sample.

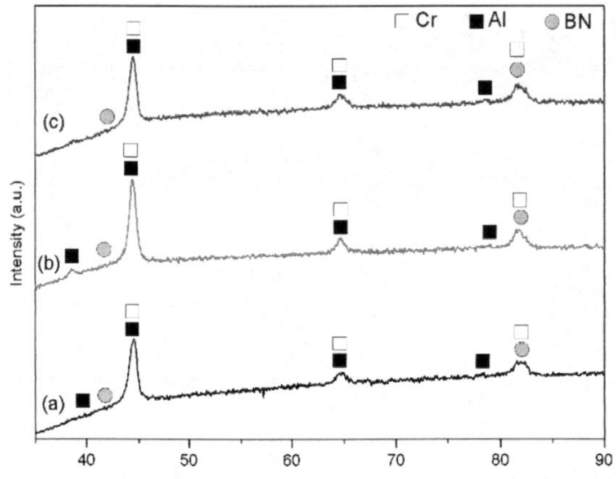

Figure 3. X-ray diffraction patterns of (a) Cr49.5-Al49.5-BN1, (b) Cr48.5-Al48.5-BN3, and (c) Cr47.5-Al47.5-BN5 coating before heat treatment

Figure 4 shows X-ray diffraction patterns of Cr49.5-Al49.5-BN1, Cr48.5-Al48.5-BN3, and Cr47.5-Al47.5-BN5 after heat treatment. The XRD patterns show that there is no trace of initial BN, implying the dissolution of BN species within the sample. The initial dominant phases were likely to be $AlCr_2$ and Al_8Cr_5. Adding 1 at% of BN did not give any additional phase. When the concentration of BN was increased to 3%, the Al_8Cr_5 was suppressed to some extent, yet still did not induce the formation of additional phase. Interestingly, the additional phases containing B and N element occurred when the BN concentration was increased to 5%. It is likely that the BN species experience bond cleavage during annealing. There is no single trace of Al_8Cr_5, while the new phase of Al_5Cr occurred. The BN bond cleavage is likely followed by the binding of Cr species in Al_8Cr_5, leading to the formation of Cr_2B and Cr_2N. As a result, the remaining Al-Cr phase experienced Cr deficiency which was resulted to the formation of Al_5Cr. As for $AlCr_2$, the existence of bond cleavaged-BN species seems had no effect on the stability of $AlCr_2$ phase.

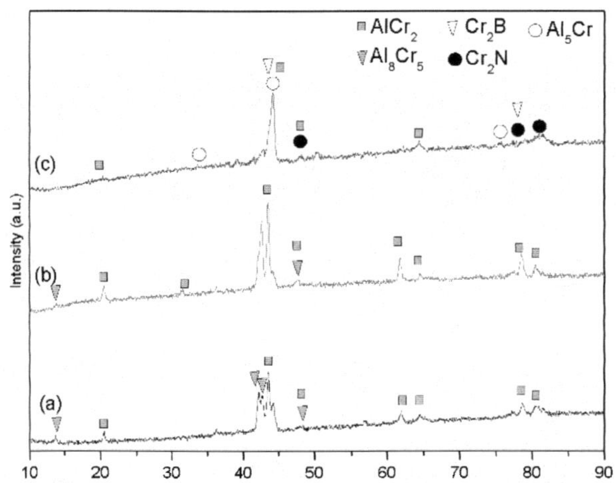

Figure 4. X-ray diffraction patterns of $Cr_{49.5}$-$Al_{49.5}$-BN_1, $Cr_{48.5}$-$Al_{48.5}$-BN_3, and $Cr_{47.5}$-$Al_{47.5}$-BN_5 after heat treatment.

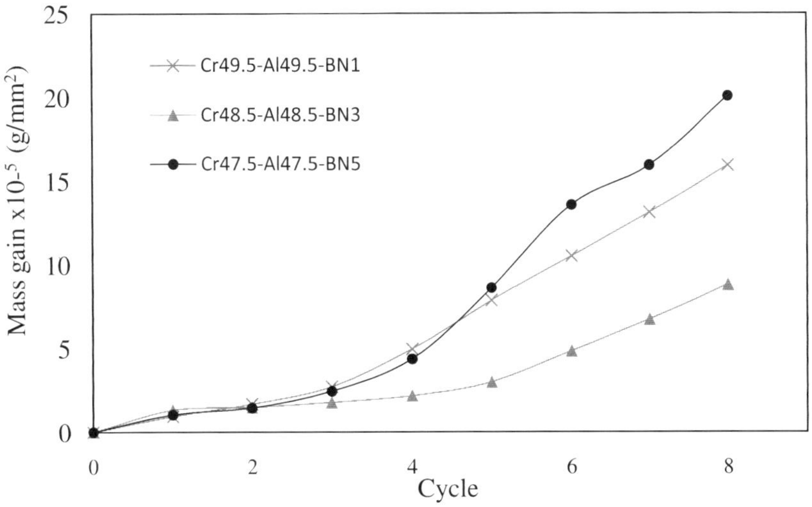

Figure 5. Cyclic oxidation curve of Cr49.5-Al49.5-BN1, Cr48.5-Al48.5-BN3, and Cr47.5-Al47.5-BN5 coating after heat treatment

Figure 5 shows cyclic oxidation curve of $Cr_{49.5}$-$Al_{49.5}$-BN_1, $Cr_{48.5}$-$Al_{48.5}$-BN_3, and $Cr_{47.5}$-$Al_{47.5}$-BN_5 after heat treatment. The mass gain in this study ~~is~~ represents the formation of oxide materials on the coating surface. The good oxidation resistance is indicated by the lowest mass gain of the materials. The most oxidation resistance was attained by Cr48.5-Al48.5-BN3.

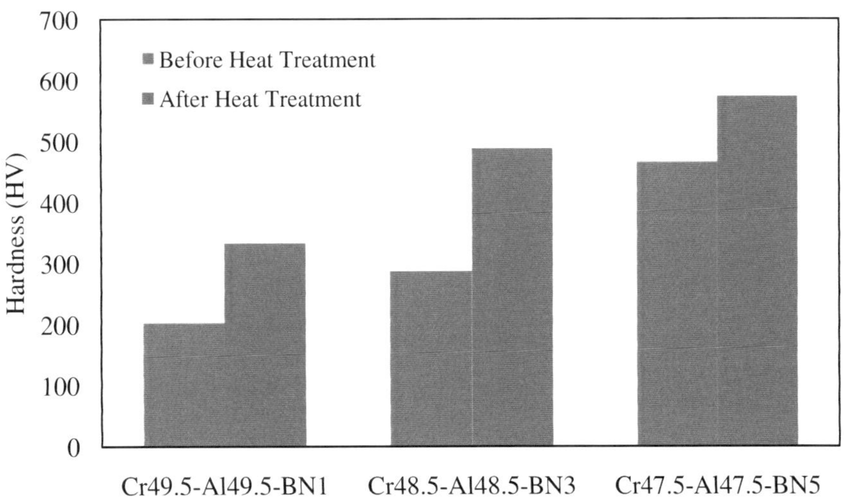

Figure 6. Hardness of Cr49.5-Al49.5-BN1, Cr48.5-Al48.5-BN3, and Cr47.5-Al47.5-BN5

Based on hardness data of Cr49.5-Al49.5-BN1, Cr48.5-Al48.5-BN3, and Cr47.5-Al47.5-BN5 (Figure 6), the hardness level is increased by heat treatment process. The more BN additions in the coating materials, the higher hardness level can be achieved.

4. Conclusion

The coating of three different Cr-Al-BN compositions has been successfully deposited on low carbon steel using mechanical alloying technique. From the observed result, it can be concluded that the void and crack were formed in all sample after heat treatment given. ThehighestBN concentration (Cr47.5-Al47.5-BN5) induces the formation of Al_5Cr, Cr_2N and Cr_2B. The increase in BN concentration was also responsible to the increase of hardness. Optimum hardness was achieved from Cr47.5-Al47.5-BN5 coatings, while the less the BN concentration the less hardness level established. However, the most oxidation resistance was attained by Cr48.5-Al48.5-BN3.

Acknowledgement

This work is partially funded by INSINAS grant (No. 058/P/RPL-LIPI/INSINAS-1/III/2018) Ministry of Research, Technology and Higher Education of the Republic of Indonesia (KemenristekDikti) and PPTI grant (title: PengembanganTeknologi High Temperature Furnace dalam Rangka Penelitian Material Lanjut dan Penguatan Inovasi Nasional Untuk Pengembangan Anoda dan Katoda).The authors would like to express our gratitude for all facilities provided by research center for physics.

References

[1] Yang W, Li Q, Liu W, Liang J, Pen, Z, and Liu B 2017 *Vacuum* **144** 207–221
[2] Zhang M, Chen C, Zhang G, Rao Y, and Ling G 2013 *Physics Procedia* **50** 206–213
[3] Park I W, Kang D S, Moore J J, Kwon S C, Rha J J, and Kim K H 2007 *Surface and Coatings Tech.* **201** (9-11)
[4] Lin J, Moore J J, Moerbe W C, Pinkas M, Mishra B, Doll G L, and Sproul W D 2010 *Int. J. of Refractory Metals and Hard Mat.* **28** 2–14
[5] Suryanarayana C 2001 *Progress in Mat Sci.* **46** 1–184
[6] Benjamin J S and Volin T E 1974 *Metallurgical Transactions* **5** 1929–1934
[7] Suryanarayana C, Ivanov E, and Boldyrev V 2001 *Mat Sci and Eng* **A304-306** 151-158
[8] Murty B S and Ranganathan S 1998 *Int Mat Reviews* **43(3)** 101–141
[9] Schwarz R, Petrich R, and Saw C 1985 *J. of Non-Crystalline Solids* **76** 281–302
[10] Goyal A and Soni P R 2018 *Mat Let* **214** 111–114
[11] Jia D, Liang B, Yang Z and Zhou Y 2018 *Progress in Mat. Sci.* **98** 1–67
[12] Guevara L, Welsh R, and Atwater M 2018 *Met.* **8 (4)** 286
[13] Wismogroho A S, Sudiro T, Didik, A, and Ciswandi 2018 *J. of Phy*, Conf. Series **985** 012042

IOP Conf. Series: Materials Science and Engineering **478** (2019) 012005 doi:10.1088/1757-899X/478/1/012005

Analysis of crucible performance for aluminum scrap casting at small and medium enterprises (SMEs) foundry

Y Hendronursito, K Isnugroho, and D C Birawidha

Research Unit for Mineral Technology – Indonesian Institute of Sciences
Jl. Ir. Sutami Km 15 Tanjung Bintang Lampung Selatan

Email: yusu016@lipi.go.id

Abstract. Performance analysis of crucible that used in aluminum casting at Small and Medium Enterprises (SMEs) Foundry have been done by non-destructive test (NDT). A non-destructive testing method is a felicitous method because it does not interfere with the production process. The crucible analysis includes the Chemical composition analysis use X-Olympus XRF to know the crucible material, fluids flow analysis and design using finite element method with Autodesk Inventor Software and Autodesk Simulation CFD. Based on the results showed that the crucible consists of 97.17% Fe, 1.68% Si, 0.6% S, 0.22% Mn, 0.26% P, and 0.08% Cr. From fluids flow simulation founded a critical point in crucible area that appropriates with the leaks occur. In this critical point, the air velocity about 8.89 m/s and the velocity in the inlet and outlet drooped 266.7 m/s, increasingly severe with fixed support the crucible only on upper hatrack while at the bottom there is no support that caused the maximum stress in this section. Finite element method divided the displacement area by a mesh average of 0.08 in four colour variations, red colour at 48.94 mm max displacement, yellow colour on 29.36 mm – 39.15 mm displacement area, green colour at 19.58 mm – 29.36 mm displacement area, and blue colour at 0 mm – 19.58 mm displacement. The crack of crucible located between red and yellow colour about 39.15 mm displacement.

1. Introduction

A crucible is a container that has a melting point higher than the material want to melted. Crucible made of a variety of materials generally available are made of clay and graphite or silicon carbide [1], but now for non-ferrous smelting especially Small and Medium Enterprises (SMEs) commonly use of metal pots from iron casting because simple in production and inexpensive [2][3]. As the main utility of non-ferrous smelting, capacity greatly influences crucible usage. For example, for 60 Kg capacity of the crucible can be used in 3 to 5 batches per day for 3 to 4 hours to smelting time per batch. The crucible failure cannot avoid in continuous use. The failure can be divided into two types i.e. crack and deformation. The crack at crucible can cause the leak of smelting fluids so disturb the production processes i.e. shrinked the volume of smelting fluids; temperature droop that impacts the failure foundry like misrun or not fulfill the model of casting. The deformation of crucible caused by the material of crucible cannot hold the high temperature during the smelting process which happens repeatedly. Fig. 1 and Fig. 2 show the example of rill crucible failure in aluminum ornament SMEs. The leak is located at the bottom side of crucible while the deformation is located in the around of crucible. Based on Irwansyah *et.al* (2010), durability or lifetime the crucible depending on the quality of aggregate which not only must high temperature resistant, but also inert or not chemically react with fused solid materials and tangled gas [4]. While in Bertha (2008), crucible failure caused by hot melt touch effect from melted material and caused penetration, corrosion and erosion melted material to crucible body [5]. The researchers explain if the deformation of crucible caused by temperature variations and over heating or the temperature operation is changed quickly and not continuous.

Crucible as a main tool in foundry, failure analysis, however without damaging the crucible so did not disturb the production, non destructive test can be used to analysis crucible (NDT) so software analysis is the best appropriate to analysis the failure of crucible.

Figure 1. Crucible in riil deformation

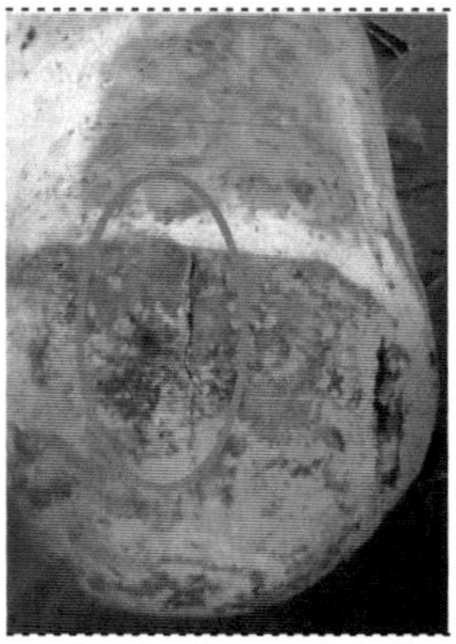

Figure 2. starting point the leak of crucible

Non destructive failure analysis uses stress analysis software can be used for predicting the workload of the design or objects like which conducted by Budimana *et.al.* (2016), analysis the safe-failed of the hydro Francis turbine blades [6]. Berkowski et al (2015), 3D modelling to examination and analysis the crack formation in a composite slab floor. Under examination, they find the cracks were caused by a reinforcement and deck plate thickness that were mismathc to the design [7]. Moreover, NASA R&D implements the same method to discovering load and forces in composite materials with the compare between simulation and experiment [8]. Furthermore, Engineering computational fluid dynamics (CFD) and Finite Element Method (FEM) can use to accurately analyze the design and safe-fail mechanically [9]–[11]. Han *et.al* (2010), used FLUENT to calculate the air force and boundary conditions the crane with 100 tons in capacity [12]. The stages of FEM are 3D models, then selecting material, and the next important step for an accurate FEM modelling is to mesh the model ineffective way so that critical regions are covered with an optimized number of elements [13].

2. Method

The method used non-destructive test (NDT) to analysis the crucible failure. The crucible failure in SMEs foundry analysis by Autodesk Inventor and Computed Fluid Dynamics (CFD) for free software, Autodesk provide two version i.e for student software & educator and free trial for 30 days. The primary data, i.e. dimensions and actual conditions obtained from UD Tami 15 located in Jl. Ir. Sutami Km. 15 Tanjung Bintang, South Lampung, Lampung Province. The measurement data such as the temperature of the burner room measuring by the Infrared Sanfix brand thermometer and air velocity measuring by anemometer type of Vane AM-4200. NDT analysis covers the chemical composition of crucible and encrustation used XRF portable merk X-Olympus, fluid flow analysis, and stress analysis. The stove and crucible drawn fit with actual dimension and assembly with real condition in UD Tami 15 workshop. The material input parameter, i.e. crucible material, furnace material used refractory brick SK 38, refractory shell used mill steel plate, air combustion from blower inputted in material data sheet of Autodesk Inventor software. The design in 3D model in the same direction as shown in Fig 3.

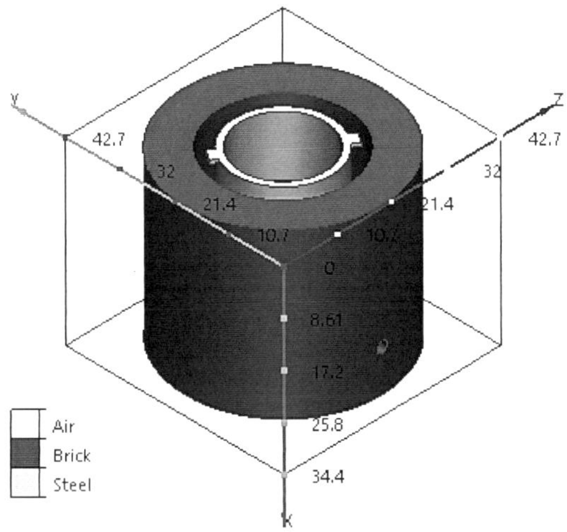

Figure 3. material selected of crucible furnace in the 3D autodesk inventor

After drawing the design, Computed fluid dynamics analysis with Autodesk Simulation CFD software. The parameter of crucible material, velocity, temperature, air pressure, inlet and outlet direction inputted as an initial parameters fluids analysis. The physical of mass, density, volume, and centre of gravity the crucible shown in Table 1.

Table 1. Physical of Crucible

Material	Iron, Cast
Density	7,25 g/cm^3
Mass	211,957 kg
Area	1731520 mm^2
Volume	29235500 mm^3
Center of Gravity	x=0 mm
	y=-246,809 mm
	z=117,886 mm

3. Result and Discussion

3.1. Chemical Composition

X-ray fluorescent (XRF) analysis the material of crucible result shown in Table 2.

Table 2. Crucible chemical composition

	Chemical Composition %								
	Fe	Si	S	Zn	Mn	P	Cu	Cr	Zr
Body	97.17	1.68	0.6		0.22	0.26		0.08	
Crust	94.22	1.68	2.84	0.62	0.25	0.20	0.14	0.04	0.015

The chemical analysis results shown the crucible made by 97.17% Fe and 1.68% Si. The crust of crucible bodies consists of 94.22% Fe, 1.68% Si and another element, such as 2.84% S, 0.62% Zn, 0.14% Cu, and 0.015% Zr. The chemical composition of Sulfur increase from 0.6 % in the body to 2.84% in the crust. The formation of 2.84% Sulfur formed from the combustion process in the outer wall of the crucible. The crucible working on high-temperature environment and this condition makes a possibility of corrosion. Sulfur (S) will react to binding Fe from the base material of crucible so will reduce the layer of the wall crucible, penetration in the wall can cause the corrosion. furthermore, the wall of crucible getting thinner and easier to deformation because cannot resist the force.

3.2. *CFD Simulation Analysis*

The air flow from the blower inlet through a pipe with dia. 1 ¼ inch with velocity $V_1$10.2 m/s and pressure P_1 1,2KPa. While the temperature obtained by burning the charcoal in combustion room. The crucible placed centre and supported by upper hatrack while at the bottom there is no support. The lines in Fig. 4 shows the fluid flow, the air from the blower entrance through the inlet pipe and formed the turbulent flow in the base of combustion room. Based on CFD analysis, show the air flow swirls two times in the base of the combustion room before up to the outlet, amongst the body of the crucible and the wall of the combustion furnace.

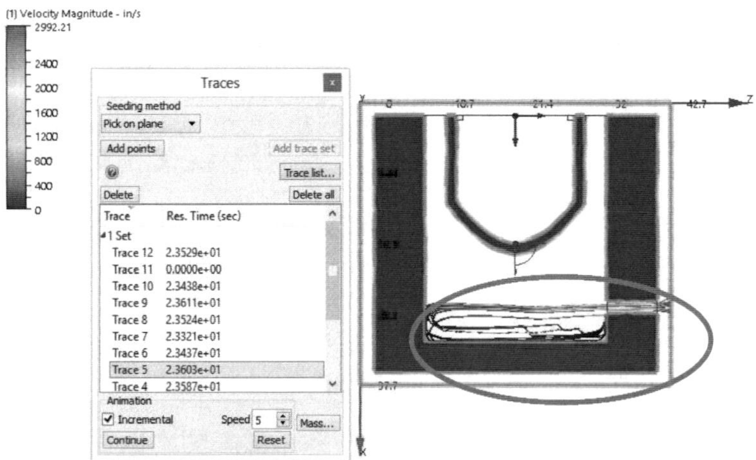

Figure 4. fluids flow in the combustion room

Fig. 5 represented the fluid flow between the crucible surface were hanged at the furnace. The first position of the air touched the outside of crucible surface and then out between the gap of the crucible wall and the wall of the furnace. The figure traces list showed the points of air forced the wall of the crucible and move leftward and eventually exit through the combustion room. This air has pressure and velocity and works continuously at high-temperature, which will be possible to stress the crucible.

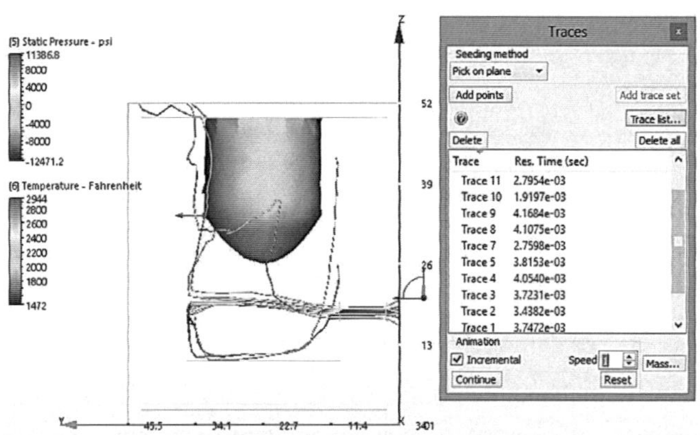

Figure 5. Traces of force air

The air velocity in the combustion room at certain points showed in Fig. 6 while the Fig. 7 showing the air velocity in 'z' direction (in 3D design, x y z coordinates).

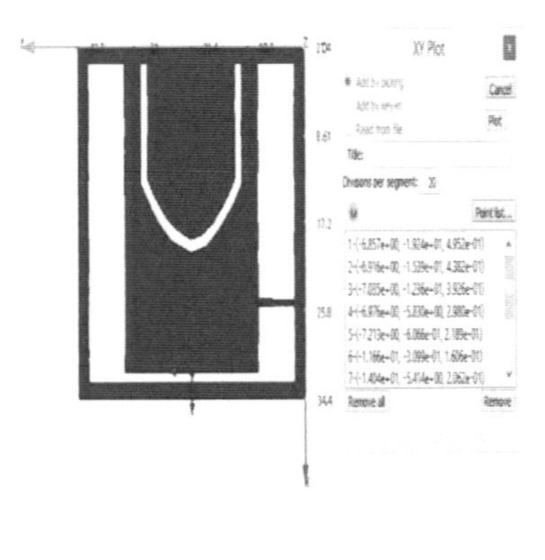

Figure 6. The air velocity at certain points

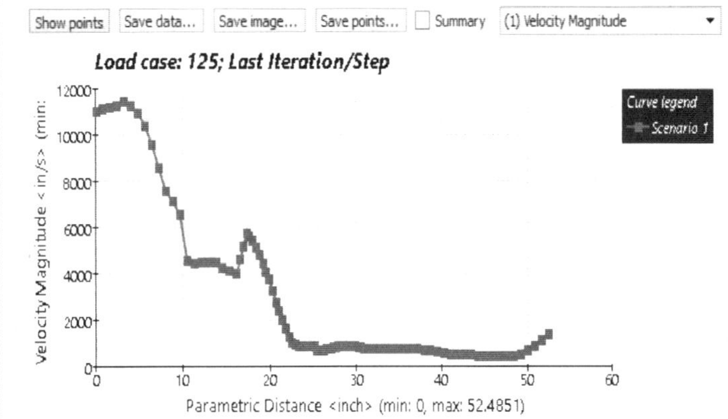

Figure 7. The air flow iteration at z coordinate

The air flow velocity at the combustion room in coordinate 0 to 10 Z direction 279.4 m/s - 114.3 m/s, or decreased velocity 165.1 m/s. This decreased happen because air vortex at the bottom of the combustion room detained by combustion wall and the fuel contact between the air flow. Whit the coordinate of 20 - 50, the air velocity 12.7 m/s. The velocity drop from inlet to outlet 266.7 m/s. The analysis of CFD finds that in coordinate 26 which is the velocity of air flow touched the outer convex body crucible is 8.89 m/s. The air continuously hit the body crucible during flow blower open in cold or hot conditions depend on the presence or absence of flame. This causes cracking point at crucible, this point is suitable with Fig. 2 where is the crack start happens in actual coditions..

3.3. *Stress analysis*

The stress analysis performed by input the parameter obtained from computational fluid dynamics analysis. The force that works in the crucible located at the bottom surface of the crucible as shown in Fig. 8. This force is the weight obtained from the crucible mass and the mass of the aluminum raw material shown in equation 1. The fixed points of the crucible located in upper hatrack. The meshing setting is shown in Table 3.

Table 3. Mesh settings

Avg. Element Size (fraction of model diameter)	0,08
Min. Element Size (fraction of avg. size)	0,2
Grading Factor	1,5
Max. Turn Angle	60 deg
Create Curved Mesh Elements	Yes

$$W = (m_0 + m_1)\ x\ g \quad (1)$$

where m_0 is mass crucible, m_1 is mass raw material, and g is gravity force (9.81 m/s^2).

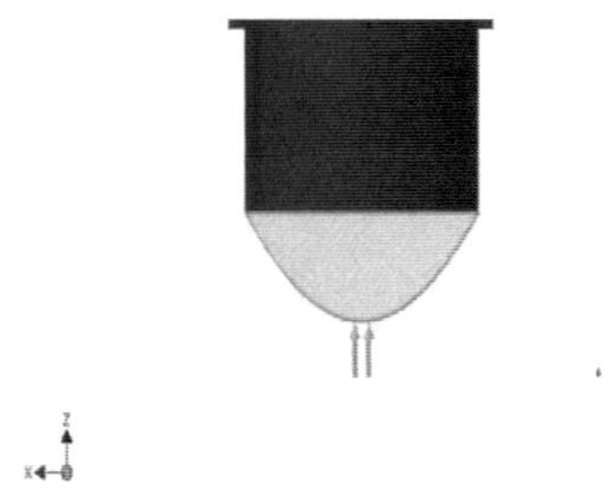

Figure 8. Stress analysis input parameters.

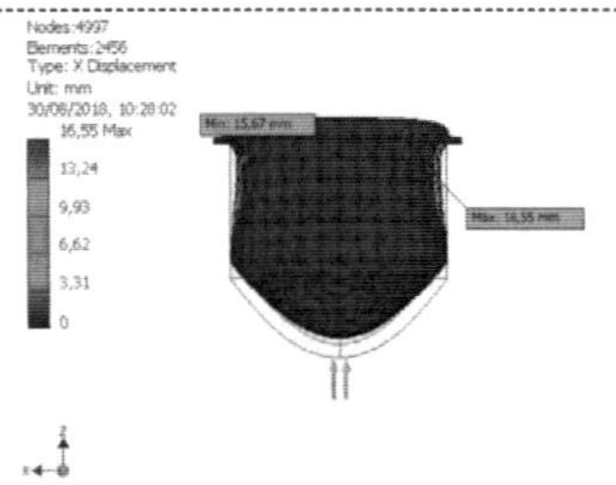

Figure 8 (a). Stress Analysis Result in X Displacement (in 3D design)

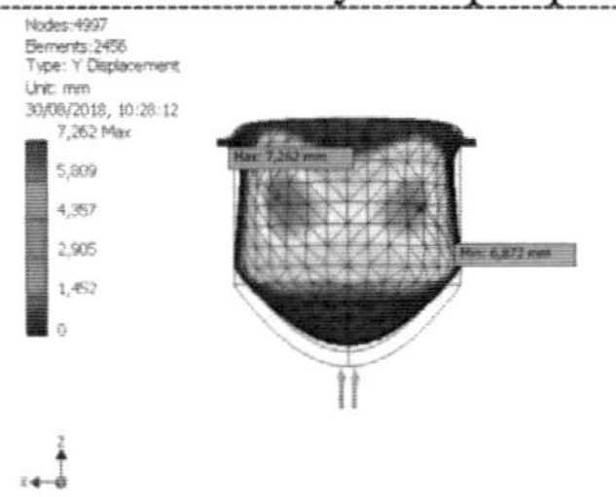

Figure 8 (b). Stress Analysis Result in Y Displacement (in 3D design)

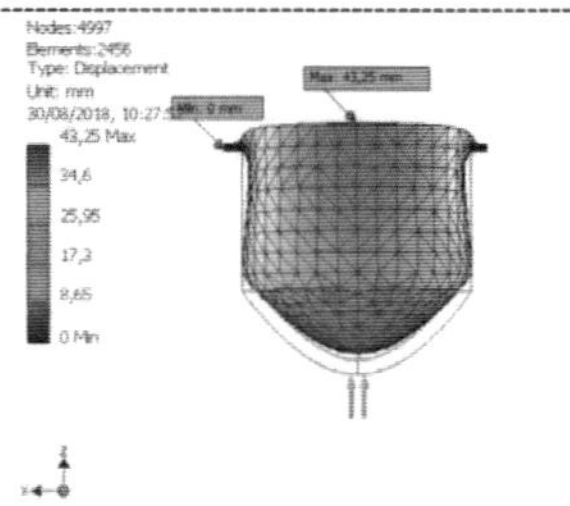

Figure 8 (c). Stress Analysis Result in Z Displacement (in 3D design)

The simulation of deformation crucible same with the real deformation crucible, the deformation crucible in z direction so crucible shown oval shaped. Similarly, on the side of crucible no longer straight, but curved in the x and y coordinates. This analysis is same with riil crucible that used in Small and Medium Enterprises UD Tami 15 shawn in Fig. 1. The stress analysis result shows the high displacement in an area with the red colour shown in Fig. 8, maximum displacement 43.25 mm located inside of the outer concave crucible. The yellow area with displacement 25.95 mm – 34.6 mm, the green area with displacement 17.3 mm – 25.95 mm, and the dark blue and light blue with displacement 17.3 mm – 0 mm. The failure of the crucible is located in red colour and yellow with displacement about 34.6 mm. In the riil crucible deformation, the displacement about 10 – 30 mm located inside of the outer concave crucible, that can see in Fig. 1 in actual condition.

4. Conclusion

The analysis of crucible performance used Computed Fluid Dynamics and Stress Analysis Software shown the same phenomena to the defect of crucible in actual condisition. Analysis software used more effective because non destructive the crucible and recomended to changes the crucible furnace design.

Acknowledgments

This study was supported by the Research Unit For Mineral Processing - Indonesian Institute of Sciences. The author would like to thank Mr Edi general Manager of UD. Tami 15 located in Tanjung Bintang Lampung Selatan, which give permission and data to complete this paper.

References

[1] Gingery and Da J 2012 *Building a Gas Fired Crucible Furnace* Rogersville: David j. Gingery Publishing, LLC.
[2] Nukman, Mataram A, and Yani I 2015 *Mechanical* **6** 6
[3] Sundari E 2011 *Jurnal Austenit* **3** 17
[4] Ferli S I, Juliandri, Iwan H, Soewanto R, and Rifki S 2010 *Journal of Indonesia Zeolites* **9** 25
[5] Bartha P 2004 *Intercera.* **33** 14
[6] Bentang A B, Djoko S, Indra D, Muhammad A, and Firman B J 2016 *Adv. Mech. Eng.* **8** 1
[7] Piotr B, Grzegorz D, Jacek G, Krzysztof S, and Zbigniew W 2013 *J. Civ. Eng. Manag.* **19** 529
[8] Venkat A, Satvir A, Hamid K, Arunkumar S, and Philip B 2016 *Proc. Am. Soc. Compos. Tech. Conf.* 31
[9] Kiran K B S, Venkata R R M, and Kumar K 2015 *International Journal of Engineering Research & Technology* **4** 165
[10] Christophe D, Thi C V, and Francois G 2015 *Int. J. Fluid Mach. Syst.* **8** 132
[11] Fan H G, Yang H X, Li F C, and Chen N X 2014 *IOP Conf. Ser. Earth Environ. Sci.* **22** 032050
[12] Dae-Suk H, Sung-Won Y, Hyun-Sik Y, Myung-Hyung K, SAng-Hyun K, and Jae-Myung L 2011 *Autom. Constr.* **20** 368
[13] Navid A, Domenic F, and Mark M T 2016 *42nd International Symposium for Testing and Failure Analysis* (Texas, USA) 6

The effect of flux's towards Mg reduction from aluminium beverage cans

A Juniarsih, S Oediyani, and A P Zain

Metallurgical Engineering, Universitas Sultan Ageng Tirtayasa, Cilegon, Indonesia

Email: andinnie@untirta.ac.id

Abstract. Aluminium beverage cans scrap recycling is expected to fulfil aluminium needs in Indonesia, reduce production cost, and pollution generated. Fluxing is one of the methods that can be used to reduce impurities in aluminium alloys such as magnesium, flux also can protect molten metal reacted with oxygen. In this research, aluminium beverage cans scrap used to consist of carbonated, isotonic, and refreshing beverage cans and NaCl+KCl as flux. The holding time of salt fluxing process was 60, 120, 180 and 240 minutes and 5, 10 and 15% variation of flux mass from total scrap mass. Ingot from the recycling process tested to know the percentages of Mg reduction and acquisition of Al using XRF analysis and metallographic. The result showed that the highest percentage of Mg reduction was 77.83% on the addition of the sample of 15% flux for holding time 120 minutes, but Al reached 97.49% for holding time 240 minutes on same flux. On the addition of 10%, flux for holding time 180 minutes was reached 98.82% for yield, 71.84% for recovery, and recycling efficiency 70.53%.

1. Introduction

Aluminium needs in Indonesia in 2016 is in the range of 899.000 tons. Meanwhile, from its supply aspect, the availability of supply in the country is still very limited. One of the single supplier factories of primary aluminium (ingot), with a production of 250.000 tons a year, is only able to supply the domestic needs of 103 thousand tons a year and most of the rest must be exported to Japan because it is tied to a cooperation agreement. Therefore, there is an aluminium supply-demand gap between country's needs and demand which outstrips supply, so that shortage of demand obtained through imports [1]. With the need which continues to increase and the company's production of primary aluminium is not able to meet those needs, a process to address the needs of the domestic aluminium is needed. A method that can be used is the process of recycling of beverage cans made of aluminium by smelting. Beverage cans that have become waste cannot be degraded by the soil and the danger posed by waste beverage cans is not only have an impact on the ground, but the human beings could be affected from beverage cans. Some of the consequences are cancer of the lungs and liver, caused by flakes of waste beverage cans [2, 3].

Beverage cans' scrap is the largest source of raw materials for recycling aluminium scrap. Beverage cans' scraps are easier to be collected than aluminium scrap from another packaging, as can be found both in the residential as well as trash in landfills. Smelting of aluminium scrap is intended to increase the levels of aluminium and one of the ways used is to use salt layer. The expected outcome of the aluminium scrap of refining process is removing impurity metal from aluminium. The process that can be done in smelting aluminium scrap is fluxing. Fluxing process on scrap smelting process will bind impurity metal so that it was separated from aluminium[4].

Fluxing process using salt flux on the aluminium scrap is one of the techniques to remove impurities that exist in aluminium. The Salt compound of the smelting process has several roles, a passive role is to protect the metal from oxidation and active role is to remove impurities from molten aluminium. Flux is often in the form of chlorides and fluoride salt with additives, most of the flux is based on a mixture of NaCl and KCl [5, 6]. The purpose of the research of aluminium beverage cans' scrap recycling cans is to increase Al's levels up to 99% and reduce Mg's levels up to 50%.

2. Experimental Design

2.1. Materials

Aluminium beverage cans' *scrap* that was used consisted of three kinds of beverages which are carbonated, isotonic, and refreshing beverage cans. The smelting process used induction furnaces with smelting process temperature is 750°C. In this research, the process of recycling was done to reduce the levels of Mg, used flux salt compound which was KCl-NaCl with the composition of the 50% KCl and 50% NaCl.

2.2. Refining Process

The refining process was done by adding KCl-NaCl flux into the smelting of beverage cans' scrap to bind the impurity metal. Flux in the smelting was added with variation mass of 5, 10 and 15% of the total scrap mass. Holding time was given to the fluxing process to see the effect of time towards the impurity metal's reduction and an increase of Al's acquisition. Holding time given during the fluxing process were 60, 120, 180 and 240 minutes. The result of scrap's smelting was moulded with a length of 2.5 cm, wide of 2.5 cm, and high of 1 cm

2.3. Characterization of Aluminium Beverage Cans' Scrap Recycling

The composition of scrap remelting could be seen by using X-Ray Fluorescence (XRF) testing with WDXRF Spectrometer model. XRD testing was used to determine Al's acquisition and Mg's reduction from beverage cans' scrap recycling's result. The Microstructure of remelting could be formed by using a proper etching solution and the results could be seen by using an Optical Microscope.

3. Result and Discussion

3.1. Influence of Fluxing Process' Holding Time towards Reduction Percentage of Magnesium's Levels

Figure 1 shows that the addition of flux mass of 5, 10 and 15%, the longer the holding time was used, the more increasing the reduction percentage of the Mg's levels. However, there is an anomaly in the addition of 15% flux with holding time of 180 minutes. The chemical reaction that occurs will affect a wide range of value on the outcome of the process of melting the aluminium beverage cans' scrap because the reaction between chemical is capable of effecting on levels of impurity metal's decreasing and capable of reducing levels of valuable metal.

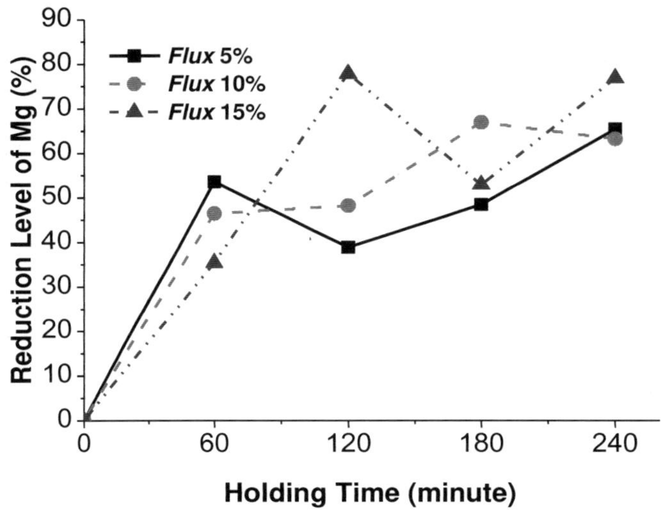

Figure 1. The Influence of Fluxing' Holding Time towards Reduction Percentage of Magnesium's Levels.

The increase of Mg's levels reduction from the holding time which was done, that increase occurred because Mg metal that reacted with oxygen and then forming oxide MgO compound and bind to Al_2O_3 which is shown in Eq. 3.Oxide compound produced from Eq. 4 reacted with the flux and H_2O that came from the air around smelting area so as to separate Mg metal from Al metal, to form $MgCl_2$ compound in form of *dross* which floats on the surface of molten metal, this reaction is shown in Eq. 5 and Eq. 6, according to Estéfano Aparecido Vieira, etc (2012) who stated that removal of element Mg in the aluminium alloy could be made by reacting the chloride compound or oxidation on the surface of the molten metal [7].

$$2Al + 1{,}5O_2 \rightarrow Al_2O_3 \qquad \Delta G_{1023} = -1.353{,}325\ KJ \tag{1}$$

$$Mg + 0{,}5O_2 \rightarrow MgO \qquad \Delta G_{1023} = -490{,}821\ KJ \tag{2}$$

$$MgO + Al_2O_3 \rightarrow MgAl_2O_4 \qquad \Delta G_{1023} = 328{,}082\ KJ \tag{3}$$

$$2Al + Mg + O_2 \rightarrow MgAl_2O_4 \qquad \Delta G_{1023} = -1516{,}064\ KJ \tag{4}$$

$$MgAl_2O_4 + 2KCl + H_2O \rightarrow MgCl_2 + Al_2O_3 + 2KOH \qquad \Delta G_{1023} = -15{,}969\ KJ \tag{5}$$

$$MgAl_2O_4 + 2NaCl + H_2O \rightarrow MgCl_2 + Al_2O_3 + 2NaOH \qquad \Delta G_{1023} = -69{,}782\ KJ \tag{6}$$

3.2. The Influence of Flux Mass and Fluxing's Holding Time towards Characteristic of the Al's Acquisition

Figure 2 shows that the longer the holding time of fluxing,more increasing Al's acquisition. This situation is caused due to the longer time spent in the process of fluxing so that the binding process of flux towards the impurity metal will increase [6, 8, 9]. However, there is an anomaly that occurs in

Figure 2 the anomaly that formed caused by chemical reaction. The chemical reaction that formed shown in Eq. 1 and Eq. 4. The reaction between the oxide and Al metal can cause a bond to form Al oxide, so that Al will be on the surface of the molten metal and become *dross*, resulting in decreased levels of valuable metal. Another circumstance that caused anomaly is a different chemical composition between the lid and body of the cans. The standard used for the lid of aluminium beverage cans is AA5182 with Mg composition ranges from 4-5%. For cans' body, its standard is AA3004 with Mg composition ranges from 0.8 to 1.3% Mg and Mn levels of 1-1.5% [2, 10-16].

The influence of adding flux mass also affects the results of Al's acquisition from research undertaken. Figure 3 shows the influence of flux mass on Al's acquisition. Gokhan Ozer, etc (2013) stated that the influence of flux mass was at its maximum in flux addition of 5%, less or more than 5% addition of flux produced yields of Al is smaller than the addition of 5% [7]. From this research, it can be seen the influence of flux mass to Al's acquisition in Figure. 3, for the result of adding flux of 10-15% it increased Al's acquisition, and it can be associated with the function of fluxing's holding time which affected flux's maximum ability to bind impurity metal. This situation came from the amount of flux which was higher so the use of much more flux that can be maximized. It is different from using 60 minutes holding time where the more addition of flux, more decreasing Al's acquisition. That situation was caused by flux which was not in its maximum ability to bind impurity metal because given time is less.

Characteristic of Al's acquisition of recycled beverage cans can be seen in the form of the microstructure shown in Figure 4. According to Figure. 4 which is about the shape of the microstructure of remelting beverage cans' scrap, there is a phase difference between the sample of after and before treatment. Based on Figure 4 (a), (b) and (d), it can be seen that the dominant phase formed is the Mg_2Si phase. Mg_2Si dominant phase held in Al-Mg alloys with Mg composition of 2-5%, the form of Mg_2Si phase is like Chinese writing or fishbone shape [12,17,18]. The results of XRF testing which is shown in Table 1 describes that the composition of Mg on Figure 4 (a) and (d) is dominant for the formation of Mg_2Si phase and it is proved by the result of metallographic which shows the microstructure of Figure 4 (a) and (d) that phase that is formed is a Mg_2Si dominant phase.

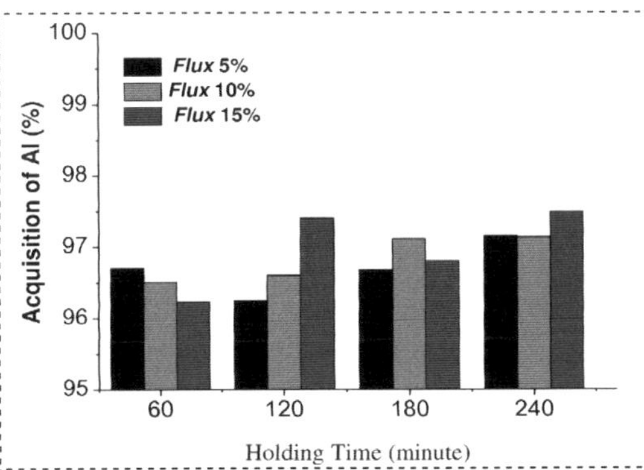

Figure 2. The Influence of Fluxing's Holding Time towards Al's Acquisition

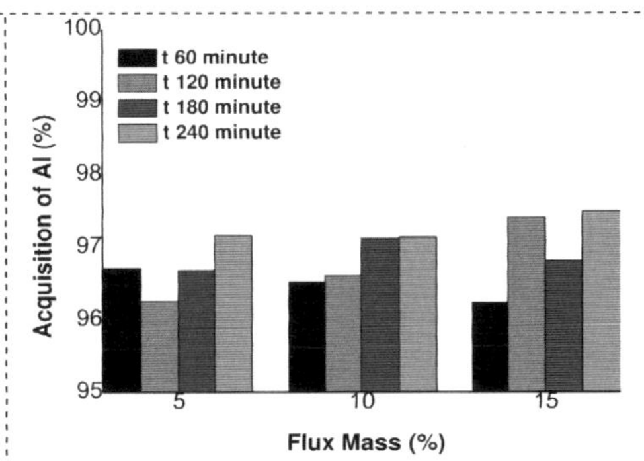

Figure 3. The Influence of Mass Flux on Al's Acquisition

Figure 4 samples (b) the dominant phase which is formed is intermetallic phase Al_6 (Fe, Mn) and a little phase of Mg_2Si. Al_6 phase (Fe, Mn) formed because Mn content is higher than Mg content. This situation is caused by the treatment process conducted with the aim of eliminating the Mg content. Increased levels of Mn produce eutectic fractions intermetallic. That situation happened during freezing occurs very slowly and produced intermetallic Al_6 phase (Fe, Mn) as the main phase [18].

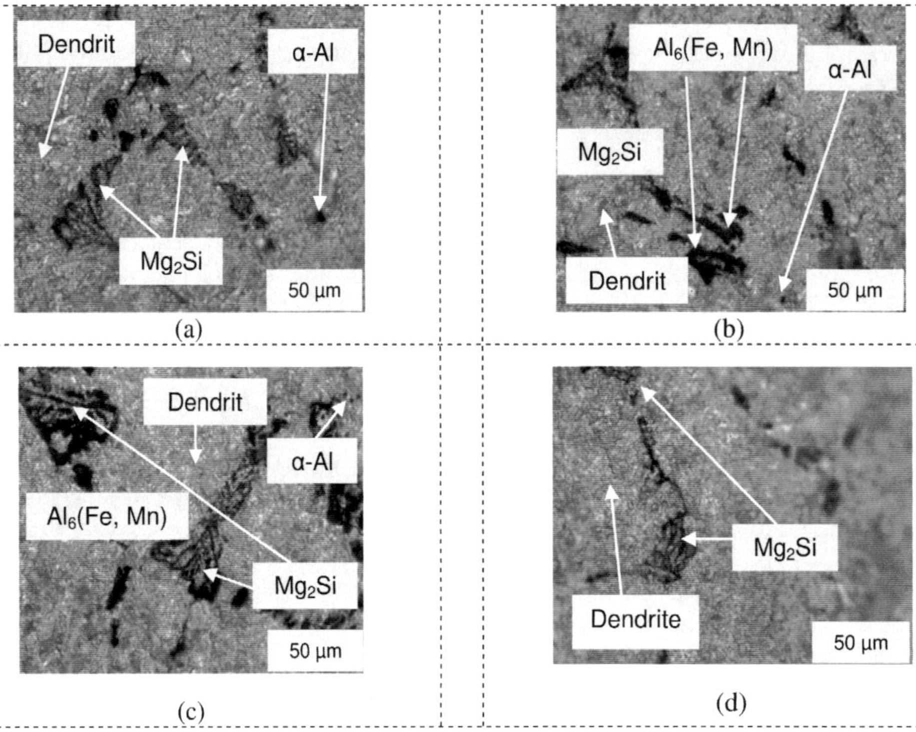

Figure 4. Microstructure (a) F15H1 (Minimum), (b) F15H2 (Maximum Reduction of Mg), (c) F15H4 (Maximum Al's acquisition) and (d) Non-Treatment. 500x magnification

Table 1. Percent Change in the Elements of Recycled Aluminium Beverage Cans' Scrap

No	Sample	Mg (%)	Al (%)
1	*Non-Treatment* 1	2,71	95,74
2	*Non-Treatment* 2	3,49	95,11
3	*Non-Treatment* 3	3,45	95,18
4	$F_5^a - H_1^b$	1,49	96,71
5	$F_5 - H_2$	1,97	96,25
6	$F_5 - H_3$	1,66	96,67
7	$F_5 - H_4$	1,11	97,15
8	$F_{10} - H_1$	1,72	96,51
9	$F_{10} - H_2$	1,67	96,61
10	$F_{10} - H_3$	1,06	97,11
11	$F_{10} - H_4$	1,18	97,13
12	$F_{15} - H_1$	2,08	96,23
13	$F_{15} - H_2$	0,71	97,41
14	$F_{15} - H_3$	1,52	96,81
15	$F_{15} - H_4$	0,74	97,49

[a] F = Addition of flux mass (%)
[b] H= The given holding time of fluxing (hour)

In Figure 4 (c), it also has an Al6 phase (Fe, Mn), but the phase which is more dominant is a Mg_2Si phase, this situation was caused due to the build up of Mg levels in some places led to the formation of Mg_2Si phase. From the overall pictures of microstructure on each sample, it can be seen that there are so many dendrites formed. Those dendrites appeared as a result of the Coran process. The ingot is research's result formed by casting therefore it is right that dendrites are formed from each of casting's result [17].

4. Conclusion

Based on the research that has been conducted, it was found that the addition of flux mass and the use of fluxing's holding time can improve Al's acquisition and reduce the levels of Mg on aluminium beverage cans' scrap's alloy, because the longer the holding time is given, the higher Al's acquisition and the bigger the reduction of Mg's levels as well as with the addition of flux mass which is used. The highest percentage of Mg reduction was 77.83% on the addition of the sample of 15% flux for holding time 120 minutes, but Al reached 97.49% for holding time 240 minutes on same flux. On the addition of 10%, flux for holding time 180 minutes was reached 98.82% for yield, 71.84% for recovery, and recycling efficiency 70.53%. Further, can be investigated about the effect of different types of fluxes on Al recovery and reduction of impurities also the addition of the stirring process during initial melting of the sample or during the fluxing process, to determine the effect of stirring on Al acquisition and Mg reduction

Acknowledgments

Authors are wishing to acknowledge to PT. Krakatau Steel by support in the conducting of the study.

References

[1] Qomarsono and Hilman 2014 *Prospek PT Inalum Pasca Pengambilalihan oleh Pemerintah*
[2] Material Safety Data Sheet 2011 2. *Used Beverage Container Scrap*, Version 1. Tri-Arrow Aluminium Inc
[3] Boin U M J and Bertram M 2005 *Melting Standardized Aluminium Scrap: A Mass Balance Model for Europe* European Aluminium Association and Organisation of European Aluminium Refiners and Remelters. Brussels, Belgium
[4] Gilstad G 2013 *Life Cycle Assessment of Secondary Aluminium Refining* Norwegian University of Science and Technology
[5] Tsunekawa M, Yoo K, Hiroyoshi N, and Ito M 2005 *Resources Processing* **52** 32-38
[6] Utigard TA 2001 *Properties of Fluxes Used in Molten Aluminium Processing* University of Toronto, Ontario, Canada
[7] Estéfano A V 2012 *Use of Chlorine to Remove Magnesium from Molten Aluminium* Department of Metallurgical and Materials Engineering Brazil
[8] Lifeng Z 2011 *Removal of Impurity Element from Molten Aluminium: A Review* Department of Science and Technology. Rolla, Missouri, USA
[9] Michael B 2015 *Fused Refining Fluxes and Their Reaction Kinetics. Aluminium Smelting Industry* MQP ltd
[10] Kiffaya A A and Layla M H B 2008 *Recycling of Aluminium Beverage Cans Materials* Engineering Dept Al-Mustansiriya University Baghdad, Iraq
[11] George E T and MacKenzie D S 2003 *Handbook of Aluminium Vol.1, Physical Metallurgy and Processes* Marcel Dekker, Inc. New York – Basel
[12] ASM Metals Handbook 2004 Volume 2 *Properties and Selection: Nonferrous Alloys and Special-Purpose Materials*

IOP Conf. Series: Materials Science and Engineering **478** (2019) 012006 doi:10.1088/1757-899X/478/1/012006

[13] Ron C 1994 *TALAT (Training in Aluminium Application Technology) Lecture 1501, Aluminium: Physical Properties, Characteristics and Alloys*. Alcan, Banbury
[14] Eric W 1994 *TALAT (Training in Aluminium Application Technology) Lecture 3710, Case Study on Can Making* Alcan Deutschland GmbH, Göttingen
[15] Gokhan O 2013 *The Effects of Proces Parameters on the Recycling Efficiency of Used Aluminium Beverage Cans (UCBs)* Istanbul, Turki
[16] Mahmoud A R 2003 *Preparation of Aluminium-magnesium Alloys and Some Valuable Salts from Used Beverage Cans* Cairo, Mesir
[17] ASM Metals Handbook 2004 Volume 9 *Metallography and Microstructures*
[18] Yulin L 2016 *Effect of Mn and Fe on the Formation of Fe⁻ and Mn⁻ Rich Intermetallics in Al-5Mg-Mn Alloys Solidified Under Near-Rapid Cooling* Shenyang Aerospace University, Shenyang, China

IOP Conf. Series: Materials Science and Engineering **478** (2019) 012007 doi:10.1088/1757-899X/478/1/012007

Effect of flux addition and reductant type in smelting process of Indonesian limonite ore in electric arc furnace

R Andika[1], W Astuti[2], Syafriadi[1], and F Nurjaman[2]

[1]Department of Physics, Lampung University
[2]Research Unit for Mineral Technology, Indonesian Institute of Sciences (LIPI)
Jl. Ir. Sutami Km. 15, Tanjung Bintang, Lampung Selatan, Lampung Province

Email: widi004@lipi.go.id

Abstract. Nickel in the form of nickel laterite ore is one of the most prominent types of mining products in Southeast Sulawesi, Indonesia. One of the utilization of laterite nickel ore is as raw material for ferronickel and NPI production. In this research, the NPI was produced from laterite nickel ore originating from Sulawesi Island in electric arc furnace (EAF). This study aims to determine the effect of the amount of flux (limestone) and type of reducing agent (coal, coke, and shell charcoal) on the content of Ni and Fe in the NPI product and metal yield. Laterite nickel ore used in this study is a type of limonite from Pomaala (Sulawesi Island). The reducing agents used are coke, coal, and coconut shell charcoal. Limestone was used as a flux to adjust the basicity of material during the smelting process of lateritic nickel ore. Reduction and melting process of limonite nickel ore to produce NPI were conducted in small EAF with capacity of 30 kg/heat. The dried pellets are reduced and melted in the EAF with added reducing agents and limestone. Based on the results of the study it can be concluded that the results of the flux addition effect obtained that the optimum Ni and Fe values were obtained on basicity of 1. The results of the reductant type effect showed that coke was the greatest reductant in this study due to the highest Ni content produced when coke used as reductant. In addition, the XRD results showed that the FeNi phase has been formed.

1. Introduction

Indonesia has a nickel resource of around 2,633 million tons with nickel reserves of 577 million tons [1]. Nickel is one of the most prominent types of mining products in Sulawesi Island, Indonesia. In addition, the largest export commodity in Indonesia is nickel with an export value of around 36.57 million dollars or 53.30 percent of the total export value [2].

Nickel is one of the most important metals and has many applications in industry. Many alloy materials are made based on nickel because they have structural strength in the creep, fatigue, and surface stability (corrosion oxidation) process [2]. As much as 62% of nickel is used as raw material for stainless steel production. Around 18% nickel is also used as raw material for the manufacture of alloy steels and non-ferrous alloys [3].

Nickel ores are classified into two namely laterite nickel ores and nickel sulfide ores. Laterite nickel ores are produced from weathering of ultramafic rocks with high iron and magnesium content and are usually found in tropical and sub-tropical regions [4]. Nickel sulfide ores are formed by volcanic or hydrothermal processes and contain lots of copper and cobalt and contain few precious metals such as gold, platinum, palladium, and rhodium [5]. Nickel sulfide ores are found in Russia and Canada [6]. During this time, as much as 60% of the need for nickel is commercially supplied from sulfide ores. In fact, around 70% of the world's nickel reserves are trapped in the form of laterite. Laterite ores have a low nickel content, so they are rarely used as the main source of nickel because special treatment is needed to increase nickel contents in these ores [3]. Basically, laterite rocks can be divided into two layers, namely limonite and saprolite layers. In each layer, there are various types of minerals. The main minerals that compose the limonite layer are goethite [FeO(OH)] and hematite

(Fe_2O_3), whereas the saprolite layer is mainly composed of serpentine minerals ($(Ni,Mg)SiO_3.nH_2O)$] [7].

One of the utilization of laterite nickel ore is as raw material for ferronickel production. Nickel content in ferronickel ranges from 20-40%. Ferronickel is commonly used to make stainless steel. Besides ferronickel, laterite nickel ore is also used in making nickel pig iron (NPI) that usually contains 1.5-25% Ni [8]. Research on the production of NPI from laterite nickel ore has been carried out by several researchers [8] that used coal as reductant. Reductions were carried out at temperatures of 900°C, 950°C, 1000°C, and 1100°C for 1 hour. The results showed that the NPI obtained had Ni content of around 1% and Fe content of around 11%.

In the current study, the NPI was produced from laterite nickel ore originating from Sulawesi Island in electric arc furnace (EAF). This study aims to determine the effect of the amount of flux (limestone) and type of reducing agent (coal, coke, and shell charcoal) on the content of Ni and Fe in the NPI product and metal yield.

2. Materials and method

2.1. Materials

Laterite nickel ore used in this study is a type of limonite from Pomalaa (Sulawesi Island). The reducing agents used are coke, coal, and coconut shell charcoal. The lateritic nickel ores were refined and sieved until the size of <74 μm was obtained. These samples, then, were analyzed using X-Ray Fluorescence (XRF, Epsilon 3XLE, PANalaytical, Netherland) and X-Ray Diffraction (XRD, X'Pert 3 Powder, PANalytical, Netherland). XRF analysis aims to determine the chemical elements contained in the sample and XRD analysis was required to determine the dominant phase in laterite nickel ore. Reductants were analyzed using a proximate method to determine the carbon, ash, and volatile matter content of each reducing agent. The ore and reducing agents were ground to 100% passing screen aperture of 74 μm to increase the contact between them before they were mixed and agglomerated into pellets with diameter 10-20 mm. Limestone was used as a flux to adjust the basicity of material during the smelting process of lateritic nickel ore.

2.2. Experimental method

Reduction and melting process of limonite nickel ore to produce NPI were conducted in small EAF with capacity of 30 kg/heat. The dried pellets are reduced and melted in the EAF with added reducing agents and limestone as flux. The addition of limestone is used to separate metal and smelting slag. The amount of limestone is calculated based on the basicity used. The effect of basicity on the recovery of nickel and iron was conducted by adjusting the composition of limestone in the raw material with the basicity used in this experiment is 0.8; 1; and 1.4 that are equal with the limestone amount of 0, 0.9, and 3.12 kg. The effect of reductant types was also studied by varying of coke, coal, and coconut shell charcoal as reducing agents. The composition of raw material in this smelting process is listed in Table 1. The furnace used is an electric arc furnace (EAF) with proceesing temperature at around 1400°C for 30-40 minutes. The pouring temperature of slag and hot metal was 1246-1578°C. Metal that produced from the experiments were analyzed using optical emission spectrometry (OES, Shimadzu, Japan) to get their chemical composition. Recovery of metal particularly iron (Fe) and nickel (Ni) as well as the yield of metal can be calculated from the iron and nickel content in the NPI produced.

Table 1. The composition of raw material

Trial No.	Laterite ore	Reducing agents	Reductant amount (kg)	Limestone (kg)	Basicity
1	30	Coke	7	-	0.85
2	30	Coke	7	0.90	1

3	30	Coke	7	3.12	1.4
4	30	Coal	7	-	0.85
5	30	Coconut Shell Charcoal	7	-	0.85

3. Result and Discussion

3.1. Characterization of limonite ore and reducing agents

Laterite nickel ore was analyzed using XRF to find out the elements contained therein, particularly the Ni and Fe contents that was shown in Table 2. The results of the analysis showed that laterite nickel ore contained 1.26% Ni which was a type of low-grade laterite nickel [9], while Fe content was 43.1%. Laterite nickel ores are known to have quite high Fe elements. To investigate the dominant phase in laterite nickel ore, analysis using XRD was carried out and the results can be seen in Figure 1. Figure 1 shows that the dominant compounds found in laterite nickel ore are goethite (FeOOH). This is consistent with previous research [10] which shows that mineral phase in laterite nickel ore are dominated by goethite (FeOOH). In addition, there is another main mineral contained, namely quartz (SiO2). Nickel is generally associated with goethite so that the XRD patterns do not show the presence of nickel compounds as an independent compound [11].

Table 2. Chemical composition of limonite nickel ore from Pomalaa (Sulawesi Island, Indonesia)

Elements	Si	Ni	Fe	Mg	Ca	Al
wt. %	3.23	1.26	43.10	6.29	0.34	2.83

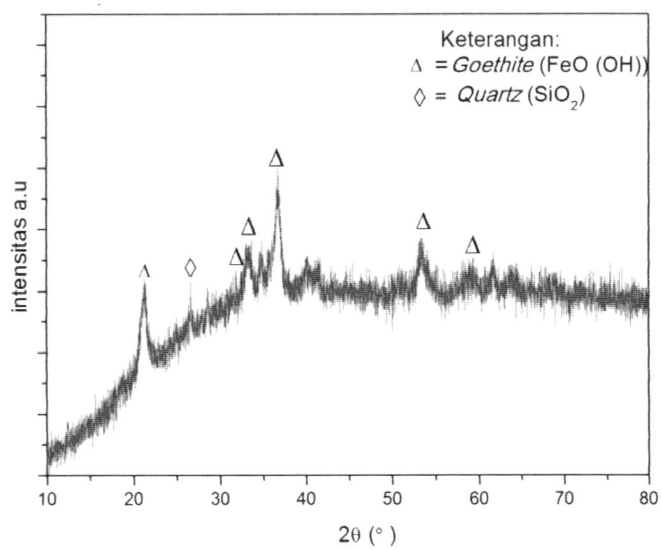

Figure 1. X-ray diffraction pattern of limonite nickel ore from Pomalaa (Sulawesi Island, Indonesia)

The reducing agents used in this study were coke, coal, and coconut-shell charcoal. Each reducing agent has a different carbon content and other compositions. To determine the carbon content of each reducing agent, a proximate analysis was conducted that can be seen in Table 3. It is known that coke has the highest carbon value compared to other types of reducing agents. The coke should be a good reductant because of its high fixed carbon and volatile matter, low ash, and relatively low harmful element (S, P) content. The coke is common reducing agent in the reduction process of laterite ore. However, another important thing is to discuss the feasibility of the other reductant to be used as a substitute in the reduction process.

Table 3. Proximate analysis of reducing agents

Parameter	Moisture (wt. %)	Volatile matter (wt. %)	Ash (wt. %)	Fixed carbon (wt. %)
Coke	0.20	9.65	27.18	62.97
Coal	0.25	48.5	14.00	37.25
Coconut Shell Charcoal	0.15	18.62	26.21	55.02

3.2. The effect of flux additions

The flux is usually used to separate metal and smelting slag. Limestone (CaO) is usually applied as a flux. As for nickel laterite, which consists largely of the magnesium silicate, the influence of CaO on generation of liquid phase was calculated thermodynamically. CaO was added to modify the quaternary basicity-{$CaO+MgO)/(Al_2O_3+SiO_2)$}. The influence of the basicity on the NPI products was studied in this study using basicities of 0.85, 1, and 1.4.

Basicity value is arranged by adding limestone to the material where the amount of limestone added is calculated according to the Equation 1 and based on the elemental composition of laterite nickel ore.

$$B = \frac{CaO + MgO}{SiO_2 + Al_2O_3} \quad (1)$$

In this experiment, limestone was applied as a flux. The impurities contained in laterite nickel ore will be discharged into the slag. Every experiment in this section used as much as 7 kg of coke as reducing agent. With the basicity of 0.85, no limestone was added where the basicity value was only calculated based on the content of SiO_2, MgO, CaO, and Al_2O_3 in the laterite nickel ore. Whereas on basicity of 1 and 1.4, limestone added was 0.9 and 3.12 kg. The results are shown in Figure 2, in which CaO could definitely enhance the formation of low melting temperature components and generate more liquid phase, particularly before 1300°C. The increase of basicity can significantly reduce the viscosity of molten slag above 1400°C, which is beneficial for the separation between slag and metal.

The weight of the NPI metal produced during the smelting process at each basicity value is 8.6 kg, 6.3 kg and 7.8 kg respectively. On the basicity of 0.85, the highest weight of the metal was obtained. The Ni contents of the NPI products that have been analyzed using OES results were 3.64%, 4%, and 3.86%, respectively for basicity of 0.85, 1, and 1.4. Iron (Fe) is also the main element in the NPI sample. The values of Fe in metals were 82.81%, 89.00%, and 87.3%, respectively for basicity of 0.85, 1, and 1.4. The highest Ni and Fe levels in metal NPI produced were obtained from experiments using basicity 1 or flux addition of 0.9 kg. From the results of this experiment, it was seen that with the flux addition of 0.9 kg, although the weight of the metal obtained was the least, but the NPI metal produced contained the greatest content of Fe and Ni. The relationship between flux addition and recovery values can be seen in Table 4.

Table 4. Effect of flux addition on recovery of Ni and Fe

No	Flux addition (kg)	Metal produced (kg)	Ni in metal (%)	*Recovery of* Ni (%)	*Metal yield* (%)	Fe in raw (%)	Fe in metal (%)	*Recovery* of Fe (%)
1	0	8.6	3.64	82.81	28.66	43.10	82.81	55.06

2	0.9	6.3	4.00	66.66	21.00	43.10	89.00	43.35
3	3.12	7.8	3.86	79.65	26.00	43.10	87.30	52.65

The addition of flux can promote the reduction of nickel and iron and optimize the dynamic condition of laterite nickel ore. The dynamic condition will affect the aggregation and the expansion of nickel and iron as well as the separation of slag and metal. It creates the increase of nickel and iron grades. However, when the flux addition was over than the optimum, there was so much liquid phase generated, resulting in the decrease of the contact area between ore and reductant. Nickel and iron particles will be difficult to aggregate and grow up that lead the entrance of more fine particles into the slag and increase the difficulty of separation. Finally, grades and recoveries of the nickel and iron declined as a result.

3.3. The effect of reducing agent types

Another important thing is to discuss the feasibility of the reductant to be used as a substitute in the reduction roasting process. Experiments to study the effect of reducing agents, namely coke, coal, and coconut shell charcoal are presented in Figure 2. Each smelting process uses each reducing agent with a weight of 7 kg and without flux. Metal NPI produced from the smelting process are 8.6 kg, 8 kg and 9 kg respectively when coke, coal and coconut sheel charcoal was used with the Ni compositions were 3.64%, 3.57%, and 2.37%. The NPI using coconut-shell charcoal as a reductant has the lowest Ni content, while the highest Ni content is obtained in coke reducing agents. Iron contained in the metal were 82.81%, 89.40%, and 89.99%, and with the reductants of coal, coke, and coconut-shell charcoal respectively.

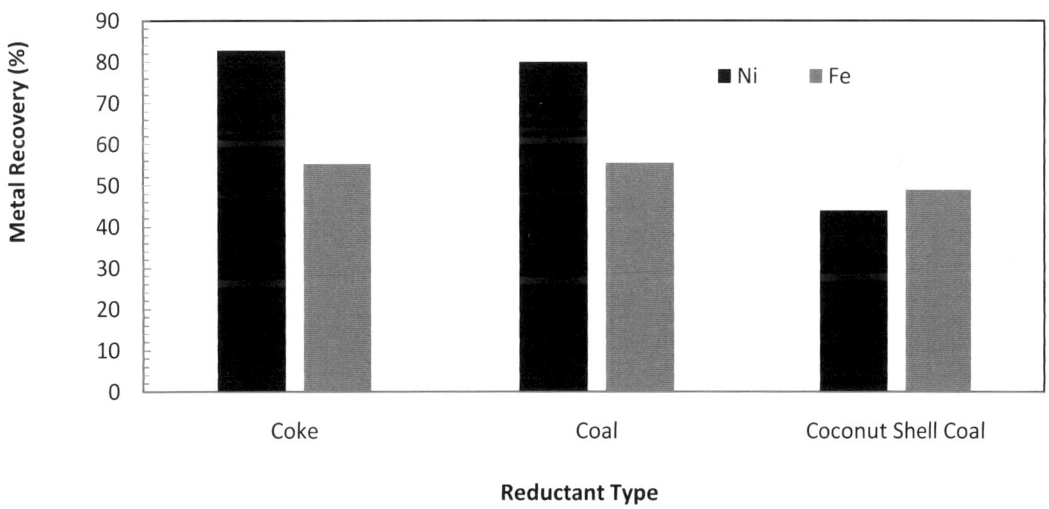

Figure 2. Effect of reductant type on the recovery of Ni and Fe

Figure 2 gives the comparation between using coconut shell charcoal and anthracite coal. Between two reductants, there are no significant differences in terms of magnetite, hematite, olivine, and quartz formation. Therefore, coconut crust charcoal is a potential substance that can be used as a substitute in the process. Mentioning the fact that the final condition of both reductants are mineralogically similat, it can be seen that the effect of the ash content the reductants (see table 3) is insignificant due to the lower concentration of ash in both reductants (see table 2).

From the results, the type of reducing agent used in Ni recovery calculation obtained 82.81%, 75.55% and 56.42% with the highest recovery value in coke. On the contrary, from the calculation of Fe recovery, 55.06%, 55.3%, and 62.63% were obtained with the highest Fe recovery in coconut shell charcoal. The results of the calculation of metal yield were obtained by 28.66%, 26.66%, and 30%, with the lowest metal yield obtained in coal, while the highest metal yield was in charcoal shells.

3.4. Characerization of NPI product

Phase identification from the results of the laterite nickel ore smelting process was carried out using XRD. NPI metal testing using XRD will provide qualitative results for each compound contained in the NPI metal. The test was carried out with a 2θ angle position from 20° to 80° and using a wavelength of 1.54056 Å. The test results that can be seen in Figure 3shows that the dominant phase in the metal product of smelting process is Fe and FeNi. It can be concluded that using electric arc furnace, Indonesian nickel laterite ore could be treated by pyrometallurgical method to produce high grade NPI that contained iron and iron nickel.

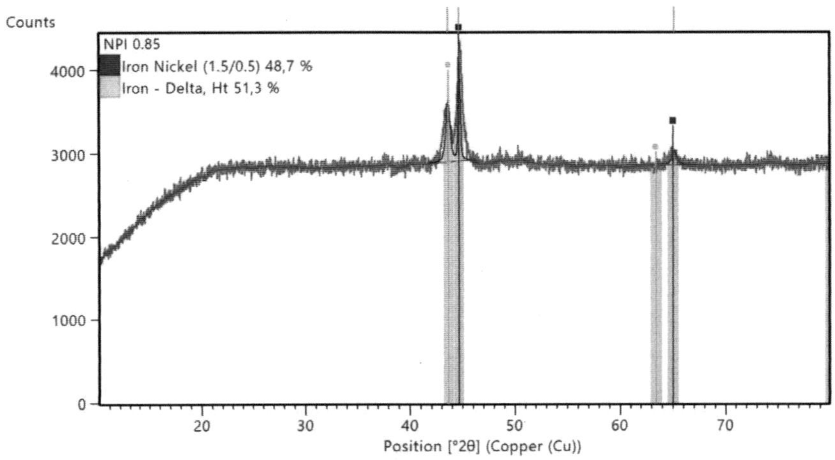

Figure 3. XRD pattern of NPI metal

4. Conclusion

Based on the results of the study it can be concluded that the optimum condition that produced NPI with the greatest Ni (4%) and Fe (89%) was the basicity of 1 and coke as reductant. Using electric arc furnace, Indonesian nickel laterite ore could be treated by pyrometallurgical method to produce high grade NPI that contained iron and iron nickel.

Acknowledgement

We would like thanks to Ministry of Research, Technology and Higher Education of The Republic of Indonesia through the scheme of Incentive Research Program for The National Inovation System with the grant number 04/INS-2/PPK/E/E4/2017 and also acknowledge to Indonesian Institute of Sciences.

References

[1] Virman, Hartiningsi E, Patiung R, and Massinai A M 2014 *Prosiding Seminar Nasional Sains Dan Pendidikan Sains IX* **5(1)** 1-6
[2] Sujiono E H, Diantoro M 2014 *Jurnal Pendidikan Fisika Indonesia* **10(2)** 163–167
[3] Kyle J 2010 Nickel Laterite Processing Technologies *(Thesis)* Murdoch University 1-37
[4] Astuti W, Zulhan Z, Shofi A, Isnugroho K, Nurjaman F, Prasetyo E 2012 *Prosiding InSINas* 66-71
[5] Pournaderi S 2014 Optimization of Ferronickel Production FrSivrihisar Nickel Laterite Ore *(Thesis)* Middle East Technical University 1-195

[6] Rochani S and Saleh N 2013 Teknologi Pengolahan dan Pemurnian Nikel *Pusat penelitian dan Pengembangan Teknologi* 1-9
[7] Yildirim H H, Morcali A, and Turan 2013 *Nickel is mainly used as f erronickel in stainless steel production industry Its usage as low grade ferronickel or nickel pig iron (NPI) in stainless steel production is a relatively new trend to reduce the cost* Istanbul Technical University 237-244
[8] Budi A and Prasetiyo P 2011 *Metalurgi: Majalah Ilmu dan Teknologi* **26** 123–130
[9] Lu J, Liu S, Shangguan J, Du W, Pan F, and Yang S 2013 *Minerals Engineering* **49** 154–164
[10] Anggarda D 2017 Bahan Umpan Mini Blast Furnace *(Skripsi)* Institut Teknologi Sepuluh November 1-57
[11] Astuti W, Hirajima T, Sasaki K, and Okibe N 2016 *Hydrometallurgy* **161** 138-151

Performance of Kulon Progo low grade manganese ore leaching using acetic acid and its selectivity

R D Anggraeni[1], A Prasetya[1], H T B M Petrus[1], R B Cahyono[1], S Sumardi[2], and W Astuti[2]

[1]Department of Chemical Engineering (Sustainable Mineral Processing Research Group), Universitas Gadjah Mada 55281 Yogyakarta, Indonesia
[2]Research Unit for Mineral Technology, Indonesian Institute of Sciences, Jl. Ir. Sutami Km. 15, Tanjung Bintang, Lampung Selatan, Lampung, Indonesia

Email:aguspras@ugm.ac.id

Abstract. The need of manganese to support human life has led to the exploitation of low-grade manganese ore. However, the appropriate technique has not been established and still under development. Having total amount of 120 thousand metric tons, Indonesia has huge potential of low grade manganese ore deposited in several areas such as Kulon Progo, Yogyakarta. In this study, leaching of low grade manganese ore was conducted using acetic acid which isenvironmentally friendly due to its ease to decompose naturally. Ore characterization was performed by X-Ray Diffraction (XRD), X-Ray fluorescence (XRF), and Scanning Electron Microscopy-Energy Dispersive X-Ray Spectroscopy (SEM-EDS) showing that 7.16 % is manganese while 32 % of the ore is dominated with calcite. Effects of leaching temperature and particle size on Kulon Progo low grade manganese ore were investigated. Particle sizes under study were less than 200 mesh and between 70 mesh and 100 mesh. The pH and temperature of experiment were varied on 0.7, 1.7, and 3; and 30 ^{o}C, 60 ^{o}C and 80 ^{o}C, subsequently. It was obtained that the ore particle size has no significant effect on Mn recovery. The manganese recovery of 76.08 % was obtained, containing lower recovery of calcium and iron which were 10%. The findings showed acetic acid is more selective towards manganese than to calcium and iron.

1. Introduction

Among other metals, manganese has important role for human life other than iron, copper, and nickel. According to Hegelstain (2009) almost 90% of world's manganese is used for iron and steel industry [1]. Manganese increase the ductility in steel, so that it is not easily broken. In addition to metallurgical purposes, manganese metal is also used for the production of chemical compounds such as $KMnO_4$ used for disinfectants, $MnSO_4$ for animal feed, manufacturing materials, addictive substances in gasoline, and manganese dioxide used as dry components that work for depolarizators [2]. Those varieties of applications lead to the rapid increase of manganese metals consumption every year. Although there are big amount of manganese metal, the deposit of high grade manganese become more limited. Therefore, it is necessary to explore and develop some techniques for processing deposits with low grade manganese. Manganese in Indonesia especially in Kulon Progo, can be a potential low grade manganese ore resources for full filing the industries needs. The acidity of a lot of inorganic acid used as leaching reagent was generally high, so it is expected that their selectivity is not much, the probability of corrosion effect to occur is high, and the controlling pH is more difficult [3]. Extraction of heavy metals by inorganic acids and complexing agent has severe drawbacks. Otherwise, for inorganic acid, fast dissolution abilities may be an advantage. Organic acids have high selectivity although the dissolving abilities of them are weak. They could be an attractive extracting agent because the extraction can be performed under mildly acidic conditions (pH=3-5), and they are biologically degradable [4]. Additionally, in industrial processes, organic acids can cause a lesser corrosion effect. Even, the corrosion effect can be controlled by adding chemical substances called inhibitors such as benzoic acid and salicylic acid [5].

Wang et al. (2017) investigated the hydrometallurgical process and kinetics of leaching manganese from semi-oxidized manganese ore with sucrose. He found the optimum ratio of sulphuric acid with sucrose about 5 mol/L of sulphuric acid concentration with 0.035 mol/L of sucrose concentration on 60 minutes and 363.2 K of leaching temperature can reach leaching efficiencies about 91.4-96.9 % Mn [6]. However, this experiment still used an inorganic acid for the leachant.

As indicated above, in the previous studies, inorganic leachant was generally used in manganese leaching. But in this study, the extraction of manganese from Kulon Progo low grade manganese ore was investigated using organic acid under atmospheric pressure to reduce the risk of environmental damage caused by the leaching process. The effects of temperature, particle size, and acetic acid pH on leaching of Kulon Progo low grade manganese ore were determined.

2. Materials and Method

The low grade manganese ore used in present study was supplied from the source in Kulon Progo, Yogyakarta. First of all the ore were crushed and sieved with various sizes of -200 mesh and -70 until + 100 mesh. The elemental contents in ores were analyzed using x-ray fluorescence (XRF). Table 1 showed that the element that dominates the Kulon Progo manganese rock is calcium (Ca), whereas the manganese contained in the sample is 7.16%. In addition to XRF testing, an x-ray diffraction (XRD) analysis was also performed to identify the mineral crystal structure contained in the sample. Figure 1 shows that the main constituent mineral phase of manganese Kulon Progo is Calcite ($CaCO_3$). The XRD test results further strengthen XRF test results where iron is the component that dominates the sample.

Table 1. Composition of low grade manganese ore with XRF and XRD instrument

Element	*% Mass*	*Mineral*	*Mineral Name*	*% Mass of mineral*
Ca	*32*	$CaCO_3$	*Calcite*	*66.14*
Mn	*7.16*	MnO_2	*Pyrolusite*	*7.03*
Si	*0.72*	SiO_2	*Quartz*	*1.29*
Zn	*0.53*	*ZnS*	*Sphalerite*	*0.46*
Fe	*0.14*	*FeS*	*Sphalerite*	*0.13*
Al	*0.3*	Al_2O_3	*Alumina*	*1.67*
Ti	*0.05*	TiO_2	*Rutile*	*0.05*
Cr	*0.00114*	Cr_2O_3	*Chromite*	*0.0038*
O	*59*			

IOP Conf. Series: Materials Science and Engineering **478** (2019) 012008 doi:10.1088/1757-899X/478/1/012008

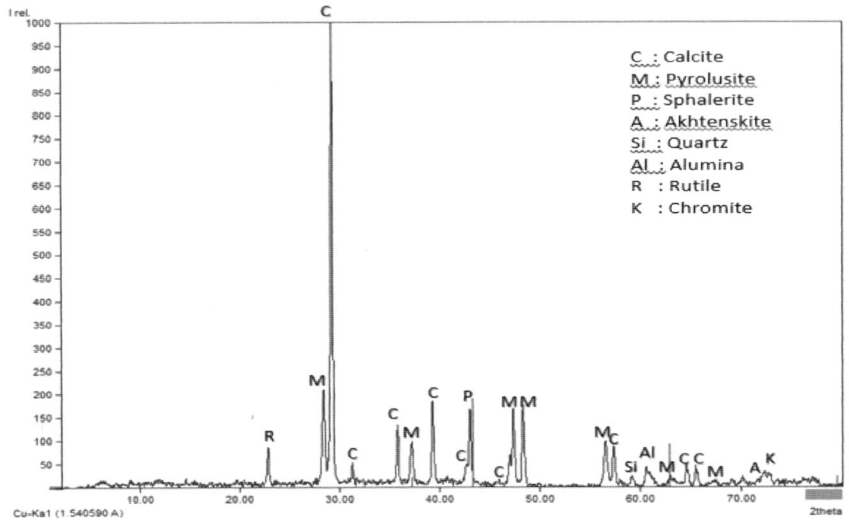

Figure 1. XRD pattern of raw manganese ore

The leaching process was carried out in 500 mL three-necked round-bottomed glass reactor at atmospheric pressure. A mechanical stirrer was used to stir the reactor contents, and a water heater batch was employed to maintain the reaction medium at a given temperature. To prevent loss of the reactants and products by evaporation, a cooler was attached to the reactor. The particle characteristics and the experimental conditions are summarized here. In the leaching experiments 300 ml of CH_3COOH solution was first put into the reactor, and after it reached the desired temperature, a given 60 gram of the ore sample was added to the solution while 400 rpm stirring started. Samples were taken at 5 mL with intervals of 5, 15, 30, 60, 120 and 240 min. The sample which taken was a suspension solution and was separated by aqueous phase and solid phase by syringe with millipore. At the end of the time reaction of 240 minutes the stirring was stopped, and all of the reactor contents were filtered off. The amount of Mn, Ca, and Fe recovery content of the filtrate was determined by AAS methods. The parameters and ranges used in the experiments are given in Table 2.

Table 2. Parameters and range employed in experiments

Parameters	Selected Parameter Values
Acetic Acid pH	0.7; 1.7; 3
Reaction Temperature (^oC)	30; 60; 90
Particle Size (mesh)	-70+100; -200

3. Result and Discussion

3.1. Characterization Kulon Progo Low Grade Manganese Ore after Leaching

Low grade manganese ore used in the present work was obtained from Kulon Progo. Comparison of both before leaching samples and after leaching samples were characterized by powder X-Ray Diffraction (XRD) to define the differences between the mineralogical composition before and after leaching. Figure 2 shows that there were deacreased intensity of pyrolusite peaks after leaching.It was a proof that manganese have recovered in fluid phase.

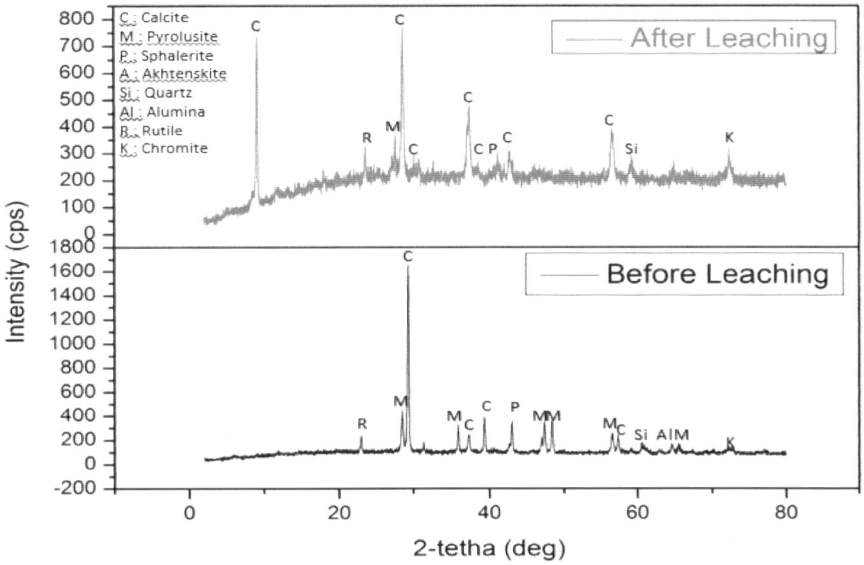

Figure 2. XRD pattern of before and after leaching of Kulon Progo manganese ore using acetic acid (temperature of 80 °C, pH of 0.7, particle size of -200 mesh, stirring speed at 400 rpm, and solid-to-liquid ratio at 20 % (w/v))

Both of before and after leaching samples were characterized by powder Scanning Electron Microscopy-Energy Dispersive Spectroscopy (SEM-EDS) mapping to define the difference of the element composition mapping before and after leaching. From Figure 3, it is observed that manganese element was decreased after leaching. It also could be seen that the large amount of manganese element were positioned on the surface of the ore.

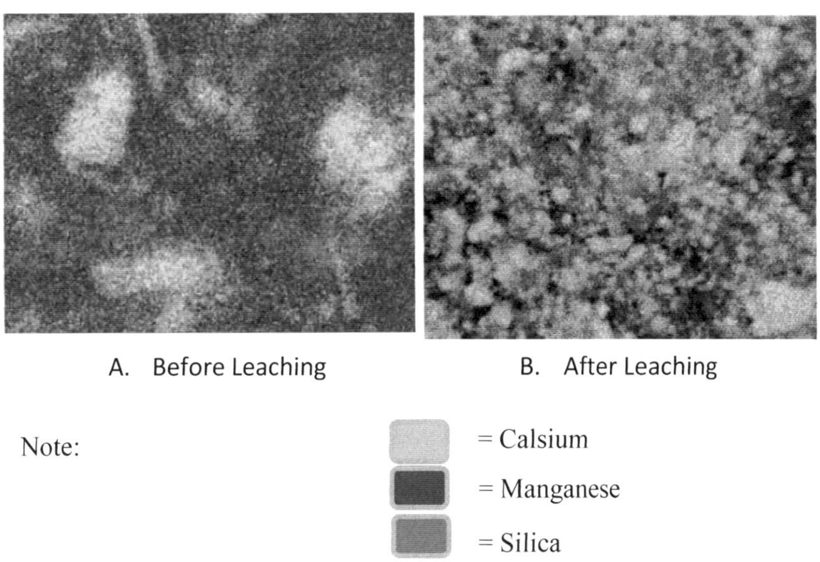

Figure 3. SEM-EDS analysis of Kulon Progo Low Grade Manganese Ore (temperature of 80 °C, pH of 0.7, particle size of -200 mesh, stirring speed at 400 rpm and solid-to-liquid ratio at 20 % (w/v))

IOP Conf. Series: Materials Science and Engineering **478** (2019) 012008 doi:10.1088/1757-899X/478/1/012008

Figure 4 showed that the recovery of manganese in acetic acid leaching was higher than other element such as calcium and iron. It was represented that acetic acid was more selective to manganese element. This condition operation experiment are temperature of 80 °C, pH of 0.7, particle size of -200 mesh, stirring speed at 400 rpm and solid-to-liquid ratio at 20 % (w/v). In Figure 4, it also can be seen the comparison between ore leaching with organic acid such as acetic acid and ore leaching with sulphuric acid. It was represented that acetic acid was more selective to recover manganese than sulfuric acid. The ore leaching with sulfuric acid acid could recover all of element in the ore. So that sulfuric acid could not selective with one elements especially manganese. The selective leaching agent was necessary to decrease load consumption of H^+ concentration which would attack to recover element because the leaching agent can be selective with one specific desired element. Decreasing load consumption of H^+ concentration made the process become more effective and efficient. It also was represented that manganese ore leaching using acetic acid could recover manganese of 76.08 %, higher than manganese ore leaching using sulfuric acid which recover of 72.07 %. It was because ligands in organic acid are stronger than ligand in inorganic acid. In order to recover an element to body of fluid, the reaction always produce complex compound. The complex compound was formed by interaction transition element with strong ligand. According to Fessenden (1982) the strength of ligand [7] are in the following order:

$$I^-<Br^-<S^{2-}<SCN<SO_2^{2-}<Cl^-<NO_2^-<OH^-<COO^-<C_2O_4^{2-}<H_2O<NCS^-$$
$$<CH_3CN<NH_3<en<bipy<phen<NO_2^-<PPh_3<CN^-<CO$$

It was represented that ligand COO^- is stronger than SO_2^{2-} in order that COO^- ligand can recover many of desired elements than SO_2^{2-} ligand.

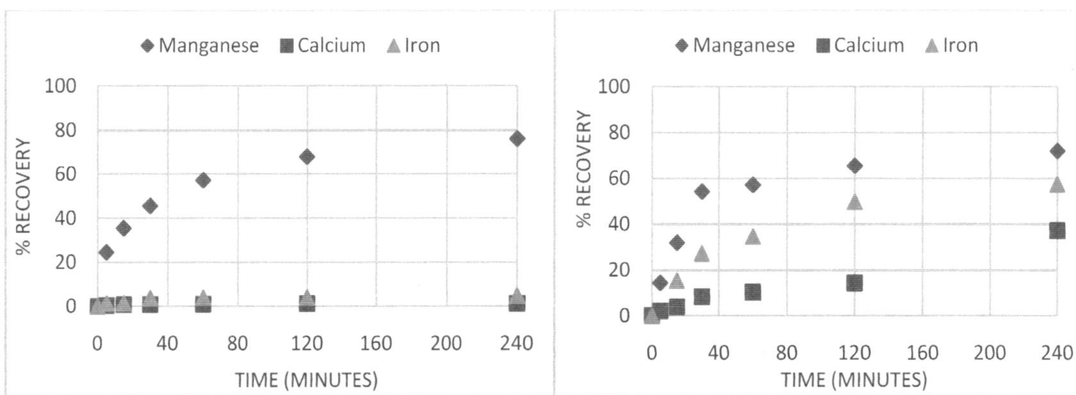

a. Leaching with acetic acid **b. Leaching with sulfuric acid**

Figure 4. Comparison of Kulon Progo manganese ore leaching with organic acid and inorganicacid (temperature of 80 °C, pH of 0.7, particle size of -200 mesh, stirring speed at 400 rpm, and solid-to-liquid ratio at 20 % (w/v))

3.2. The effects of leaching parameters

3.2.1 The effect of reaction temperature

Experiments on the effect of temperature were carried out in the temperature with range of 30-80 °C, maintaining the solid-to-liquid ratio at 20 % (w/v); stirring speed at 400 rpm, the average particle size of -200 mesh, and acid pH at 0.7. As seen from the experimental results given in Figure 5, the leaching rate increases regularly as the temperature increases. Figure 5 showed that in different of temperature, recovery of manganese, calcium and iron were increase with the increase of temperature. It showed recovery of manganese in different temperature is higher than recovery of calcium and iron. As observed, the leaching rate of Mn increases from 48.43 % to 76.08 % as leaching temperature increases from 30 °C to 80 °C. The results obtained are illustrated in Figure 5. As observed, the leaching rate of Ca increases from 1.08 % to 1.38 % as leaching temperature increases from 30 °C to 80 °C. The leaching rate of Fe increases from 3.14 % to 4.83 % as leaching temperature increases from 30 °C to 80 °C. It was adjusted that acetic acid is selective with manganese ion. This behavior can be explained by the fact that high leaching temperature can accelerate reaction rate between Kulon Progo manganese ore and acetic acid. Therefore, the leaching temperature of 80 °C was chosen as the optimum condition in the next experiment. The same results was obtained by Yuksel Abali et al. (2007) which showed that the leaching rate increases gradually as the temperature increases [7].

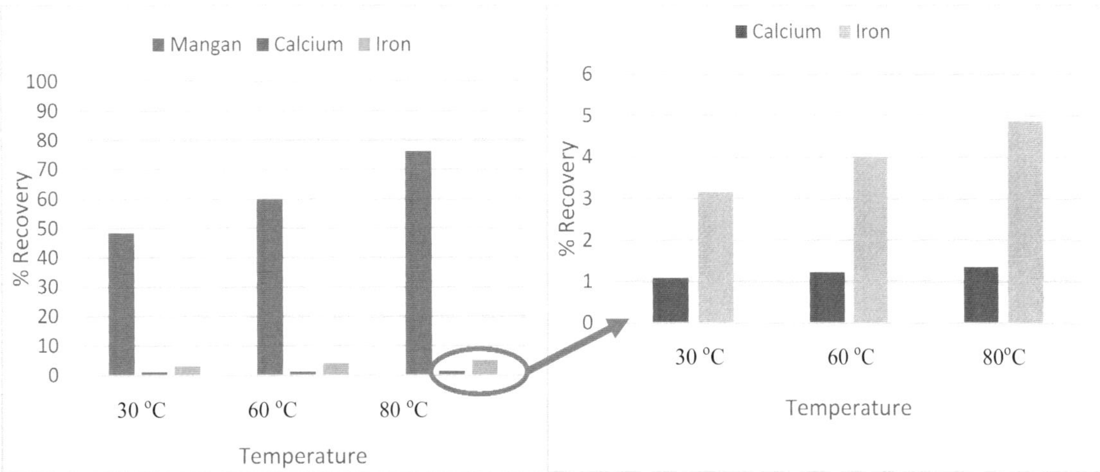

Figure 5. Effect of temperature on the leaching in the range of 30-80 °C in some elements (solid-to-liquid ratio at 20 % (w/v); stirring speed at 400 rpm, the average particle size at -200 mesh, and acid pH at 0.7)

3.2.2 The effect of pH

The effect of acidity level on the leaching was studied using the values of pH in the range of 0.7-3. For these experiments the value of the other parameters maintained constant during the reaction were: leaching temperature of 80°C, stirring speed at 400 rpm, the average particle size of -200 mesh, and solid-to-liquid ratio of 20 % (w/v). The results obtained by these experiments are shown in Figure 6. Leaching rate was increased by decreasing the acid pH. Smaller of acid pH then H^+ concentration was increased or amount of hydrogen ion which from dissosiation step of acetic acid was increased. According to Yuksel Abali et al. (2007), the reaction occured in dissociation step [8] was:

$$2CH_3COOH \leftrightarrow 2CH_3COO^- + 2H^+ \quad (1)$$

Increased of the hydrogen ion amount would involve the increased activity in proton attack step. The reaction of proton attack step [9] was:

$$MO_2 + 2H^+ \rightarrow M^{2+} + 2H_2O + \tfrac{1}{2}O_2 \quad (2)$$

Hydrogen Ion (H^+) which formed attacks pyrolusite mineral on the ore. Increased of the hydrogen (H^+) ion amount which reacted with pyrolusite mineral make manganese ion detached from the ore and increase the amount of proton attack step. Because of that, the recoveries of manganese on leaching were increased. The results are summarized in Figure 6. The leaching rates of Mn, Ca, and Fe in Figure 6 are enhanced with the increase of the pH of acetic acid from 3 to 0.7. However, the leaching rate of Mn is almost unchanged when the pH is over 0.7, indicating that leaching reaction nearly reaches equilibrium. A pH of acetic acid of 0.7 corresponds to the maximal leaching rate of Mnbeing 76.08 %, leaching rate of Ca being only 1.38 % and leaching rate of Fe being only 4.83 %.

It was represented that acetic acid in same of pH is selective with manganese ion. Thus, the most favorable pH of acetic acid is 0.7 in this work.

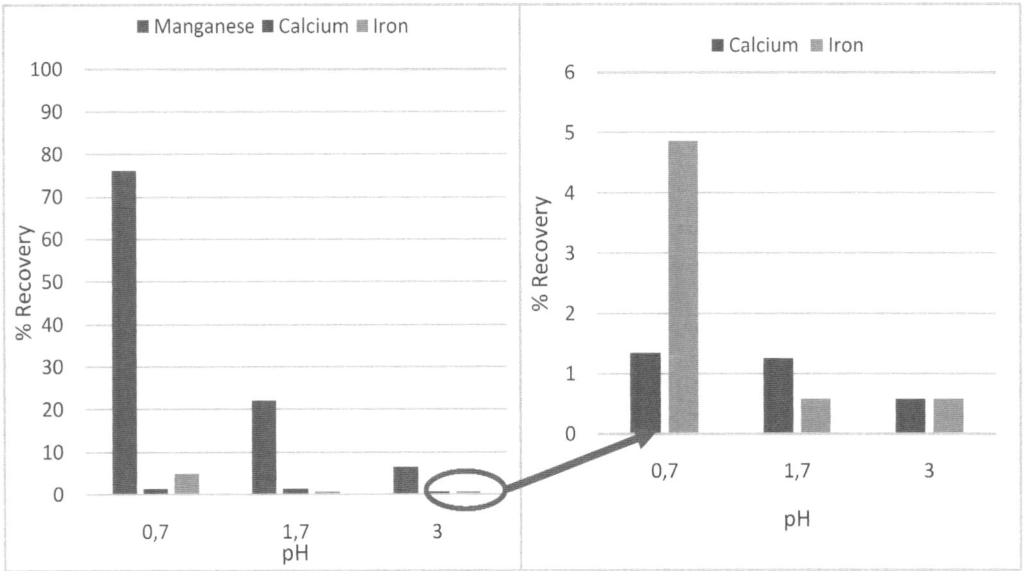

Figure 6. Effect of acid pH on the leaching in the range of 0.7-3 (leaching temperature 80 °C, stirring speed 400 rpm, the average particle size at -200 mesh and solid-to-liquid ratio of 20 % (w/v))

3.2.3 The effect of particle size

The effect of particle size on the leaching rate was investigated using two different particle size fractions in the range of (-70+100)- (-200) mesh and keeping other parameters constant: temperature at 80 °C, stirring speed 400 rpm, acid pH of 0.7 and solid-to-liquid ratio of 20 % (w/v). There was no effect particle size ore with results manganese, calcium and iron recovery of leaching. The results are summarized in Figure. 7. Showed that recovery of manganese, calcium, and iron were extracted from Kulon Progo manganese ore when 0.7 pH of acetic acid but only 76.08 % of manganese recovery, 1.38 % of calcium recovery, and 4.83 % of recovery of Iron. The results adjusted that acetic acid selective with manganese ion.

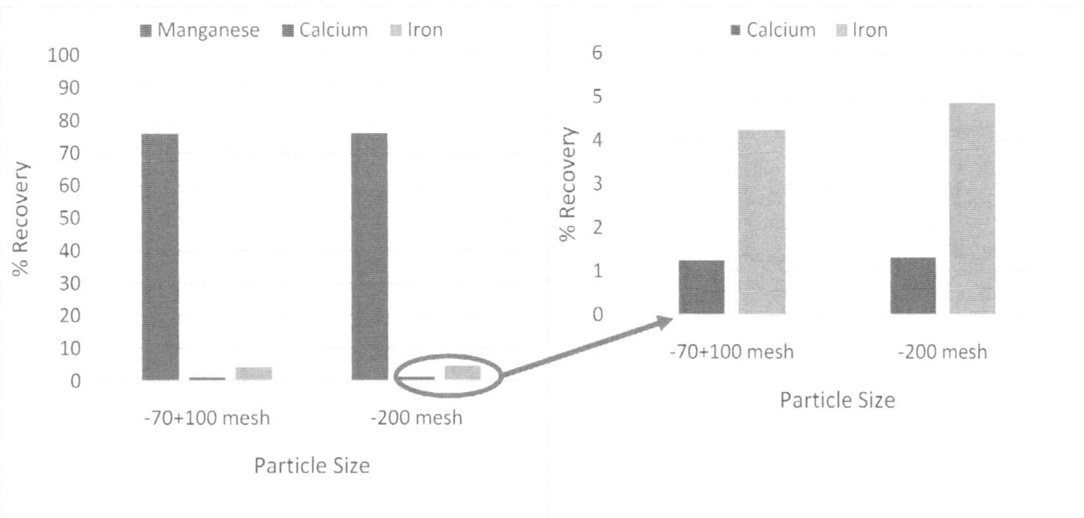

Figure 7. Effect of particle size onl (temperature at 80 °C, stirring speed 400 rpm, acid pH of 0.7 and solid-to-liquid ratio of 20 % (w/v))

3.2.4 Dissolution behavior of different metal ions

Kulon Progo manganese ore is assumed to be incorporated in Calcite, Sphalerite, Chromite, Alumina, Rutile, and Quartz in Kulon Progo; therefore, manganese dissolution must always coinide with the dissolution of the other metals contained in these minerals, particularly calcium and iron. Sphalerite particles (0.7 m^2/g) have smaller surface area than calcite (3.82 m^2/g) and pyrolusite (1223.9 m^2/g) particles [10], [11], [12]. The data in Table 1 showed that calcite is the first dominant mineral in this ore, pyrolusite is second dominant mineral in this ore, and sphalerite is minor mineral in this ore. From Figure 3, we obtained that pyrolusite are on the outside layer in this ore. Because of pyrolusite have a big surface area and located on the outside layer in sample ore, it makes manganese ion easy to dissolve in acetic acid. However, sphalerite have small surface area so that it is very difficult to dissolve in atmospheric leaching even with acid solutions of small pH of acetic acid. Calcite have slightly moderate surface area, so it is relatively difficult to dissolve in atmospheric leaching even with acid solutions of small pH of acetic acid. From Figure 5, 6, and 7 it is observed that manganese recovery is much higher than calcium and iron. Accordingly, it can be predicted that the manganese originates in separated mineral with other mineral and not associated with other element in same mineral. In conclusion, it is very easy to dissolvethan calcium and iron.

4. Conclusions

The leaching of Kulon Progo low grade manganese ore was investigated, with the parameters having the ranges: temperature of 30-80 °C; acid pH of 0.7-3, and average particle size of (-70+100) and (-200) mesh. It was observed that the leaching of Kulon Progo lowgrade manganese ore increased with increasing reaction temperature and decreasing acid pH. It was obtained that the ore particle size has no effect on Mn recovery. The recovery of 76.08 % was obtained with significant low extraction of iron and calsium below 10%. The finding showed the manganese selectivity form its impurities when acetic acid applied.

Acknowledgements

The authors would like to expressed our gratitude to Department of Chemical Engineering, Universitas Gadjah Mada and Indonesian Institute of Sciences (LIPI) for providing experimental facilities. The support from Sustainable Mineral Processing Research Group are gratefully acknowledged.

References

[1] Hagelstein K 2009 *J. Environ. Manag.* **90** 3736-3740
[2] Lu J, Dreisinger D and Gluck T 2014 *Hydrometallurgy* **141** 105-116
[3] Housmanns S, Laufenberg G and Kunz B 1996 Desalination **104** 95-98
[4] Veeken A H M and Hamelers H V M 1999 *Water Science and Technology* **400** 129-136
[5] Bilgic S 2002 *Material Chemistry and Physics* **76** 52-58
[6] Wang Y, Jin S, Lv Y, Zhang Y, Su H 2017 *Minerals* **7** 27
[7] Fessenden 1982 Kimia Organik Jilid 2 Edisi Ketiga Erlangga Jakarta
[8] Abali, Yuksel., Kelesoglu, Seda., Kaymak, Julide 2007 *C.B.U. Journal of Science* **3(1)** 81-88
[9] McDonald R G and Whittington B I 2008 *Hydrometallurgy* **91** 56-69
[10] Bang J H, Jang Y N, Kim W B, Song K S, Jeon C W, Chae S C, Lee S W, Park S J, and Lee M G 2012 *Chemical Engineering Journal* **198-199** 254-260
[11] Birkner N and Navrotsky A 2012 American *Mineralogist* **97** 1291-1298
[12] Stanton, Mark R, Taylor, Cliff D, Gemery-Hill, Pamela A, and Shank III W C 2006 *American Society of Mining and Reclamation (ASMR)* 2090 - 2097

The kinetic study of limonitic low grade nickel ore leaching using hydrochloric acid

H S E A Gustiana[1], I M Bendiyasa[1], Agus Prasetya[1], F R Mufakhir[2], W Astuti[2] and H T B M Petrus[1]

[1]Department of Chemical Engineering (Sustainable Mineral Processing Research Group), Universitas Gadjah Mada 55281 Yogyakarta, Indonesia
[2]Research Unit for Mineral Technology, Indonesian Institute of Sciences (LIPI) Jl. Ir. Sutami Km. 15, Tanjung Bintang, Lampung Selatan, Indonesia

Email: bayupetrus@ugm.ac.id

Abstract. The leaching kinetic ore from Pomalaa (Southeast Sulawesi, Indonesia) using hydrochloric acid solution was investigated in a batch reactor. Variables studied were temperature, pH and particle size. To evaluate the kinetic parameters, the experimental data were fitted with a shrinking core model of a constant particle size. The experimental data fit very well with the proposed model and the overall rate is controlled by dissolution rate of nickel through product layer. The activation energy (E_{De}) for the dissolution of nickel is 22.5 kJ/mol. The maximum nickel recovery is 41.5% at pH 1, reaction temperature of 90°C, particle size of -200 mesh, solid to liquid ratio at 0.2 (w/w) and stirring speed of 400 rpm for 240 minutes.

1. Introduction

Nickel is one of strategic metals in many industrial and metallurgical applications. The use of nickel in the preparation of stainless steel and non-ferrous alloys is necessary because it can improve the impact strength, corrosion resistance and other electrical, heat and magnetic properties [1-3]. Nickel reserves in the world are mostly dominated by laterite ores (72%). However, the world's nickel production, around 54% of it, is resulted from nickel sulphide ores [4]. Indonesia is one of the countries that has an abundant amount of laterite ore reserves with the total amount of 15% of total world laterite ore reserves [5].Around 70% of the total lateritic Ni reserves worldwide are limonitic laterites, which are rich in Fe but low in Mg [6,7]. The dominant mineral in most laterites is goethite (FeOOH) [1,9].

Hydrometallurgical processes are the main methods applied for the extraction of Ni from limonitic ores. Two existing hydrometallurgical processing currently used are Caron process and high pressure acid leaching (HPAL) [10,11]. The Caron process involves three steps which are drying, calcination and ammonia leaching, while HPAL needs high pressure and temperature within autoclaves. Due to such as high capital expenditure and energy consumption make these two processes less attractive. Therefore, McDonald and Whittington [12], Basturkcu et al. [13] and Wanta et al. [14] studied an atmospheric leaching method to extract Ni from limonitic ores.

One of the recent development in the extraction of nickel from laterite ores in atmospheric pressure is due to leachant which can be regenerated, such as hydrochloric acid. An atmospheric leaching has been performed by a few researchers such as Rice [15], Agacayak et al. [16] and Astuti et al. [17].

In this present study, the kinetic leaching of a limonitic ore using hydrochloric acid solvent was investigated. Variables investigated were temperature, pH, and ore particle size.

2. Materials and Methods

2.1. Materials

Nickel laterite was limonite type and mined from Pomalaa, Southeast Sulawesi Province, Indonesia. The chemical components of the samples were identified using X-ray fluorescence (XRF; PANalytical Epsilon 3XLE). The nickel content of the sample was 2.65% (Table 1).

The mineral phases of the raw ore samples were identified by X-ray diffraction (XRD; X'Pert 3 Powder, PANalytical) using Cu Kα radiation, at an accelerating voltage and applied current of 40 kV and 40 mA, respectively, in the 2θ range 5° to 80°. The result of XRD analysis is shown in Figure 1. It shows that goethite is the main minerals in limonite ore from Pomalaa. XRD test results further strengthen the results of XRF testing where iron is a component that dominates the sample.

As for the leachant reagent, hydrochloric acid (Merck) was used and deionized water was used for all the analytical preparations.

Table 1. Chemical analysis of the laterite ore sample

Component	Fe	Ni	Si	Al	Cr	Mn	Mg	LOI (Lost on Ignition)
Content wt.%	80.637	2.651	5.148	4.953	2.512	1.374	1.286	1.439

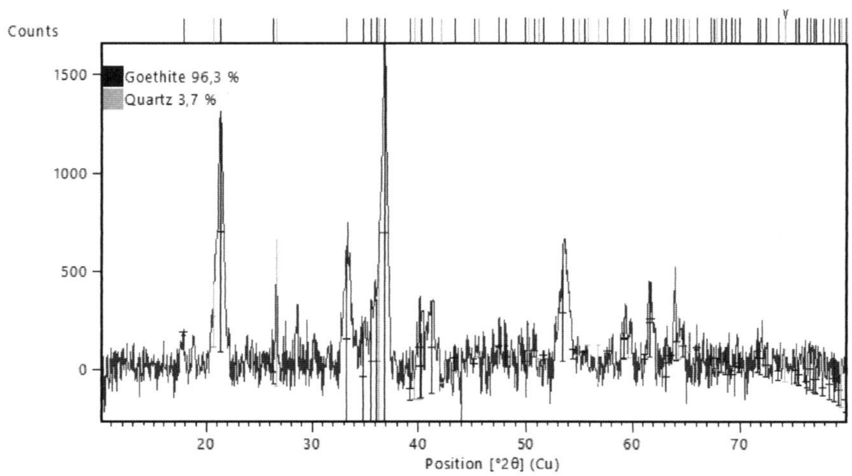

Figure 1. XRD pattern of the laterite ore sample

2.2. Experimental design

The experiments were carried out in a 500 mL three-neck flask reactor equipped with stirrer, temperature control unit and condenser. Tree hundred milliliters (300 mL) of hydrochloric acid solution (pH = 1 or 2) was poured into the tree-neck flask and heated using a waterbath to a desired temperature (30°C, 60°C or 90°C). Sixty grams (60 gr) of nickel laterite sample (particle size : -70 + 100 mesh or -200 mesh) was then put into the reactor. The stirrer speed was set constant at 400 rpm during the leaching process. Each run of the leaching process was performed for 4 hours. Samples of the leach suspension were drawn at predetermined intervals and then the solid phase was separated from the liquid phase using a centrifuge at 1,000 rpm for 10 minutes. Then, the element in the liquid phase was analysed using the atomic absorption spectrometer (AAS).

3. Results and discussion

3.1. Leaching studies

3.1.1. Effect of particle size

Effect of particle size on nickel recovery is shown in Figure 2. It can be seen from Figure 2 that nickel recovery increases with decreasing particle size. For the first 30 minutes, nickel recovery increases rapidly, followed by insignificant change due to the attained equilibrium condition. For particle size of -200 mesh nickel recovery is about twice than that of -70+100 mesh. This phenomenon is caused by the increase of surface area in the smaller particle sizes per unit weight. This condition will surely enhance the mass transfer process of leaching [18,19].

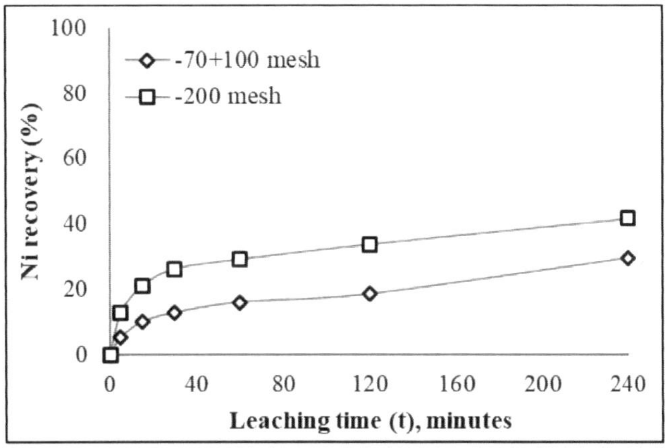

Figure 2. Effect of praticle size on the nickel recovery (pH = 1; stirring speed = 400 rpm; temperature = 90°C and solid/liquid ratio = 0.2 w/w)

3.1.2. Effect of pH

The effect of hydrochloric acid pH on nickel extraction is shown in Figure 3. Figure 3 exhibits that nickel recovery increases with decreasing pH. At pH 1, nickel recovery is about 3 to 4 times than that at pH 2. pH will increase the affinity of nickel in the ore so that nickel recovery decreases with increasing pH. This is in accordance with experimental results conducted by Yan et al. [20]. Nickel recovery increases at the first 30 minute and starts to show insignificant increase due to approaching its equilibrium.

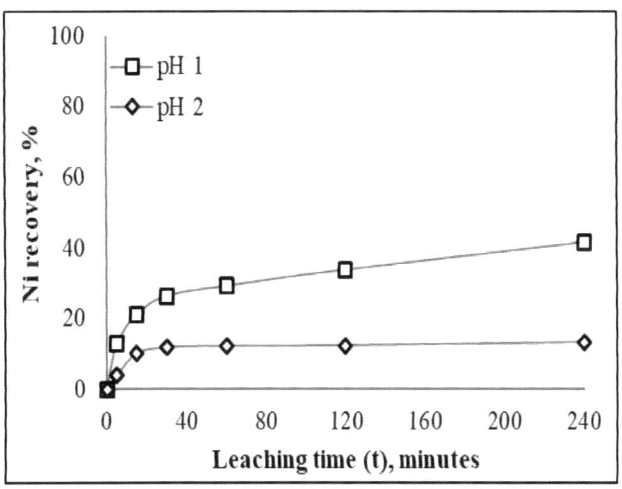

Figure 3. Effect of hydrochloric acid pH on the nickel recovery (particle size = -200 mesh; stirring speed = 400 rpm; temperature = 90°C and solid/liquid ratio = 0.2 w/w)

3.1.3. Effect of temperature

The leaching temperature also plays a significant role in metal extraction. Figure 4 shows the effect of leaching temperature on nickel recovery. Nickel recovery increases very significantly from the beginning to 15 minutes and then increasing slowly, because it is reaching the equilibrium condition. It can be seen from Figure 4 that nickel recovery increases with the increase of temperature. The optimum nickel recovery obtained is approximately 40% leaching temperature of 90°C for 4 hours. This is caused by the increase of both rate of chemical reaction and rate of diffusion at higher leaching temperature [21].

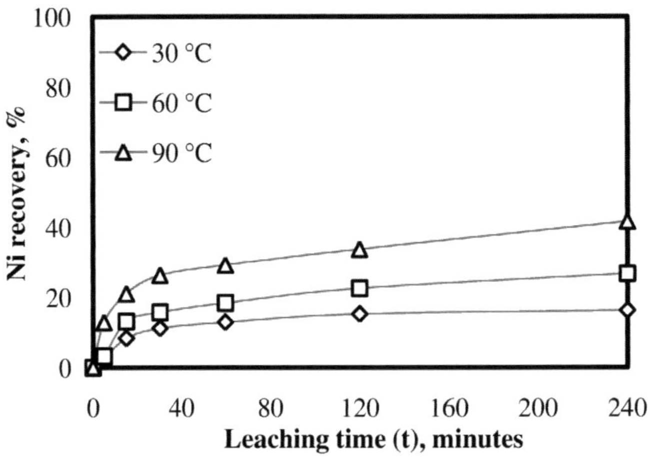

Figure 4. Effect of temperature on the nickel recovery (particle size = -200 mesh; pH = 1; stirring speed = 400 rpm and solid/liquid ratio = 0.2 w/w)

3.2. Kinetic model analysis

3.2.1. The selection of kinetic model

Nickel extraction from limonite ore can be represented by the general reaction of aA(aq) + bB (s) → products. In a heterogenous reaction system, the reaction rate can be controlled by one or a combination of the following steps :diffusion through the fluid film, diffusion through the ash (or solid product) layer on the particle surface or the chemical reaction at the surface of reacted particles [24].Several studies, like the studies conducted by Wei et al. [22], Agacayak et al. [19] and Parhi et al. [23] used shrinking core model to describe the physical phenomenon of the leaching process.In this study, the diffusion through film layer is assumed not to control the process because of high speed of stirring and high concentration of hydrochloric acid applied in the experiment. Therefore, diffusion through the ash layer and chemical reaction models were further investigated in this work.

For diffusion through the product layer, the correlation between t and x is calculated using Equation (1) and D_e is calculated using Equation (2) [24].

$$\frac{t}{\tau} = 1 - 3(1-x)^{\frac{2}{3}} + 2(1-x) \quad (1)$$

$$\tau = \frac{\rho_B R^2}{6bD_eC_{Al}} \quad (2)$$

While for the reaction controlling regime, the correlation between t and x calculated using Equation (3) and k_s is calculated using Equation (4) [24].

IOP Conf. Series: Materials Science and Engineering **478** (2019) 012009 doi:10.1088/1757-899X/478/1/012009

$$\frac{t}{\tau} = 1 - (1 - x)^{\frac{1}{3}} \quad (3)$$

$$\tau = \frac{\rho_B}{k_s C_{Al}} \quad (4)$$

Where x is fraction of nickel leached, t is the reaction time, τ is the overall time to reach unreacted core at the radius = 0, ρ_B is the density of the ore, R is the radius of the solid particle at t = 0, b is the stoichiometric coefficient of the solid reactant, D_e is the diffusion coefficient in the porous product layer, C_{Al} is the concentration of A in the bulk of solution, $k_s$is the chemical kinetic constant. Equations (1), (2), (3) and (4) are applied to evaluate the kinetic parameters that contain in Equations (2) and (4). The calculated parameters are shown in Table 2.

Table 2. D_e, k_s and correlation coeffcients for different temperatures

Temperature (°C)	Apparent rate constants for Ni		Correlation coefficients (R^2)	
	D_e (cm^2/min)	k_s (/min)	Diffusion Through Product Layer	Chemical Reaction
30	7.32 x 10^{-8}	4.48 x 10^{-2}	0.79	0.61
60	1.83 x 10^{-7}	6.72 x 10^{-2}	0.92	0.72
90	1.65 x 10^{-6}	1.12 x 10^{-1}	0.98	0.72
	0.36 x 10^{-6}		1	

From Table 2, it can be seen that the results indicate that the recovery of Ni is controlled by diffusion in the porous product layer. $1\text{-}3(1\text{-}x)^{2/3}+2(1\text{-}x)$ versus t plotting will give a straight line with a slope of 1/τ, as depicted in Figure 5. By knowing the value of τ , the value of D_e can be evaluated using Equation (2). At experimental temperature of 90°C, the data shows two regions that give two different values of slopes. It means that there are two different values of D_e. This is subjected to the internal structure changes of the ore. Around 30 minute starting from the beginning of experimental, the changing of pore size would be very fast. This is indicated by the changing value of De. The higher the slope, the higher the D_e is. In general, the experimental results are in good agreement with the results reported by Agacayak et al. [19], Wei et al. [22] and Parhi et al. [23]. The decreasing rate of difusion in the product layer may be due to the smaller porosity. That is resulted by the formed products are still remained in the ash layer [25].

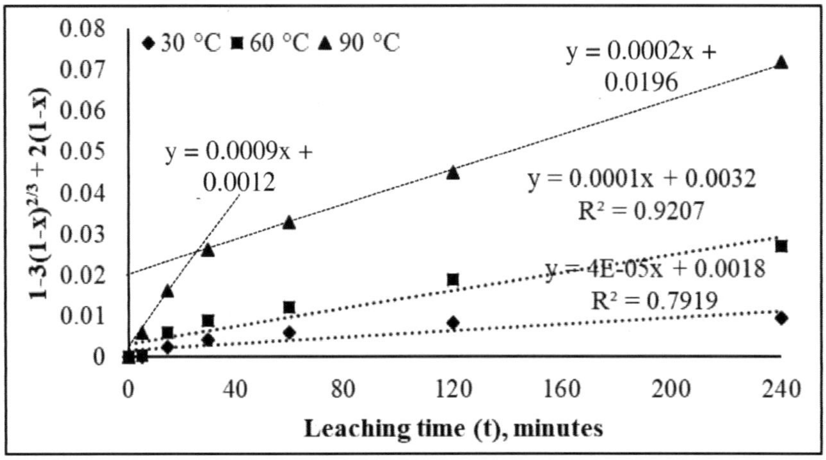

Figure 5. $1-3(1-x)^{\frac{2}{3}}+2(1-x)$ vs time for leaching of nickel from limonite ore at different temperature

To evaluate whether the chemical reaction phenomenon control the system, plotting $1-(1-x)^{1/3}$ versus t should give a straight line with a slope of $1/\tau$. Figure 6 shows that experimental data do not fit the model related to the chemical reaction control.

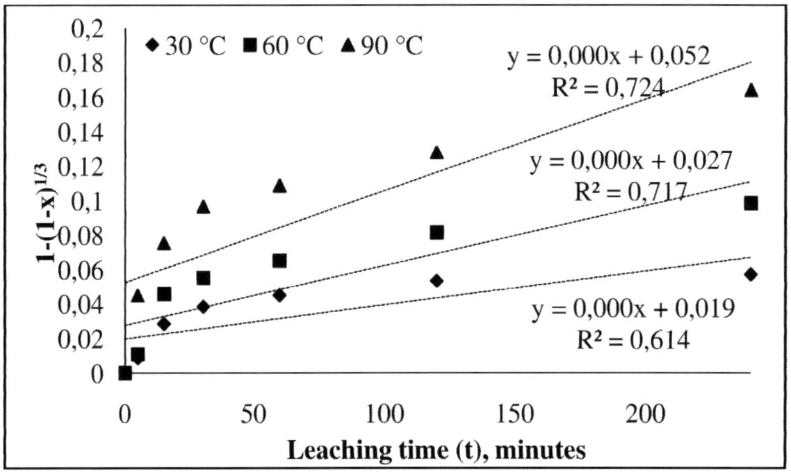

Figure 6. $1-(1-x)^{1/3}$ vs time for leaching of nickel from limonite ore at different temperature

3.2.2. Calculation of diffusion activation energy. The diffusion activation energy can be expressed in accordance with Equation (5).

$$D_e = A\,exp\left(-\frac{E_{,D_e}}{RT}\right) \tag{5}$$

Equation (5) can be linearized as follows:

$$\ln D_e = \ln A - \frac{E_{,D_e}}{RT} \tag{6}$$

Plotting ln De versus 1/T gives a straight line with a slope of (-E/R), as exhibited in Figure 7. The slope in Figure 7 is -2.7117. Therefore the values of E/R = 2.7117, so that E = 2.7117 (8.413 kJ/mol.K) = 22.5 kJ/mol. This value is so closed to Wei, et al. [22] result. Table 3 shows the activation energy results researchers and this study.

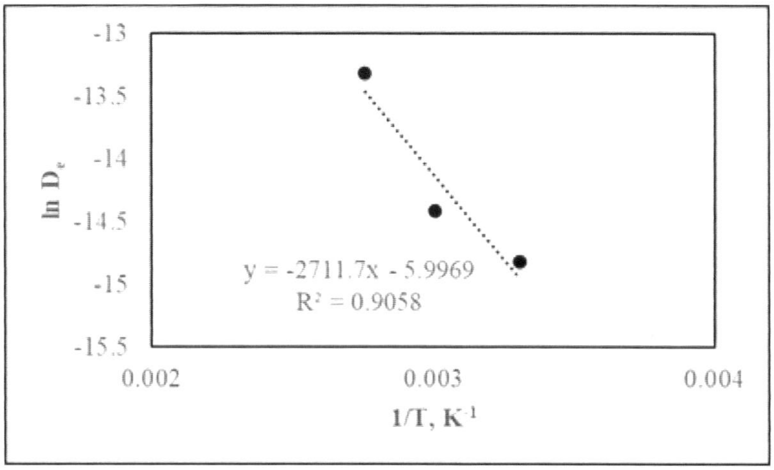

Figure 7. ln De versus 1/T

Table 3. The study activation energy

Researchers	Activation Energy, kJ/mol	Rate controlled by
Wei et al. [22]	23.279	Diffusion through product layer
Agacayak and Zedef [19]	68.66	Diffusion through product layer
Parhi et al. [23]	17	Diffusion through product layer
This study	22.5	Diffusion through product layer

4. Conclusions

The leaching kinetic of Ni from low grade limonite ore was affected by pH, particle size andtemperature. The lower pH and the smaller particle size, the higher the nickel recovery is. On the other hand, when the temperature decreases, nickel recovery is decreasing. Optimum nickel recovery is obtained at pH1, temperature of 90°C and particle size of -200 mesh, in which the recovery of nickel is 41.5%.

Shrinking core is the most appropriate model to describe the kinetic of nickel leaching from low grade limonite ore usingof hydrochloric acid solution. The kinetic analysis shows that the nickel dissolution from low grade limonite ore is controlled by diffusion through product layer. The activation diffusion energy for conducted temperature in this study in between 30°C-90°C is 22.5 kJ/mol.

Acknowledgements

The authors would like to expressed our gratitude to Department of Chemical Engineering, Universitas Gadjah Mada and Indonesian Institute of Sciences (LIPI) for providing experimental facilities. The support from Sustainable Mineral Processing Research Group are gratefully acknowledged.

References

[1] Guo X, Li D, Park K, Tian Q, and Wu Z, 2009 *Hydrometallurgy* **99** 144-150
[2] Jephcott and Brendan 2016 *Nickel Industry Analysis 2016* Golden Dragon Capital
[3] Kuck P H 2012 *Nickel. U.S. Geological Survey* Mineral Commodity Summaries
[4] Dalvi A D, Bacon W G, dan Osborne R C 2004 *The Past and The Future of Nickel Laterites PDAC 2004* International Convention, Trade Show & Investors Exchange
[5] Rochani S and Saleh N 2013 Teknologi Pengolahan dan Pemurnian Nikel *Pusat Penelitian dan*

Pengembangan Teknologi Mineral dan Batubara **11** 1-9
[6] King M G 2005 *Journal of Metals* 35-39
[7] Sudol S 2005 *Canadian Mining Journal*
[8] Elias M 2002 *Nickel laterite deposits – geological overview, resources and exploitation. Centre for ore deposit research* University of Tasmania **Special Publication 4** 205-220
[9] Gustiana H S E A, Bendiyasa I M, Petrus H T B M, Mufakir F R, and Astuti W 2018 Pelindian nikel dari bijih limonitlow-grade Pomalaa menggunakan pelarut asam asetat *Seminar Nasional Teknik Kimia Kejuangan*
[10] Moskalyk R R and Alfantazi A M 2002 *Minerals Engineering* **15** 593-605
[11] Mudd G M 2010 *Ore Geology Reviews* **38** 9-26
[12] McDonald R G and Whittington B I 2008 *Hydrometallurgy* **91** 56-59
[13] Basturkcu H, Acarkan N, and Gock E 2017 *International Journal of Mineral Processing* **163** 1-8
[14] Wanta K C, Perdana I, and Petrus H T B M 2017 *IOP Conf. Series: Materials Science and Engineering* **162** 012018
[15] Rice N M 2016 *Minerals Engineering* **88** 28-52
[16] Agacayak T, Zedef V, and Aydogan S 2011 *SGEM 2011 Conference Proceesings/ISSN 1314-2704* **Vol. 1** 1155-1162
[17] Astuti W, Sasaki K, Hirajima T, and Okibe N 2016 *Minerals Engineering* **85** 1-16
[18] Donmez B, Demir F, and Lacin O 2009 *Journal of Industrial and Engineering Chemistry* **15** 865-869.
[19] Agacayak T and Zedef V 2012 *Acta Montanistica Slovaca : Ročnik 17* **Cislo 1** 33-41
[20] Yan Y, Gao J, Wu J, and Li B 2014 *Effect of inorganic and organic acids on heavy metals leaching in contamined sediment* China University of Mining and Technology Press ISBN 978-7-5646-2437-8
[21] Yu-chun Z, Wen-ning M, Yan L, and Qian X 2010 *Transactions of Nonferrous Metals Society of China* **20** s65-s70
[22] Wei L, Mo-tang T, Chao-bo T, Jing H, Sheng-hai Y, and Jian-guang Y 2010 *Transactions of Nonferrous Metals Society of China* **20** 910-917
[23] Parhi P K, Park K H, and Senanayake G 2013 *Journal of Industrial and Engineering Chemistry* **19** 589-594
[24] Levenspiel O 1999 *Chemical Reaction Engineering* New York : John Wiley & Sons
[25] Fogler S H 1999 *Elements of Chemical Engineering* 3th ed London : Prentice Hall International

Application of electro analytical chemistry for the corrosion process on the stainless steel electrodes in the acidic thiourea

A I Nazri[1], Sudibyo[2], N I Basir[1], and M N Murat[1]

[1]School of Chemical Engineering, Universiti Sains Malaysia, Engineering Campus, Pulau Pinang, Malaysia
[2]Research unit for mineral technology, Indonesian institute of sciences, Lampung, Indonesia

Email: chnazri@usm.my

Abstract. A cyclic voltammetry study is one of electro analytical chemistry methods which able to study the corrosion of material. In this research, the cyclic voltammetry has been carried out in order to determine the voltage range that contributes to the initiation of the corrosion on the surface of stainless steel in the acidic thiourea solution. In this study, the knowledge of corrosion on stainless steel is useful in the optimisation process for gold recovery from electronic waste based on electrodeposition technique. Since stainless steel material has been used as both anode and cathode in the experimental setup, under certain applied voltage, corrosion might occur and hinder the gold recovery process by damaging the electrodes. With the aid of cyclic voltammetry study, the suitable applied voltage for electrodeposition process has been investigated and found to be approximately 1.4 V for two-electrode setup for a system with sulfuric acid concentration of 0.1 M in the presence of 0.1 M thiourea. The data obtained can be used as a basis to determine a suitable pre-set applied voltage in the electrodeposition process such that corrosion on the electrodes can be avoided.

1. Introduction

Corrosion is a process involving the destructive attack of a metal by chemical or electrochemical reaction with its environment. In electrochemical cell, corrosion most likely to occur at the anode via metal oxidation processes. Additionally, introducing electricity flow through the cell might also increase the rate of corrosion of metal electrode in the presence of corroding agent. The usage of coatings, inhibitors, passivators and suitable alloyed materials could reduce the amount of metal being corroded [1].

Stainless steel is one of the long listed alloyed materials available in the market. Generally, stainless steels are widely used in industrial applications due to their strength, corrosion resistance, mechanical workability and excellent electrical and thermal conductivities [2]. Also, it is affordable compared to other nobler metals such as platinum and tantalum. High corrosion resistance of stainless steels is primarily attributed to the passive oxide film formed on its surface and the resistance of this passive film is determined by the environmental conditions in which it being exposed to [3].

Eventhough stainless steel has high corrosion resistance under normal application, however, corrosion of stainless steel can still occur under certain conditions, such as in the presence of acid, or when certain voltage being applied. Thus, in this paper, CV experiments has been done in order to study the corrosion process on the stainless steel electrode in the acidic thiourea solution as well as determining the limiting voltage in order to minimize the effect of corrosion on the electrodes. For the application in electrodeposition process, knowledge of suitable applied voltage is very useful so that the corrosion of the electrodes can be avoided. This in turn can increase the deposition yield of the overall process.

Cyclic Voltammetry (CV) technique has widely being used in electrochemical studies especially in gold electroleaching and electrodeposition processes using acidic thiourea solution [4]-[8]. Gold recovery via thiourea solution has received many attentions from a number of researchers since it was first recognized by Moir in 1906 and then pioneered by Plaskin and Kozukhova in 1941 [9]. By implementing CV in this study, further understanding on the reaction mechanisms involved in the process could be grasped. Also, a voltage region which separately represents electroleaching process and electrodeposition process could also be determined.

This paper discusses the CV studies for further understanding of corrosion behaviour on stainless steel electrodes. The understanding of corrosion behaviour is useful to optimise the electrodeposition process of gold recovery from electronic waste (e-waste), specifically when using a stainless reactor [10]. On the point of optimising the electrodeposition process, CV study on the redox process of gold in the acidic thiourea solution as well as determining the appropriate applied voltage range to enhance the gold leaching and deposition was also conducted earlier [11]. The area of study related to this research is electrochemical study, for enhancing the electrodeposition process and avoiding corrosion of the electrodes.

2. Materials and methods

2.1. Chemicals and solutions
Thiourea (99%, Acros) and sulfuric acid (95% - 98%, Sigma-Aldrich) were used without further purification. All solutions were prepared by mixing desired amount of reagent together with deionized water purified in a Millipore Milli-Q system (18 MΩ cm) and were used immediately after preparation [7].

2.2. Apparatus
Cyclic Voltammetry (CV) experiments were carried out using a potentiostat (CH Instrument, model: 600E) by measuring electrochemical reading of the system. CV experiments were carried out in a simple three-electrode cell using a glass beaker. Two stainless steel plates (about 15cm^2each) were used as working and counter electrodes and all the potentials measured in the three-electrode cell were reported relative to a saturated calomel electrode (SCE) purchased fromCH Instrument. The saturated calomel electrode acted as the reference electrode to the system. For CV experiments with two-electrode cells, the reference electrode is omitted.

2.3. Initial preparation of electrodes
Prior to each experiment, both the working and counter electrodes were carefully ground and polished with ultra-fine P2000 grit abrasive paper. After polishing, the electrodes were rinsed with deionized water and left to dry in oven (about 50 °C).

2.4. Experimental setup
Figure 1 shows the experimental setup for the CV experiments to study the corrosion of stainless steel electrode in acidic thiourea solution. The prepared stainless electrodes were connected to both positive and negative terminals of the potentiostat whilst the SCE was connected to another terminal for reference electrode (RE). The electrode connected to the positive terminal are then referred to as working electrode (WE) and the electrode connected to the negative terminal are referred to as counter electrode (CE).

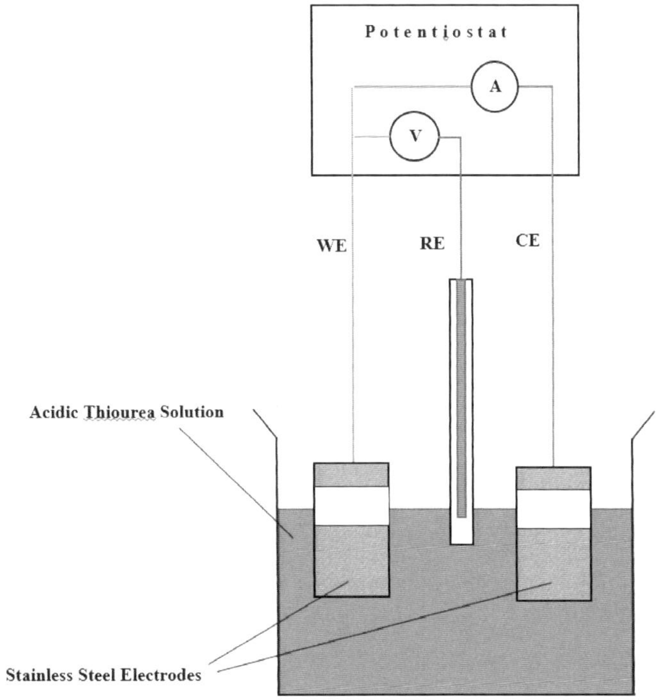

Figure 1. Experimental setup for 3-electrode system of CV study on corrosion of stainless steel electrode in acidic thiourea solution.

All the electrodes including the reference SCE were immersed into the freshly prepared acidic thiourea solution and positioned as shown in Figure 1 by which the RE was positioned in between the WE and the CE. Once ready, the experiment was started and data collected by potentiostat were recorded and logged to the computer connected to the potentiostat. The data were then interpreted by using the appropriate software included by the manufacturer.

As for the experimental work using the 2-electrode system, the configuration in Figure 1 was slightly altered in which the SCE was disconnected from the RE output terminal. The output terminal for RE was then connected directly withthe CE, thus forming the 2-electrode system with both stainless steel as the anode (WE) and cathode (CE). This configuration was provided by the manufacturer, CH Instrument through the manuals.

2.5. Experimental method
The CV experiments were carried out in predetermined voltage range with variation concentration of thiourea and sulfuric acid. All experiments were carried out under room temperature (25 °C) with CV scan rate of 100 mV/s.

3. Result and Discussion
In this section, results regarding system behavior on stainless steel electrode in acidic thiourea solution are discussed based on cyclic voltammogram. Results on the limiting voltage that could be applied based on the comparison between the three-electrode system and a two-electrode system are also presented.

3.1. System behaviour on stainless steel electrode in acidic thiourea solution

Figure 2 (A) and Figure 2 (B) show cyclic voltammograms of stainless steel electrodes in acidic thiourea solution from applied voltage region of -0.8 V to 1.8 V using different amount of concentration for both sulfuric acid (H_2SO_4) and thiourea (TU) reagents in the presence of each other. Note that Fig. 2 (A) represents condition in which sulfuric acid concentration were varied between 0.1 M, 0.3 M and 0.5 M while thiourea concentration were fixed at 0.1 M. Meanwhile, Fig. 2 (B) represents condition in which TU concentration were varied between 0.1 M, 0.3 M and 0.5 M while H2SO4 concentration were fixed at 0.1 M.

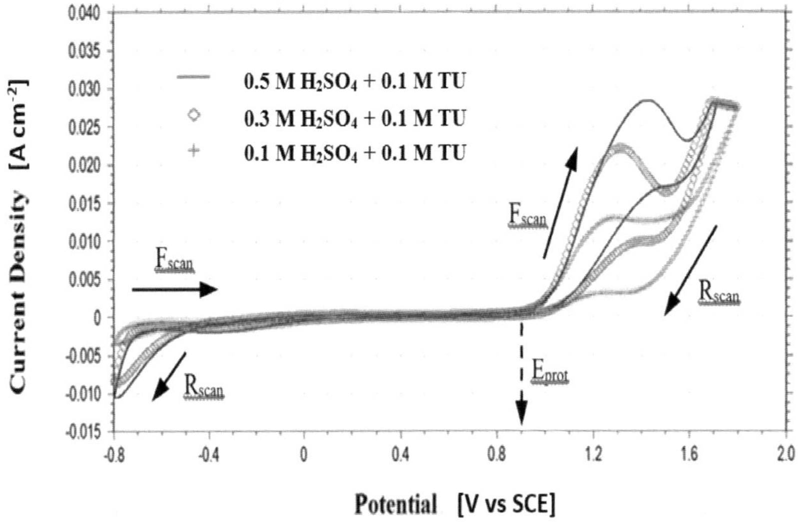

Figure 2. Cyclic voltammogram of stainless steel electrodes in the acidic thiourea solution for fixed concentraton of thiourea.

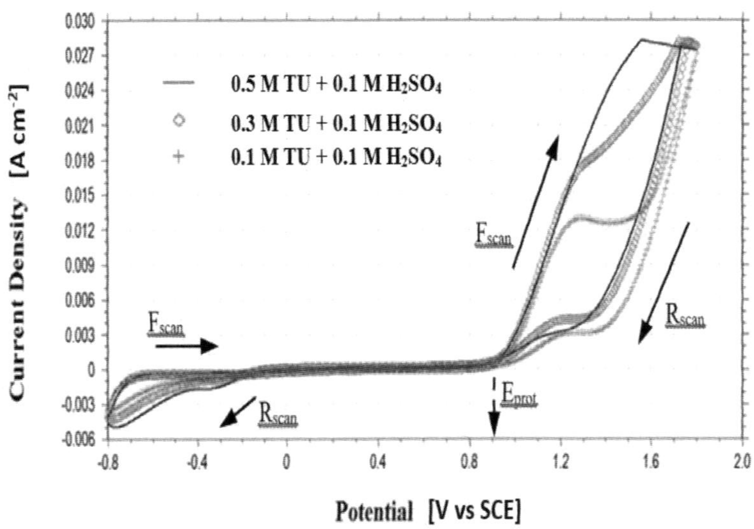

Figure 3. Cyclic voltammogram of stainless steel electrodes in the acidic thiourea solution for fixed concentration of sulfuric acid.

It can be seen in either voltammogram in Figure 2 or Figure 3 that there are basically two anodic peaks and one cathodic peak exist during both forward (Fscan) and reverse (Rscan) scan. Both anodic peaks are related to corrosion process in which the one during the Fscan represents the breakdown of the passive film and the initiation of pitting corrosion. The one during the Rscan represents the activation of the electrode. Note that the voltage in which the current density reached zero during the activation process in Rscan indicates the protection potential (Eprot). It can be concluded that there are no pitting corrosion occurs at the potential below Eprot and pit nucleation tends to start at the potential above Eprot. On the other hand, the cathodic peak represent the onset of hydrogen evolution reaction. The origins of all three peaks are agreed with what has been reported by Albrimi and his colleagues in their publication [2].

Generally, there are no significant differences between both voltammogram in Fig. 2 and Fig. 3. It can be seen in both figures that as the concentration increased, the anodic peak also increased in current density as well as shifted towards more anodic voltage. In the presence of both TU and H2SO4 in the solution, it was found that the results from both figures show the same peak behaviours no matter which reagent was beingmanipulated, thus making it impossible to directly determine which reagent was responsible for catalysing the corrosion process. Due to that reason, further experiments need to be carried out in order to determine which reagent catalysed the corrosion of the stainless steel electrode.

3.2. Influence of sulfuric acid on the system

The experimental results involved the CV studies on stainless steel electrodes at different concentration of sulfuric acid in the absence of thiourea is shown in Figure 4.

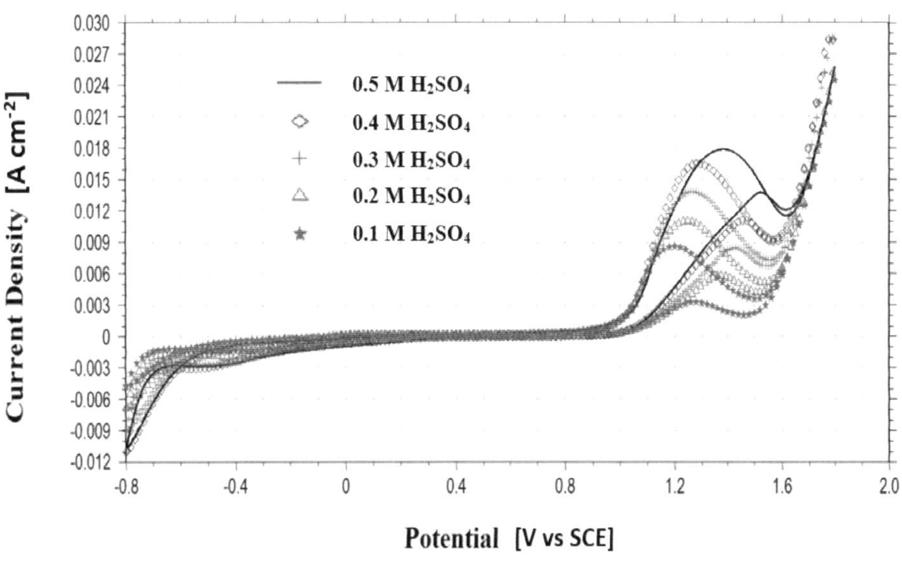

Figure 4. Cyclic voltammogram of stainless steel electrodes using different concentration of sulfuric acid in the absence of thiourea.

As shown in Figure 4, the increasing of sulfuric acid concentration from 0.1 M up to 0.5 M tends to increase the current density for both anodic peaks during Fscan and Rscan. Note that the behaviour of this voltammogram is quite the same with what being reported in Figure 2. This phenomenon indicates that sulfuric acid in the solution catalysed the corrosion of stainless steel electrodes. Note that, the voltage in which the nucleation of pit corrosion tends to occur, Eprot for all concentration of sulfuric acid maintained the same i.e. approximately 0.9 VSCE.

In order to further prove the facts that sulfuric acid acts as the catalyst for the corrosion process of stainless steel electrodes, another set of experiment involving manipulation of thiourea in the absence of sulfuric acid has been done and discussed in the next section.

3.3. Influence of thiourea on the system

The influence of various concentration of thiourea on stainless steel electrodes in the absence of sulfuric acid was shown in Figure 5.

In the absence of sulfuric acid, the behaviour recorded in voltammogram shown in Figure 5 is different to what has been reported in the previous sections. The increment of thiourea concentration up to 0.5 M does increase the current density but in a very small values compared to what being recorded when using sulfuric acid. Also, there are no visible current plateaus during the whole scans. This indicates that thiourea did not significantly affect the corrosion of stainless steel electrodes as much as sulfuric acid.

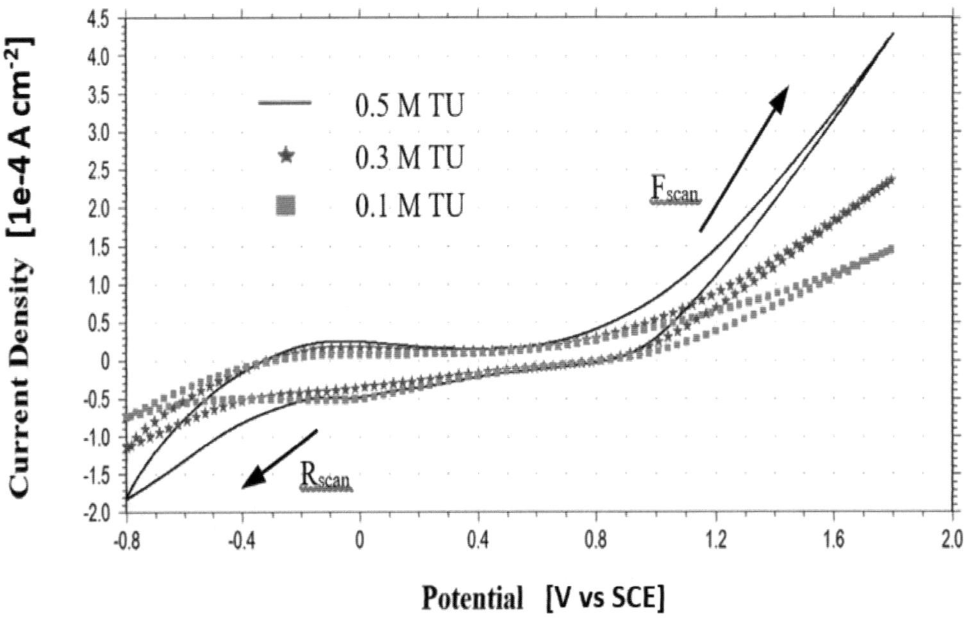

Figure 5. Cyclic voltammogram of stainless steel electrodes using different concentration of TU in the absence of sulfuric acid.

3.4. Limiting Applied Voltage

As the actual application of gold recovery is based on two-electrode system connected with a direct currect (DC) power supply, it is more beneficial to have the data gained from the three-electrode system CV studies being modified and readily used for two-electrode system arrangement. Thus, CV experiment by using 2-electrode system setup described in section II (D) has been done by implementing the best parameters used in the CV experiment using the 3-electrode system which provided the lowest possible corrosion toward the stainless steel electrode. Repetition of experiments was carried out in order to make sure the uniformity of the result and the average results data has been used to develop the trends for CV study of both 3- and 2-electrode system.

A series of experiment comparing the applied voltage of corrosion process for two different system i.e. three-electrode system and two-electrode system with the same parameters i.e. 0.1 M H_2SO_4 + 0.1 M TU has been done and the result is shown in Figure 6.

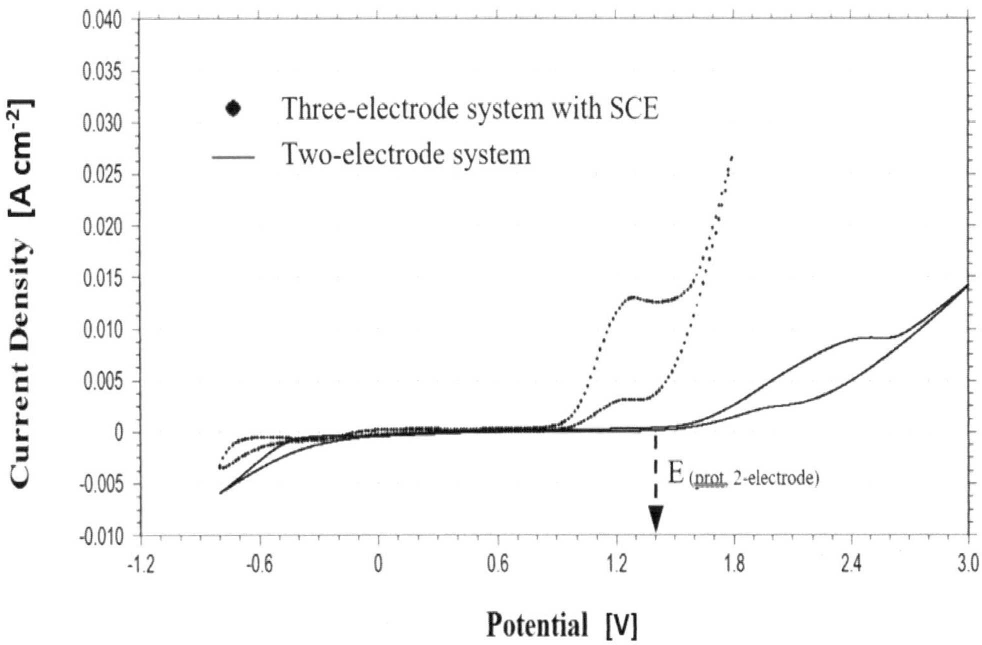

Figure 6. Cyclic voltammogram of stainless steel electrodes using (•) three-electrode system with saturated calomel electrode (SCE) as reference electrode, and (—) two-electrode system.

From the figure, the anodic current plateau from three-electrode system shifted toward more positive voltage as the system changed to a two-electrode system whilst maintaining the general peaks behaviour. It is acceptable as the calculation from (1) which represent electrochemical reaction for SCE requires the data gained from SCE-based reference electrode needs to be added with 0.241 V (CH Instrument SCE model: CHI150) in order to change the values to normal hydrogen electro (NHE) instead of SCE [12]. The values gained from CV ($E_{(prot,\ 2\text{-}electrode)}$ = approx. 1.4 V) however differs to the values from (1) i.e. 1.141 V due to differences in experimental setup etc.

$$Hg_2Cl_2 + 2e = 2Hg + 2Cl^- \quad , E° = 0.241\ V_{NHE,\ 25\ °C} \qquad (1)$$

Besides that, the current density recorded from two-electrode system for the corrosion plateau is generally lower than the corrosion plateau by the three-electrode setup. This indicates that corrosion effects on the stainless steel electrode could be further reduced if direct use of two-electrode system is implemented.

4. Conclusion

Series of experiments were conducted to investigate the unwanted process of corrosion of stainless steel electrodes in acidic thiourea solution. The investigation was performed based on Cyclic Voltammetry (CV) technique, with the aid of an electrochemical device, namely potentiostat, to measure the electrochemical reading of the system. From the experiments, it was found that thiourea ligand has no significant contribution towards the corrosion process of stainless steel electrodes. In fact, it was found that sulfuric acid was the one plays major role in the corrosion process. In addition to that, the chances of the stainless steel electrodes to corrode increase proportionally with the concentration of sulfuric acid being used. On the other hand, it was found that the two-electrode system in actual process is less prone to corrosion compared to the three-electrode system, which is used specifically in CV studies. Last but not least, the limiting applied voltage was found to be approximately 0.9 V_{SCE} for three-electrode system or approximately 1.4 V for two-electrode system.

Acknowledgement
Authors are grateful to CREST-USM grant 304/PJKIMIA/6050265/C121, the company in collaboration i.e. Scrap Computer Trading (SCT) Sdn. Bhd., the MYBRAIN15 scholarship for their financial assistance and to all technicians for their technical supports.

References
[1] Revie R W and Uhlig H H 2008 *Corrosion and corrosion contro*. 4thed.United States of America: John Wiley & Sons
[2] Albrimi Y A 2011 *International Journal of Electrochemical Science* **6** 4614-4627
[3] Iliyasu I, Yawas D S, and Aku S Y 2012 *Advances in Applied Science Research* **3 (6)** 3909-3915
[4] Bolzan A E, Piatti R C V and Arvia A J 2003 *Journal of Electroanalytical Chemistry* **552** 19-34
[5] Gomez H, Lizama H, Suarez C and Valenzuela A 2009 *Journal of the Chilean Chemical Society* **54 (4)** 439-444
[6] Shetsova O N, Bek R Y, Zelinskii A G and Vais A A 2006 *Russian Journal of Electrochemistry* **42 (3)** 239-244
[7] Yang X Y, Moats M S and Miller J D 2010 *Minerals Engineering* **23 (9)** 698-704
[8] Zhang H, Ritchie I M and La Brooy S R 2001 *Journal of the Electrochemical Society* **148 (10)** D146
[9] Hilson G and Monhemius A J 2006 Journal of Cleaner Production **14 (12–13)** 1158-1167
[10] Nazri A I, Basir N I and Murat M N 2015 *Advanced Materials Research* **1113** 187-192
[11] Nazri A I, Basir N I and Murat M N 2015 *Cyclic Voltammetry Application in Determining Oxidation and Reduction of Gold in Acidic Thiourea Solution Using Gold-Stainless Steel Paired Electrode* **5th International Conference on Environment 2015 (ICENV 2015)** Malaysia
[12] Pavlishchuk V V and Addison A W 2000 *Inorganica Chimica Acta* **298 (1)** 97-102

Adsorption of lithium in the manganese hydroxide precipitation processes

F Firdiyono[1], E Sulistiyono[1], L H Lalasari[1], M R Arwanda[2], and S Wahyuningsih[2]

[1]Research Center of Metallurgy and Materials, Indonesian Institute of Sciences, Kawasan PUSPIPTEK Gedung 470, Tangerang Selatan, 15314, Indonesia
[2]Department of Chemistry, Faculty of Mathematics and Natural Sciences, Sebelas Maret University, Surakarta

E-mail: f.firdiyono@yahoo.com

Abstract. Lithium Manganese is an important ingredient in the process of making lithium battery cathode materials. The purpose of this study was to observe the presence of lithium ions in the manganese hydroxide deposition process in mixing $MnSO_4$, LiCl, and NH_4OH. The expected result is that the adsorbed lithium ion reacts with manganese hydroxide and settles when sintered at high temperatures. In this experiment, the addition of $MnSO_4$ varied by 5, 10, 15, 20, 25 grams in 200 ml of LiCl solution with a concentration of 10 g / l and ammonia 25% by volume was added. The resulting solids were calcined at 800 ^{O}C for 24 hours. The results showed that the best adsorption of lithium ions on the addition of manganese sulfate was 15 g and 25 g. Characterization results obtained by solids have a 3.3% $LiMn_2O_4$ phase and 96.7% $Li_2Mn_2O_4$ at the addition of 15 g manganese sulfate. On the addition of 25 g of manganese sulfate, 36.6% of $LiMn_2O_4$ and 63.4% of $Li_2Mn_2O_4$ were obtained. The success of this study will make it easier for us to synthesize the manufacture of lithium battery cathode materials such as lithium manganese phosphate.

1. Introduction

Lithium is a natural product and one of the most needed materials in the present. This happened because of the rapid development of electronic equipment such as mobile phones, tablets, and portable computers, and electric cars in recent years. Industry analysts and lithium producers estimate that 28,000 tons of lithium are used throughout the year [1].

Brines are one of the most commonly used sources of lithium natural. Besides that, lithium can also be obtained by spodumene mineral, sea water, and geothermal fluid [2][3]. Among secondary battery types, lithium ion batteries have received special attention because they have a higher energy density, high capacity, slow self-discharge rate, low weight, high cell voltage, high specific energy and no effect memory [4]. The main reason for using Li-ion technology on these batteries is that lithium is the lightest metal and the most electropositive metal, thus providing a high energy density. Li-ion batteries show a stable life cycle (more than 500 cycles), can be made in various sizes and also require little maintenance [5].

Manganese is a reactive element that easily combines ions in water and air. Manganese was chosen because it is non-toxic, cheap and safe and has the ability to adsorb lithium ions which are realized in the form of Lithium Manganese Oxide (LMO) [6] or lithium manganese. It has the molecular formula of $LiMn_2O_4$ which is displayed in the spinel phase used in the cathode battery [7].

Lithium manganese can be used as cathode material in batteries that can be coupled with other materials such as lithium cobalt oxide ($LiCoO_2$), lithium nickel oxide ($LiNiO_2$) and lithium ferro phosphate (LFP). The advantage of $LiMn_2O_4$ material is a higher life cycle (500 cycles), thus facilitating the nature of rechargeable batteries. In addition, $LiMn_2O_4$ also has good structural stability during the charge-discharge process and is quite stable during intercalation. LiMn2O4 is a cathode material that is easily synthesized. It has good structural stability when compared to other types of cathode material [8].

In this research, we would investigate the lithium adsorption in a mixture of $MnSO_4$, LiCl, NH_4OH which was proven by identification using ICP-OES and XRD analysis. This analysis was used to ensure lithium adsorption during manganese hydroxide deposition.

2. Material and Methods

The research was conducted using materials such as manganese (II) sulfate, lithium chloride, and ammonia 25%. All materials have the type of pro analysis (PA) that were purchased from Merck Company. The research was conducted in four stages, namely preparation of $MnSO_4$ samples without lithium chloride, preparation of $MnSO_4$ samples with lithium chloride, calcination and characterization.

2.1. Preparation of $MnSO_4$ Samples without Lithium

The initial stage of the study was to weigh $MnSO_4$ in mass variations, namely 5, 10, 15, 20, and 25 grams. Each sample was dissolved with 200 ml of distilled water and added 25% by volume of ammonia according to its stoichiometric calculation. Then the sample is stirred for 1 hour and filtered. The residue obtained is then dried in an oven for 14 hours at a temperature of 100 ° C. Furthermore, the drying residue in the form of $Mn(OH)_2$ without lithium was weighed.

2.2. Preparation of MnSO4 Samples with Lithium

The experimental steps are the same as above but in this second experiment, each solution was added with LiCl with a concentration of 10 g/L. After stirring, filtering, and drying, the residues obtained were weighed as $Mn(OH)_2$ with lithium.

2.3. Calcination

Calcination was carried out on $Mn(OH)_2$ residues for 24 hours with a temperature of 800 °C [9].

2.4. Characterization

The sample analysis was carried out by using ICP – OES (Inductively Coupled Plasma - optical emission spectrophotometers) on the filtrate samples. The X-Ray Diffraction (XRD) analysis was carried out on residue samples.

3. Result and Discussion

3.1. Adsorption of Lithium Ions during the Manganese Hydroxide Precipitation Process

Preparation of $LiMn_2O_4$ samples could be carried out through the precipitation stage using inorganic precursors [9]. In this experiment, the mixture of lithium chloride and manganese hydroxide was carried out more intensely until precipitation of manganese oxide was obtained which acted as an adsorbent and was able to absorb lithium. The precipitated residues were then calcined to produce homogeneous $LiMn_2O_4$ crystals. The possible reactions were as follows:

Precipitation reaction:

$$MnSO_{4(s)} + 2NH_4OH_{(aq)} \longrightarrow Mn(OH)_{2(s)} + (NH_4)_2SO_{4(aq)} \quad (1)$$

Calcination reaction:

$$Mn(OH)_{2\,(s)} \longrightarrow MnO_{2(s)} + H_2O_{(l)} \quad (2)$$

$$MnO_{2\,(s)} + LiCl_{(aq)} \longrightarrow LixMnyOz_{(s)} \quad (3)$$

Samples of manganese oxide deposited with lithium had a different morphology than MnO_2 without lithium. The most prominent morphological difference was that the color of the MnO_2 sample with lithium looks darker than the sample without lithium. The results of the measurement of the weight of the MnO_2 sample without the addition of lithium and with the addition of lithium can be seen in Figure 1.

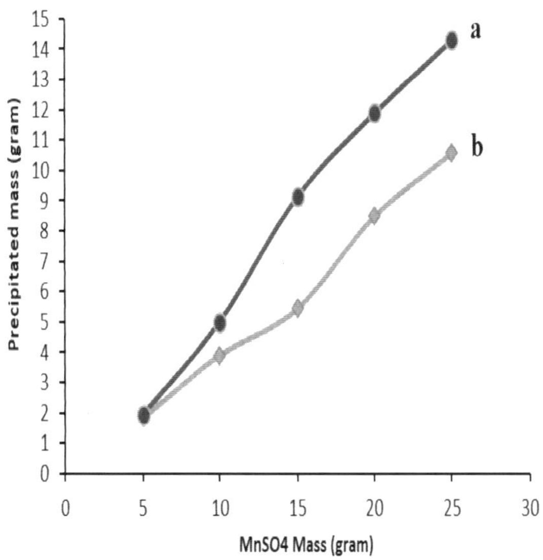

Figure 1. The change of precipitated residue mass by drying at a temperature of 100 °C in various condition : (a) The addition of LiCl; (b) Without the addition of LiCl.

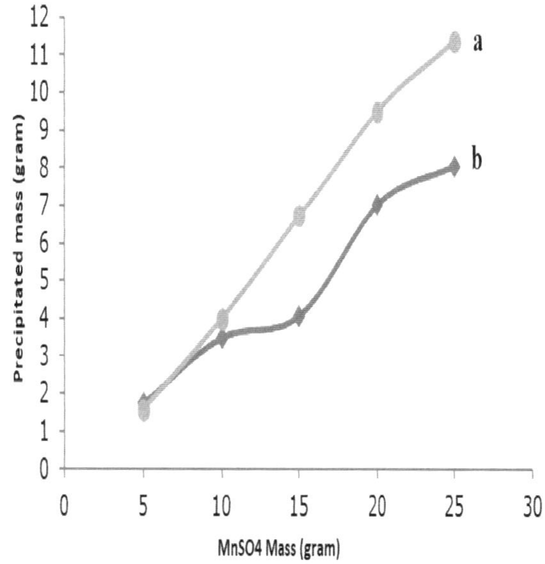

Figure 2. The change of precipitated residue mass by calcination at a temperature of 800 °C in various condition : (a) The addition of LiCl; (b) Without the addition of LiCl.

Based on Figure 1 showed that the residual mass precipitated with LiCl (Fig. 1a) had a mass greater than the residual sample precipitated without the addition of LiCl (Fig. 1b). The difference between the two types of precipitated residues was ranging from 0.11 grams to 3.72 grams which indicated the presence of lithium adsorbed when the $Mn(OH)_2$ precipitation process formed MnO_2.

Figure 2 showed the residues calcinated at a temperature of 800 °C. Figure 2 looked that the residual mass with the addition of LiCl after the calcination process of 800 °C had a tendency to be greater than the residue without the addition of LiCl. These data reinforce the presumption of lithium adsorption when the precipitation process of $Mn(OH)_2$ forms MnO_2. The difference in residual mass (Fig. 2a and Fig. 2b) is significant in both samples which are 2.65 grams for mass variations of 15 grams of $MnSO_4$ and 3.32 grams for 25 grams of $MnSO_4$ mass variation. The difference in mass can be indicated as the mass of lithium adsorbed by manganese oxide and form lithium manganese after the calcination process.

3.2. Characteristic of filtrates by ICP-OES analysis

Manganese sulfate and lithium chloride used are the main constituents in the synthesis of lithium manganese. By analyzing the amount of lithium contained in the filtrate compareted to initial solution (LiCl solution with a concentration of 10 g/L), it can also be known that the lithium was adsorbed in the manganese hydroxide precipitation process. The amount of ability of each adsorbent in absorbing lithium can be seen in the chemical composition of the precipitated solids [2]. The following results of ICP analysis on 5 filtrates can be shown in Table 1.

Table 1. The chemical composition of initial solution and filtrates in variation $MnSO_4$ mass

$MnSO_4$ Mass (gram)	Li content (ppm)
Initial solution	12571
5	9406
10	8894
15	8919
20	8197
25	7907

Table 1 shows the greater the $MnSO_4$ used, the smaller the lithium content remaining in filtrates. This means that the greater the MnSO4 used, the greater the lithium which will be adsorbed in the process of manganese hydroxide precipitation. Based on the data above, the amount of lithium adsorbed during manganese hydroxide precipitation can be seen in Figure 3.

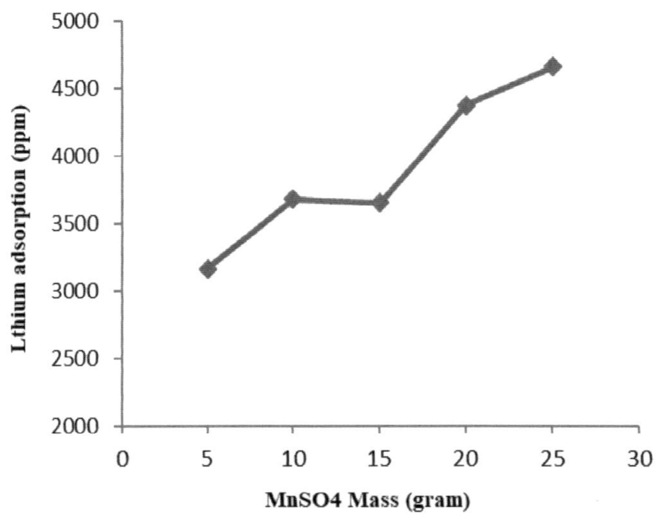

Figure 3. The adsorbed lithium (ppm) in variation $MnSO_4$ Mass

3.3. Characteristic of precipitated residues

Characterization using XRD (X-Ray Diffraction) analysis was carried out on samples of residues calcined at a temperature of 800 ° C. The results of the analysis were shown in Figure 4 and Figure 5.

The XRD analysis results in Figure 4 showed that lithium was successfully identified in the form of lithium manganese (IV) oxide ($LiMn_2O_4$) with a percentage reaching 3.3%. The lithium content identified in other forms was dominated by dilithium manganese oxide ($Li_2Mn_2O_4$) with a percentage of 96.7%. These results indicated that lithium manganese oxide was successfully formed in the $Li_2Mn_2O_4$ or $LiMn_2O_4$ phases. This result was also supported by the literature which stated that lithium can be adsorbed by manganese and will form a synthesis product of $LiMn_2O_4$ when calcined at temperatures of 800 °C [10]. The LiCl reagent which was a source of lithium was not fully effective in the formation of lithium manganese oxide when compared to using Li_2CO_3. The reaction of lithium manganese formation in this experiment was possible as follows:

$$LiCl_{(aq)} + 2MnOOH_{(s)} + {}^{1}/_{4}O_{2\cdot(g)} \longrightarrow LiMn_2O_{4(s)} + {}^{1}/_{2}H_2O_{(l)} + HCl_{(aq)} \text{..................} (4)$$

$${}^{3}/_{2}LiCl_{(aq)} + {}^{1}/_{2}LiMn_2O_{4(aq)} + O_{2\cdot(g)} \longrightarrow Li_2Mn_2O_{4(s)} + {}^{3}/_{4}Cl_{2(aq)} \text{........................} (5)$$

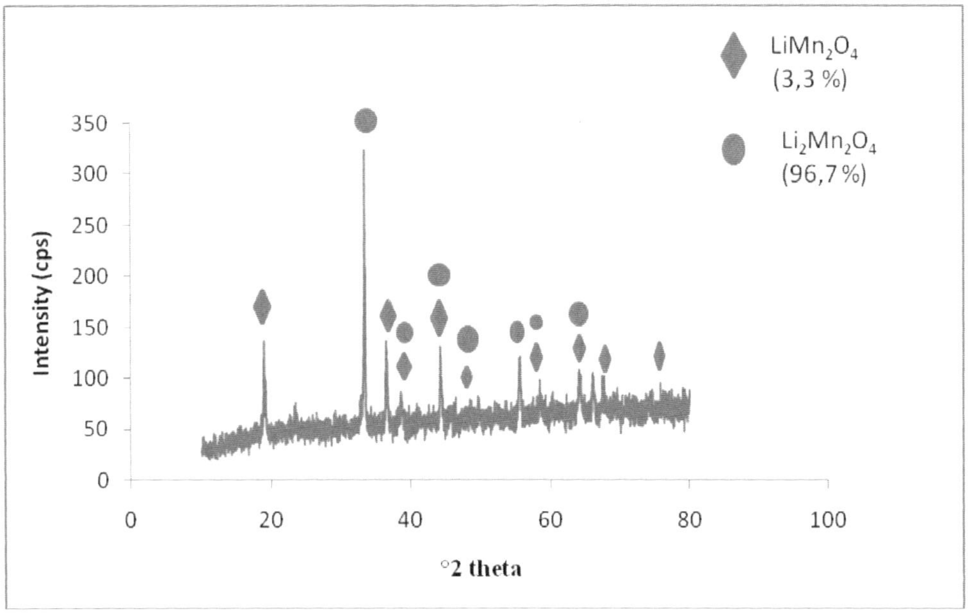

Figure 4. The XRD pattern of precipitated residues by using 15 grams of $MnSO_4$ with LiCl addition

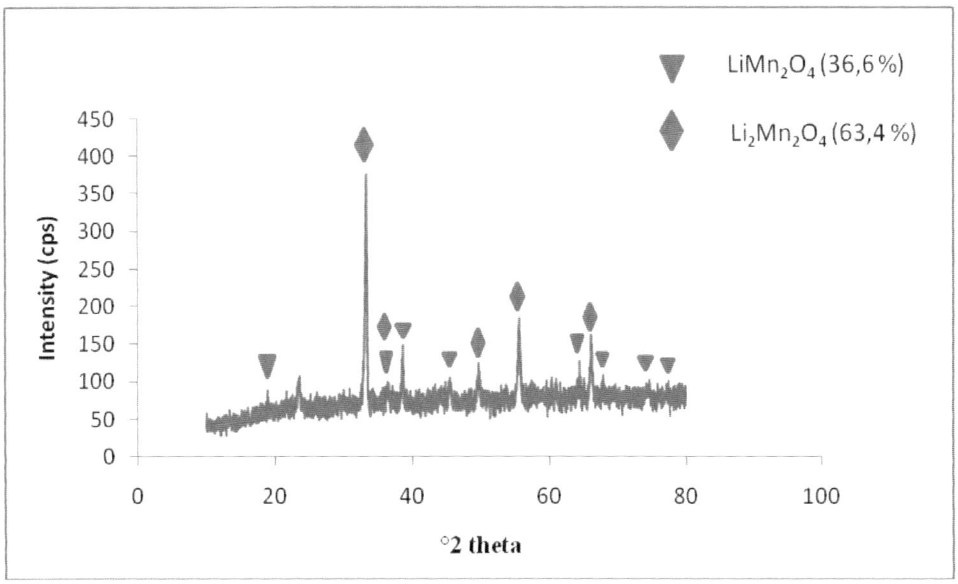

Figure 5. The XRD pattern of precipitated residues by using 25 grams of $MnSO_4$ with LiCl addition

The phase of the formation of both types of lithium manganese products namely $LiMn_2O_4$ and $Li_2Mn_2O_4$ run systematically and gradually. The formation of the $LiMn_2O_4$ product occurred first at a temperature of 650 °C according to reaction (4) then proceed with the formation of the $Li_2Mn_2O_4$ product at a higher temperature of about 800 °C according to reaction (5) 9). Reactions (4) and (5) which are crystallization processes of $LiMn_2O_4$ and $Li_2Mn_2O_4$ which are controlled by diffusion of O^{2-} ions in the samples.

The results of XRD analysis in Figure 5 showed that lithium in the dilithium manganese oxide phase ($Li_2Mn_2O_4$) is 63.4% and in the lithium manganese oxide ($LiMn_2O_4$) phase is 36.6%. The formation of $LiMn_2O_4$ and $Li_2Mn_2O_4$ products in a sample of 25 grams of $MnSO_4$ was also based on reactions (4) and (5) which were also affected by the presence of oxygen.

4. Conclusion

From the results of the experiment were obtained some conclusions, as follows: lithium can be adsorbed together with the manganese hydroxide precipitation process to form manganese oxide. The amount of lithium adsorbed was proportional to the reduction in lithium levels in $MnSO_4$, NH_4OH and LiCl solutions. The more mass of $MnSO_4$ used in the experiment indicates that more lithium was adsorbed in solution. Lithium adsorbed with MnO_2 formed lithium manganese oxide crystals with the dominant phase in the form of $Li_2Mn_2O_4$ and $LiMn_2O_4$ when dried at 100 ^oC and continued calcination at 800 ^oC.

Acknowledgments

This research was carried out at hydrometallurgy laboratorium and pyrometallurgy laboratorium in Research Center for Metallurgy and Material, Indonesian Institute of Sciences. The authors are grateful for the main financial supports from INSINAS budget 2018.

References

[1] Brian W J 2015 *2013 Minerals Yearbook* no. January U.S. Geological Survey 44.1-44.11
[2] Satriawangsa G A and Noerochim L 2014 *J. Tek. Pomits* **3(2)** 6–11
[3] Sulistiyono E, Lalasari L H, Mayangsari W, and Prasetyo A B 2018 in *Proceedings of the International Seminar on Metallurgy and Materials (ISMM 2017), AIP Conference Proceedings* **010001 (1964)** 1–6
[4] Lange J G 2012 *Improving Lithium-Ion Battery Power and Energy Densities Using Novel Cathode Architectures and Materials* University of Illinois at Urbana-Champaign
[5] Paravasthu R 2012 *Synthesis and Characterization of Lithium-Ion Cathode Materials in The System (1-x-y)LiNi1/3Mn1/3Co1/3O$_2$ · xLi$_2$MnO$_3$ · yLiCoO$_2$* 1–93
[6] Wang G 2000 *Investigations on electrode materials for lithium- ion batteries* Faculty of Engineering University of Wollongong
[7] Mikolajczak C, Kahn M, White K, and Long R T 2011 *Lithium-Ion Batteries Hazard and Use Assessment* Exponent Failure Analysis Associates, Inc, 149 Commonwealth Drive, Menlo Park, CA 94025
[8] Julien C M, Mauger A, Zaghib K, and Groult H 2014 *Inorganics* **2** 132–154
[9] Naghash A R and Lee J Y 2000 *J. Power Sources* **85** 284–293
[10] Yang X, Tang W, Kanoh H, and Ooi K 1999 *J. Mater. Chem.* **9(10)** 2683–2690

IOP Conf. Series: Materials Science and Engineering **478** (2019) 012012 doi:10.1088/1757-899X/478/1/012012

Application of Taguchi optimization on the titanium recovery from iron sands using flotation process

Sudibyo[1], A Junaedi[1] , M Amin[1], S Sumardi[1], F Nurjaman[1], F R Mufakhir[1], A S Handoko[1], Darmansyah[2], and M N Hidayat[2]

[1]Research Unit for Mineral Technology, Indonesian Institute of Sciences (LIPI), South Lampung, 35361, Indonesia
[2]Department of Chemical Engineering, Engineering Faculty, Lampung University, Bandar Lampung, 3514, Indonesia

Email: sudibyo@lipi.go.id

Abstract. Titanium is very important metal which commonly used in industrial process, automotive, electronic devices and other application. Titanium is not pure element in nature but mixed with other elements such as in irons sands. Hence, it is required to do research to gain high concentration titanium and also eliminate its impurities. One of method which able to overcome this problem is the flotation method. On this research, titanium recovery process is done from natural iron sands of Lampung using oleic acid as a collector in the flotation process. The iron sands which used for each variation are 10, 15 and 20 wt% with a size of 60, 100 and 150 mesh. Those iron sands were mixed into 0,10; 0,20; 0,30 lb/tons of collector by varying pH condition of 6, 8 and 10. After that, each variation product was characterized by XRF characterization and then calculated based on Taguchi method. From the research that had been done, it is known that the optimum values are at the content: oleic acid = 0,1 lbs/ton, ore 20 wt%, pH 8, the Particle size of 150 mesh.

1. Introduction

Titanium is a very important material which can be alloyed with other metal such as iron, aluminum, vanadium, and molybdenum, among other elements, to produce strong and lightweight alloys. The titanium alloys commonly used in aerospace, military, industrial processes, automotive, agrifood, medical prostheses, orthopedic implants, dental and endodontic instruments and files, dental implants, sporting goods, jewellery, mobile phones, and other applications [1,6].

Table 1. The analysis result of XRF from Lampung Iron Sand

Number	Compositions of Iron Sand	%	Value +/-
1	Fe	53,883	0,306
2	Le	26,185	0,370
3	Ti	3,828	0,040
4	Cl	2,028	0,071
5	Si	4,772	0,042
6	Al	2,004	0,132
7	K	0,636	0,088
8	Ca	1,344	0,011

Titanium can be produced from a natural material which available on earth such as iron sands. Titanium which produces from iron sand such as in Uni Soviet, Australia and South Africa[2]. Lampung province (Indonesia) has a lot of iron sand which contains of 3.83% titanium as listed in Table 1. One of the methods which able to increase the titanium contained in the sand is froth floatation. Froth flotation is a mineral separation process into ore from its impurities by floating ore to the surface through its bonds with froth. This process is used for some ores such as Cu, Pb, Zn, Ag, Au and Ni. The procedures are done by blowing air into mineral grains which mixed with water and frothing substances. The hydrophobic mineral grains then will be flown by air to the surface. The flotation process will be more effective if it is added to a collector substance. One of economical and effective collector for titanium floatation is oleic acid [3]. The parameters which have significant effect towards titanium floatation were studied and analyzed using Taguchi methods such as oleic acid concentration, ore concentration, pH in floatation cell and particle size. In this work, Taguchi method for four parameters and three levels only need 9 experiments [4,5]. This method needs less experiments than the full factorial methods which need 81 experiments.

2. Methodology

The sand used in this work was from iron sand from West Lampung beach which contains titanium as shown in Table 1. An oleic acid from Brataco Chemika Co. Ltd. (Jakarta, Indonesia) was used as a collector. Pine oil or surfactant Frother chemical which used in this work was purchased from Brataco Chemika Co. Ltd. (Jakarta, Indonesia). Lime (CaO) (Mutiara co. Ltd., Lampung, Indonesia) was used to adjust pH in floatation cell. In this work, iron sands were Iron sand which will be used is gained from Lampung that is separated by magnetseparator then grilled by a ball mill. Then, sieved by sieve in size of 60, 100 and 150 mesh. Flotation process was conducted in 6 Liters capacity of froth flotation cell (Quinn process equipment co., Denver, United States). The pH inside froth flotation cell was measured using Universal indicator from Merck (German). the composition of tailing and product of froth Flotation were analyzed using handheld X-ray fluorescent (XRF) analyzers from Olympus, (Massachusetts, US).

All experiments were conducted by following design experiment of the Taguchi orthogonal arrays (OA) method for four parameters and three levels as shown in Table 2. This method was used to determine the optimum operating condition and to study the effect of parameters towards a recovery of titanium [6,7]. The orthogonal arrays of the L9 type which has 9 experiments were chosen to study four parameters and three levels as shown in Table 2. The benefit using Taguchi orthogonal arrays (OA) this project only has 9 experiment, meanwhile when if this project using full factorial, its need 81 experiment. The optimum condition obtained was obtained using Signal to Noise (S/N) ratio of "larger the better" which was calculated using equation as follows:

$$S/N = -10*\log(\Sigma(1/Y2)/n)$$

where Y = responses for the given factor level combination and n = number of responses in the factor level combination [4]. All experiments were conducted for 10 minutes operations. During the experiment, the opening of the gas valve was kept constant to ensure all experiments were conducted in the same condition of the gas inlet. The froth from flotation cell was collected with an automatic scraper, which is installed in the flotation cell. At the end of an experiment, the gas inlet valve was closed and the compressor was turned off. Then concentrate sample or product of flotation and the tailing of floatation was sun-dried. Then, the drying products were analyzed by XRF and weighed using the analytical balance. The schematic diagram of froth flotation cell was used in this research was shown in Fig. 1.Percent of titanium recovery was calculated using formula as follows:

$$\frac{\text{Ti product x product weight}}{\text{tailing x tailing weight}} \text{ x } 100\%$$

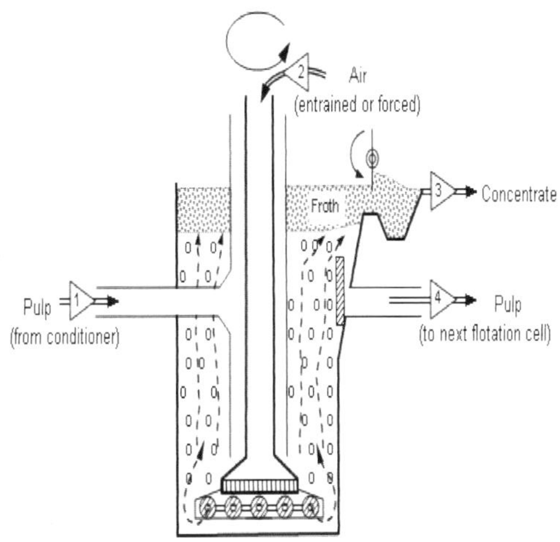

Figure 1. Diagram of Froth Flotation

Table 2. Titanium recovery from Iron Sand by Flotation Method

Run	*Oleic acid* (lb/ton)	*Ore* (wt %)	pH	Particle Size (mesh)	Ti product (%)	Ti tailing (%)	Product Weight (gram)	Tailing Weight (gram)	Recovery (%)
1	0.1	10	6	60	3,8	3,0	360	390	18,1631
2	0.1	15	8	100	5,1	4,4	700	460	36,8741
3	0.1	20	10	150	4,3	4,3	1240	340	41,1661
4	0.2	10	8	150	2,9	4,7	570	180	41,1661
5	0.2	15	10	60	3,0	3,5	980	130	10,8112
6	0.2	20	6	100	4,9	4,8	1270	280	35,9132
7	0.3	10	10	100	4,8	4,3	700	80	6,2504
8	0.3	15	8	150	4,1	4,0	700	389	27,1697
9	0.3	20	6	60	4,3	3,7	1280	270	37,1039

3. Result and Discussion

3.1. The effect of oleic acid concentrations towards titanium recovery

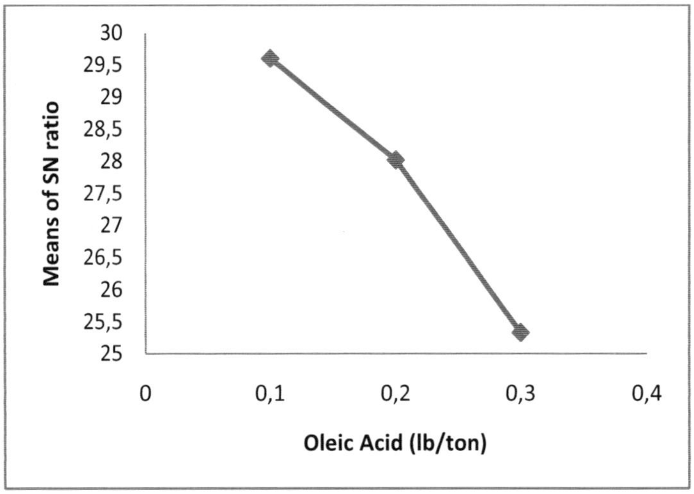

Figure 2. The effect of oleic acid content on the means of SN ratios

The effect oleic acid concentration towards titanium recovery is shown in Fig. 2. The figure shows the larger means of S/N ratio of titanium recovery for oleic acid concentration is at level 1. This result indicates that the best flotation recovery at 0.1 lb/ton of oleic acid. The means of SN ratios decreased from level 1 to level 3 for oleic acid concentration. This caused that the flotation recovery oleic acid act as a surfactant that increases the hydrophobic characteristic of the mineral. In nature, titanium commonly binds to sulfide material which has a hydrophobic characteristic. Oleic acid cause these sulfide materials become more hydrophobic hence, these hydrophobic materials easily float together with gas bubbles. However, the means of SN ratios decreases at levels 2 and 3. This decrease of percentage recovery is due to the excess of oleic acid as a surfactant will cause a lot of non-sulfide minerals which do not contain titanium will turn into hydrophobic, hence these float with gas bubbles. This phenomenon led to the decrease in the recovery percentage.

3.2. The effect of ore concentration towards titanium recovery

The effect of ore concentration towards titanium recovery is shown in Figure 3. The figure shows that the increment of ore concentrations will enhance the titanium recovery. The figure also shows that the optimum ore concentration (wt %) is at level 20 wt % of ore. This caused by more titanium particle in the mixture of sand and water, hence more titanium in the sand which able to separate using flotation process.

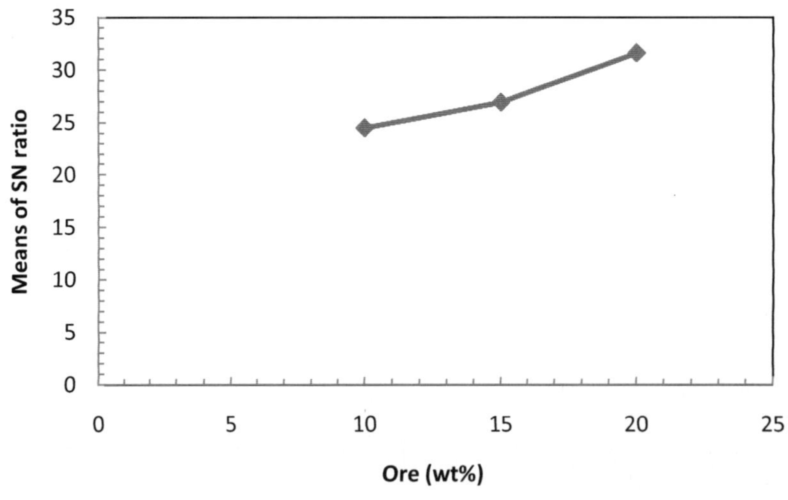

Figure 3. The effect of ore concentration (wt %) on the means of SN ratios

3.3. The effect of pH towards titanium recovery

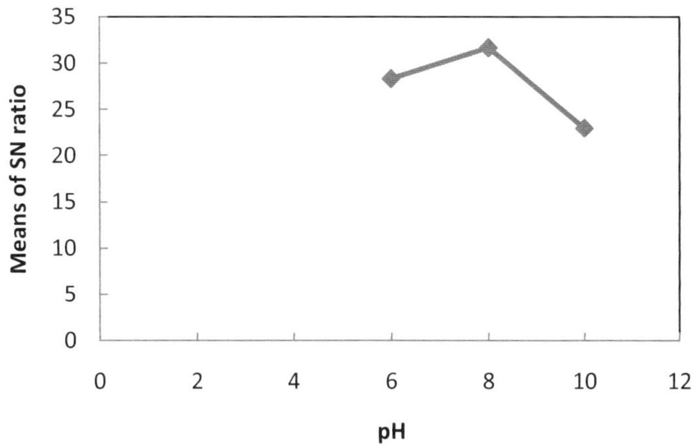

Figure 4. The effect of pH on the means of SN ratios

The correlation of pH towards titanium recovery in flotation process of iron sand is shown in Fig. 4. The figure shows that the increase in pH will increase the titanium recovery. However, at the pH of 10, the titanium recovery was decreased. This caused by the increased pH as the increase of CaO concentration which leads to the increase of impurities on the product floatation. Hence, the titanium recovery was decreased as the increase in pH. The surface of titanium ore in froth floatation process is affected by pH, at the pH of 8 the attraction of collectors to the titanium ore surface become more effective as shown in Fig. 4 [8] .

3.4. The effect of particle size towards titanium recovery

The titanium recovery on flotation process was affected by particle size as shown in Fig. 5. The figure shows that the increase of mesh or the decreased of the particle will enhance the titanium recovery on froth flotation process. This decrease of individual particle size caused the decrease of mass gravitation than these particles were lifted up by the gas bubbles easily. The smaller particle size has a good the degree of liberation which the most of valuable mineral particles has been separated with other mineral impurities. Hence, collector reagent interacts with the surface of the precious mineral particles and resulting the recovery will increase. However, if the particle size is too small it will cause high consumption of reagent or cause the poor selectivity due to mechanical entrainment as well as entrapment of ore with gangue or vice versa [9]. From the figure, it is shown that the optimum condition of flotation process for titanium is at 150 of mesh.

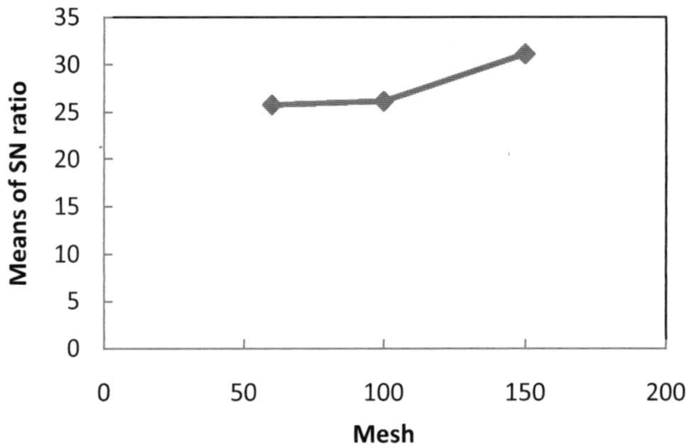

Figure 5. The effect of particle size on the means of SN ratios

3.5. Optimum Condition of Flotation

The largest titanium recovery percentage indicates the best flotation process. The titanium recovery percentage has been analyzed using Taguchi analysis "Larger the better ". The S/N ratio and mean analysis result was listed in Table 2 and 3, respectively. The data of the responses for the means and the values of the delta are shown in Table 2 and 3. Delta is the difference in values between the biggest to the smallest means or SN ratio of percentage titanium recovery. It represents the most parameter, which affects percentage gold recovery. From Table 3 and 4, pH has the biggest delta value, which means that the pH is the most significant parameter to the percentage of titanium recovery. This is followed by the ore concentration, particle size and oleic acid concentration. Based on Taguchi method which is shown in table 4.2, the optimum condition is the highest value of SN ratio of "larger the better". The optimum values are at the content: oleic acid = 0,1 lbs/ton, 20 wt% of ore, pH 8, the Particle size of 150 mesh.

Table 3. Graph of SN ratio analysis of Taguchi method by using Minitab

Level	Oleic acid	Ore	pH	Particle Size
1	29.60	24.46	28.32	25.75
2	28.02	26.90	31.67	26.12
3	25.33	31.59	22.96	31.09
Delta	4.27	7.13	8.71	5.34
Ranking	4	2	1	3

Table 4. Graph of analysis of mean recovery result from Taguchi Method by Minitab

Level	Oleic acid	Ore	pH	Particle Size
1	32.07	21.86	27.08	22.03
2	29.30	24.95	38.38	26.35
3	23.51	38.06	19.41	36.50
Delta	8.56	16.20	18.97	14.47
Ranking	4	2	1	3

4. Conclusion

The flotation process of titanium from iron sand has been conducted using Taguchi design of the experiment. Based on Taguchi analysis it is found that the pH is the most significant parameter to the percentage of titanium recovery than followed by the ore concentration, particle size and oleic acid concentration. The optimum condition also enable to determine using SN ratio of "larger the better". The optimum values are at the content: oleic acid = 0,1 lbs/ton, ore 20 wt%, pH 8, the Particle size of 150 mesh.

Acknowledgement

Financial support from Indonesian Institute of Sciences and Ministry of Research, Technology and Higher Education - the Republic of Indonesia through INSINAS research grant no. RT-2016-0227 is greatly acknowledged.

References

[1] Encyclopædia Britannica 2006 *Titanium*
[2] Carp O 2004 *Progress in Solid State Chemistry* **32** 33-177
[3] Alonso L M 2014 *Chemical Engineering.*
[4] Aulia A, Sudibyo, Bramantyo B A, Slamet S, Fika R M, Nurbaiti M A, Karna 2016 *LIPI. Indonesia 978-0-7354-1473-0*
[5] Ross T J 1996 *Taguchi Techniques for Quality Engineering: Loss Function, Orthogonal Experiments, Parameter and Tolerance Design*. New York: McGraw-Hill International Edition
[6] Sharma N K and Cudney E A 2011 *International Journal of Engineering, Science and Technology* **3(7)** 15-24
[7] Ain Z N, Rizal A S, and Abdulrazak A K 2015 *J. Engineering Science* **11** 9–16
[8] Ferihan G 2002 *Turkish J. Eng. Env. Sci.* **26** 309 – 318
[9] Rao D S, Kumar V, Bhaskar R G, and Prabhakar S 2011 *Journal of Mining and Metallurgy* **47A(1)** 37 – 49

Effect of concentration, agitation, and temperature of Pomalaa limonitic nickel ore leaching using hydrochloric acid

S Oediyani[1], U Ariyanto[1], and E Febriana[2]

[1]Metallurgical Engineering Department, Faculty of Engineering, Sultan Ageng Tirtayasa University, Cilegon 42435, Indonesia
[2]LIPI, Serpong, Indonesia

E-mail: s_oediyani@untirta.ac.id

Abstract. Pomalaa limonite nickel ore has nickel content of 1.29% and it has not been processed in Indonesia until now. In addition, The Minister of Energy and Mineral Resources Regulation No.5/2017 on adding the value of minerals. Therefore, the limonite nickel ore should be processed to produce nickel metal with the content 99.93%. The limonite nickel ore generally has low nickel content (less than 1.5%) therefore hydrometallurgy such as leaching is the suitable process. In this research is done 2 series of leaching. First, determination of leaching controlling rate by varying the temperature and time of leaching. Second, determination the optimum condition of the leaching variables (concentration of leaching agent, agitation and temperature) based on the leaching controlling rate. The amount of activation energy obtained from this research is 16.23 and 38.47 kcal/mol then it could be said that the reaction control of leaching limonite nickel ore from Pomalaa is a chemical reaction. In addition, a maximum extraction of 88.9% was achieved at 400 rpm, 6M hydrochloric acid concentration and a temperature of leaching about 90 °C.

1. Introduction

This research was motivated by the regulation of minister of energy and mineral resources on added value of local resources that mentioned in abstract section. There are two type resource of nickel such as nickel laterite and nickel oxide ores. Nickel laterite ores can be divided into two type i.e. saprolite and limonite. Saprolite nickel ore is high in nickel and magnesium (Ni> 1.6% and Mg> 20%), but low iron content (Fe <20%). In Indonesia, saprolite nickel ore has been commercially processed through pyrometallurgy to produce nickel matte and ferronickel (FeNi).

It is recorded that in 2016, Indonesia is the world's sixth nickel producer with production of 160,000 tons. Indonesia has an abundant potential of nickel resources (5.8 billion tons), spread over areas such as South Sulawesi, Southeast Sulawesi, Maluku and Papua. The limonite nickel ore is a type of laterite with characteristics having a low nickel and magnesium content (Ni <1.5% and Mg <2%), but high iron content (Fe> 50%) [1] therefore the limonite nickel is not suitable to be processed through pyrometallurgy. It could be said that, hydrometallurgy is the suitable process for limonite nickel ore.

In nature, saprolite ore lies in the innermost layer, while limonite ore is present in the upper layer. To mine the saprolite ores must first mine the limonite layer above it. One of the areas in Indonesia that has limonite nickel ore reserves is in Pomalaa, Southeast Sulawesi with an average nickel content of 1.3% and iron 28.63% [2]. Nowadays, the limonite layer is not processed because of its low nickel content (Ni <1,5) but in the presence of Ministry of ESDM regulation, it is need to do a research to process limonite nickel ore.

Leaching is a selective dissolution process that aims to dissolve precious metals from a concentrate without dissolving the impurity metal.The most influential factors in the leaching process include the concentration of leaching agent and the speed of agitation. If the reaction is controlled by a chemical reaction, the concentration of leaching agent affects the reaction whereas the speed of agitation will affect the reaction if the reaction is controlled by diffusion.

Leaching is the main stage of the hydrometallurgical process using leaching agents. the leaching technique chosen is atmospheric leaching. this process has advantages in terms of simple equipment procurement and done at atmospheric pressure [3]. In the leaching process, leaching agent plays an important role and in this research hydrochloric acid is used as leaching agents because of its ability to dissolve limonite nickel ore. Another important that should be considered in the leaching process is the reaction controlled such as diffusion or chemical controlled reaction that can be known from the amount of activation energy. Therefore, the first stage of leaching process with variations in temperature is done.

2. Methodology/experimental

In this research, the preparation processes include ore beneficiation (drying, crushing, and sizing) and atmospheric leaching method was used. Then, limonite nickel ore analyzed by using XRF to measure the chemical composition of the ore.

First, the leaching stage I is done with the variation of temperature (ambient, 50 and 90°C) and leaching time (3, 15, 30, 60, 90, 120, 180, 240, 360 and 480 minutes. Then, by collecting data from variation of leaching temperature and using Arrhenius equation it could be known the controlling leaching reaction steps.

Second, the leaching stage II is carried out with the variation of concentrations of hydrochloric acid (1, 4, and 6 M), agitation speeds (200, 400, and 600 rpm) and time of leaching (3, 180 and 480 minutes). After leaching is done, the filtration stage is generated then the filtrate and residue are produced. The filtrate was tested its chemical composition with AAS and the residue tested with XRF. Figures 1, 2, 3 and Table 1 show the limonite nickel ore, the laboratory scale of leaching limonite nickel, filtration process after leaching and the composition of limonite nickel ore that used in this research, respectively.

Figure 1. Pomalaa limonite nickel ore

Figure 2. Laboratory scale of leaching limonite

Figure 3. Filtration process

Table 1. Chemical composition of limonite nickel ore pomalaa

Substitute	%	Substitute	%	Substitute	%
SiO_2	11.20	Na_2O	0.035	Cr_2O	2.18
Al_2O_3	6.23	K_2O	0.22	NiO	1.64
Fe_2O_3	49.50	TiO_2	0.10	Co_2O_3	0.10
MnO	0,61	P_2O_5	0.014	CuO	0.014
MgO	2.20	CaO	3.86		

3. Results and discussion

3.1. Determination of rate controlling leaching

According to some researchers that there are several kinetics models are known in the leaching process such as unreacted core models [4] and Shrinking Core Models [5,6]. There are three mechanisms that can be determined by leaching process such as mass transfer from the ambient fluid through the fluid film surrounding the particle; diffusion through the inert product layer left behind as the core shrinks; chemical reaction at the surface of unreacted core. The standard Shrinking Core model is able to account. The first mechanism relatively has little impact on the leaching process [5] therefore in this research diffusion and chemical mechanisms are discussed.

On the application of shrinking core is assumed that the shape of particles is spherical. Figure 4 shows the kinetic model of leaching limonite nickel rate controlling steps such as diffusion and chemical, respectively while Table 2 shows the R^2 values obtained from linear regression equation using shrinking core model on Figure 4. Based on the data shown on Table 2, the value of R^2 equals one is a diffusion model therefore it could be said that leaching limonite nickel Pomalaa is controlled by diffusion.

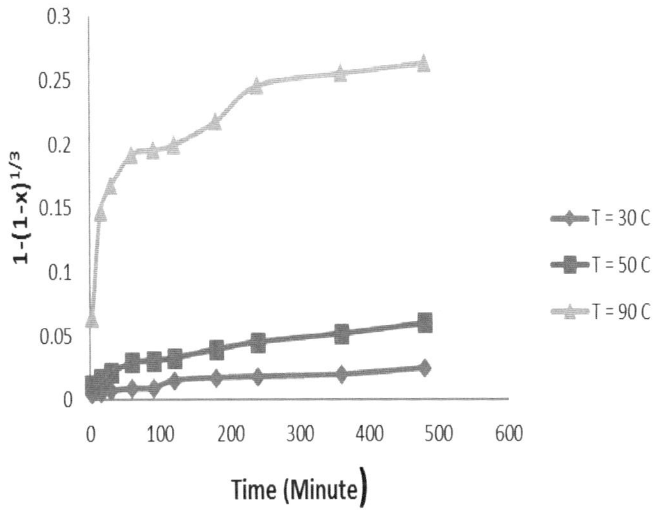

(a)

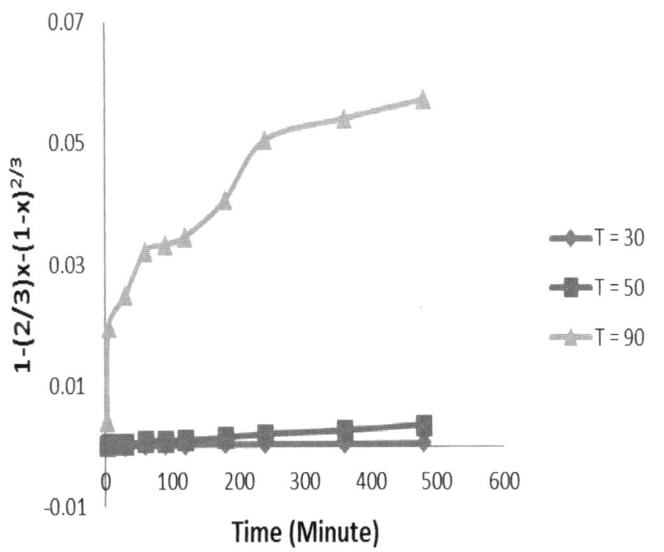

(b)

Figure 4. Shrinking Core Model for (a) Diffusion controlled reaction and (b) Chemically controlled reaction of leaching limonite nickel Pomalaa

Table 2. The Value of R^2

Temperature (°C)	R^2	
	$1-(1-x)^{1/3}$	$1-(2/3)x-(1-x)^{2/3}$
30	0.913	0.956
50	0.932	0.993
90	0.672	0.801

In addition, the amount of activation energy can be used to determine the rate controlling step of limonite nickel leaching process. The value of the activation energy can be determined from the curve (-ln K) vs (1 / T) while the value of (-ln K) is obtained from the slope of the curve shown in Figure 5. Figure 5 shows the curve (-ln K) vs (1/T).

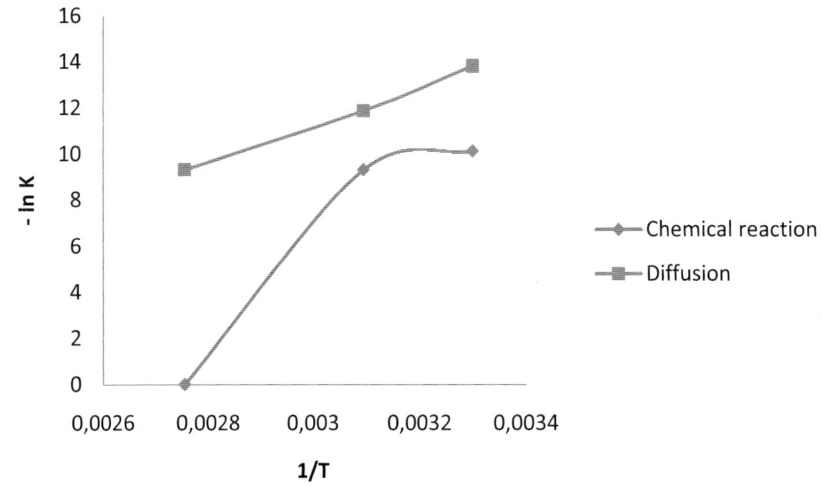

Figure 5. The Graph of (-ln K) vs (1/T)

Based on figure 5, the value of activation energy for chemical controlled reaction is 38.47 kcal / mol, while for diffusion controlled is 16.23 kcal / mol. According to Habashi, the activation energy with values greater than 10 Kcal / mol indicates that the process is controlled by chemical controlled reaction [7]. Therefore, it could be said that leaching limonite nickel ore from Pomalaa is chemically controlled reaction.

3.2. Effect of leaching temperatures on percent extraction of nickel

Figure 6 shows the effect of temperature on the percentage of nickel extraction. It could be seen from Based on Figure 6 it can be seen that the temperature significantly affects the percentage of nickel extraction, the higher the leaching temperature, the higher the percentage of nickel extraction. A maximum extraction percent value of 60.01% is achieved at 90 °C. Meanwhile, the percentage of extraction increased significantly from the temperature of 50 (16.91%) to 90 °C (60.01%). This is one of the indication that the leaching process of the limonite nickel ore of Pomalaa is chemically controlled reaction In addition, it supports the results of this research using the shrinking core model as described in point 3.1.

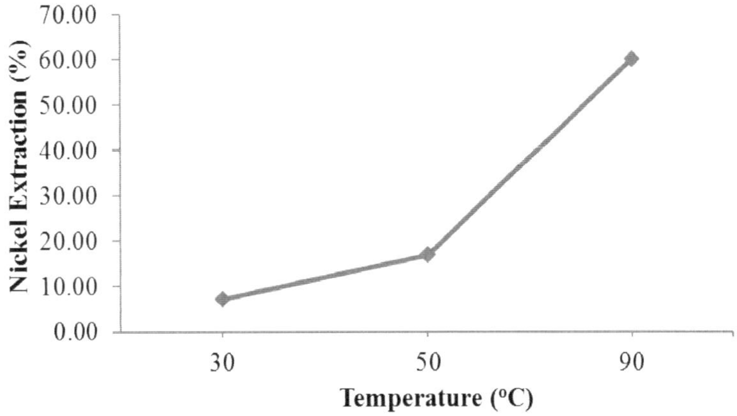

Figure 6. Effect of temperature leaching on nickel extraction

3.3. Effect of stirring rate on percent extraction of nickel

Figure 7 shows the effect of stirring rate on percentage of nickel extraction. The maximum percentage of extraction (88.90%) was obtained at a stirring rate of 400 rpm, while the minimum extraction percent was 84.10%. In addition, the percentage of nickel extraction did not change significantly with the increasing of stirring rate. According to Habashi on Kinetics of heterogeneous reactions, on increasing the rate of the rate of stirring in a solid - liquid reaction (leaching process), "an increase in the rate of dissolution sometimes takes place". This is the case when the process is diffusion controlled because the thickness of the boundary layer decreases with increased speed of stirring. In addition, the rate of dissolution also increases. On the other hand, "a chemically controlled process independent of the speed of stirring" or is affected to a limited extent by a stirring rate [7]. It could be said that the leaching limonite nickel is controlled by chemical reaction.

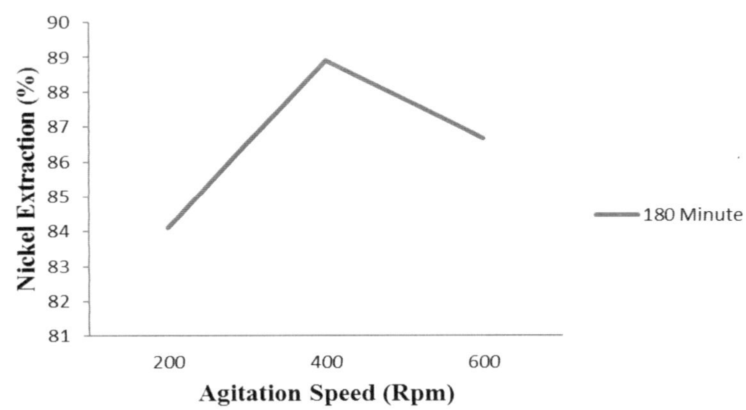

Figure 7. Effect of stirring rate (rpm) on nickel extraction

3.4. Effect of concentration of hydrochloric acid on percent extraction of nickel

Figure 8 shows the effect of concentration of hydrochloric acid on Nickel extraction. It could be said that the percentage of Nickel extraction is increasing with increasing HCl concentration. Maximum nickel extraction (86,49%) was obtained at HCl concentration 6 M, while minimum extraction percentage

(31.22%) at 1M HCl concentration. The formula to calculate the percentage of nickel extraction is stated in the following equation:

$$\%\, extraction\, of\, nickel = \frac{\left[\frac{ppm\ x\ amount\ of\ dilution}{1000}\right] x\ [volume\ of\ hydrochloric\ acid]}{(weight\ of\ limonite\ ore\ x\ \%Ni\ on\ limonite\ ore)} x\ 100\%$$

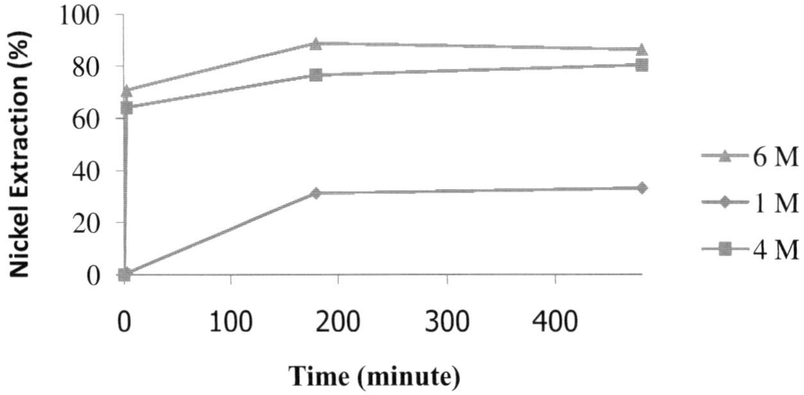

Figure 8. Effect of concentration of hydrochloric acid on nickel extraction

The increasing of percentage of nickel extraction due to the increasing of concentration of HCl is supported by the research conducted by N.M. Rice and L.W. Strong [8] which states that the increasing of hydrochloric acid concentration causes the increasing of percentage of nickel. Figure 9 shows the result of N.M. Rice and L.W. Strong [8] research.

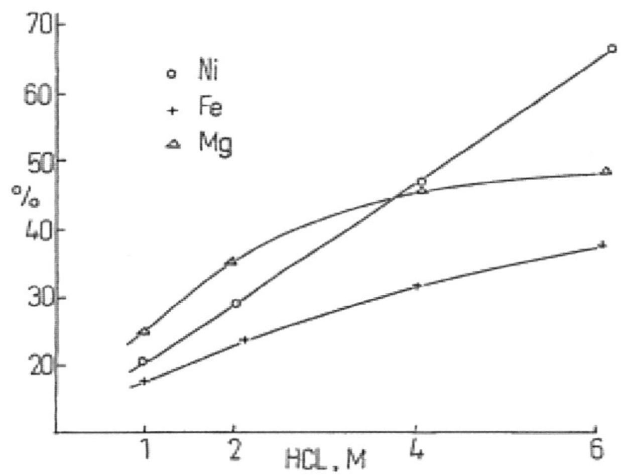

Figure 9. Effect of acid strength on nickel, iron and magnesium extraction ore [8]

4. Conclusion

Based on this research, it can be concluded that:

- According to the shrinking core kinetic model, it was found that the leaching process of nickel limonite ore from Pomalaa by using hydrochloric acid had an activation energy as much as 38.47 kcal/mole. It could be said that the leaching process was a chemically controlled reaction.
- The optimum percentage of nickel leaching process for limonite Pomalaa ore was 88.90%. This result was achieved under leaching process conditions such as 400 RPM of stirring rate, 90oC of leaching temperature and 180 minutes of leaching time.

Acknowledgement

This research was conducted by research collaboration between the Metallurgical Engineering Department of Engineering Faculty of Sultan Ageng Tirtayasa University with Department of Metallurgy LIPI, Serpong. Special thanks is given to my colleague, Mrs. Eni Febriana and also to Ulya, Agung, Taufik and Irvan, students of Metallurgical Engineering Department Untirta.

References

[1] Rudi S 2015 *Metalurgi* **30(2)** 71-80
[2] Widi A, Mohammad Z M, Siti K C 2012 *Prospek Fungal Bioleaching untuk Pengolahan Bijih Nikel Laterit di Indonesia* Prosiding Indonesian Process Metallurgy
[3] Nurhayati I C 2012 *Efek Scale-Up Pada Proses Pelindian Bijih Nikel Kadar Rendah Jalur Hidrometalurgi* Universitas Indonesia.
[4] Hem S R 1992 *Kinetics of Metallurgical Reactions* New York International Science Publisher
[5] Richard J F 2015 *Models for apparent reaction kinetics in heap leaching: a new semi-empirical approach and its comparison to shrinking core and other particle-scale model*
[6] Sohn H and Wadsworth M E 1979 *Rate Process of Extractive Metallurgy* Plenum New York
[7] Habashi F 1968 *Extractive Metallurgy Vol 1: General Principle* Gordon and Breach, Science Publishers, Inc. New York
[8] Nevill M R and Strong LW 1974 *Canadian Metallurgical Quarterly* **13 (3)**

IOP Conf. Series: Materials Science and Engineering **478** (2019) 012014 doi:10.1088/1757-899X/478/1/012014

Effect of ore mineralogy on the reductive-leaching of manganese ores

W Astuti, F R Mufakhir, E Prasetyo, A P T Yuda, S Sumardi, F Nurjaman, Y I Supriyatna, A S Handoko, and Suharto

Research Unit for Mineral Technology, Indonesian Institute of Sciences (LIPI), Jl. Ir. Sutami Km. 15, Tanjung Bintang, Lampung Selatan, Lampung, Indonesia

Email: widi004@lipi.go.id

Abstract. Manganese ores from Waykanan region (WK ore) contain pyrolusite (MnO_2) as major minerals, whereas the main minerals in manganese ore from Tanggamus region (TG ore) are iron manganese silicide. The effects on manganese extraction of leaching reagents without reductant, leaching temperature of non-reducing leaching, and reducing agents of sulfuric acid leaching were determined to investigate the leaching performances of the two manganese ores. Manganese leaching efficiency from WK ore is always higher than that from TG ore under the same leaching conditions. The mineral contents of the ores significantly affected the leaching performances and mineral dissolution behaviors of the samples.

1. Introduction

Manganese plays an important role in many fields, such as steel production, preparation of dietary additives, fertilizers, cells, and fine chemicals [1,2,3,4,5,6,7,8,9,10,11,12,13. Many efforts have applied recently to develop a commercial hydrometallurgical process to recover manganese from various manganese ores because most manganese products are produced from hydrometallurgical processes. Reductive leaching is one of the most important methods for extracting manganese from manganese ores. Some reductants such as molasses, biomass, sugar, oxalic acid, etc. have been applied in the leaching of manganese ores using acids mainly inorganic acids as the acidic medium and leaching reagents [14,2, 3, 4, 5, 6, 7,8, 9, 10, 11, 12, 13]

Indonesia has potential manganese ore resources that are distributed in some islands. They are widely distributed and spread among the islands of Sumatera, Kalimantan, West Java, Yogyakarta, East Java, and East Nusa Tenggara. The properties of manganese ores from every area may differ, depending on the sample origins; this is also the case for Indonesian manganese ores. The chemical and mineral compositions of manganese ores from different origins or even from different depths in one area can vary significantly. Therefore different leaching efficiency and leaching behavior might be occured, even though the similar method was used. Leaching rates and products will depend not only on the leaching conditions but also on the mineralogy of ores [15]. Specific mineralogical research should therefore always be conducted when a new manganese ore sample is found.

The aim of this study was to obtain a better understanding of the comparative mineral characterization of the ores, leaching performances, and mineral dissolution behavior of different Indonesian manganese ores from two different mining areas in Lampung Province, under atmospheric-reductive-leaching conditions using sulfuric acid and various reducing agents. This is important in the design of alternative techniques for treating such ores. The findings of the study can be adopted for leaching of other manganese ores from different locations. The leaching processes of manganese and iron from two different manganese ores by applying several leaching parameters were demonstrated to investigate the effect of reducing agents and leaching parameters on the leaching efficiency and leaching behaviour of each manganese ore.

2. Materials and Method

2.1. Materials

Manganese ores were originated from two mining areas i.e. Way Kanan (WK ore) dan Tanggamus (TG ore) in Lampung Province, Indonesia. The samples (WK ore and TG ore) were mineralogically and chemically characterized. X-ray fluorescence (XRF, Epsilon 3XLE, PANalaytical, Netherland) was used to determine the chemical compositions of these samples. The mineral phases of the raw ore samples were identified by X-ray diffraction (XRD, X'Pert 3 Powder, PANalytical, Netherland), using CuKα radiation, in the 2θ range 5° to 80°. Scanning electron microscopy/energy-dispersive X-ray spectroscopy (SEM/EDS; JEOL, USA) was used to determine the nickel contents of the different minerals by metal mapping.

2.2. Reductive Leaching method

To evaluate the leachability of mangan from two different manganese ores, the ores with particle size of <75 μm were acid leached (5% w/v of pulp density) by sulfuric acid (various concentrations) using various reducing agents at 150 rpm of shaker speed, 30°C of leaching temperature, and 8 hours of leaching period. Analytical-grade of all chemicals (Wako Pure Chemicals Industries, Ltd., Tokyo, Japan and Sigma-Aldrich, Germany) were used to prepare the solutions. Ultrapure water with a resistivity of 18 ΩM was used throughout this work. Reducing agents used in the experiments were molasses, glucosse, fructose, cane sugar, sodium sulfit, tannic acid, oxalic acid, citric acid, black tea, green tea, cellulose, sawdust of candlenut shell, sawdust of Acacia wood, sawdust of coconut shell, sawdust of Albizia chinensis wood, and hydrogen peroxyde. Liquid samples after leaching were analyzed by Induced Couple Plasma-Optical Emission Spectroscopy (ICP-OES; Perkin Elmer 8500, Waltham, MA, USA) to calculate metal concentrations in the pregnant leaching solution.

3. Results and Discussion

3.1. Comparison of ores characterization

XRF analyses in Table 1 show that manganese ores used have different chemical composistions. WK ore has higher manganese content (36.77%) than TG ore (12.5), while the iron content of TG ore (11.02%) is greater than that of WK ore (6.59%). In addition, Si and Al content in both samples are similar. Figure 1 shows the XRD pattern of WK ore and TG ore. From this figure, it can be seen that major minerals in WK ore are quartz and pyrolusite (MnO_2), while TG ore contains quartz and Fe_2MnSi. Manganese in both samples is incorporated in different minerals. Figures 2 and 3 present the metal mapping of each sample respectively by SEM/EDS analysis. Figure 2 shows that manganese in WK ore is closely associated with oxygen as manganese dioxide (MnO_2), whilst from figure 3, it can be seen that manganese in TG ore is incorporated with iron.

Table 1. Chemical composition of manganese ores (determined using XRF)

Sample	Elements (wt %)				
	Si	Mn	Fe	S	Al
WK ore	4.13	36.77	6.59	0.31	0.45
TG ore	6.89	12.50	11.02	0.59	0.69

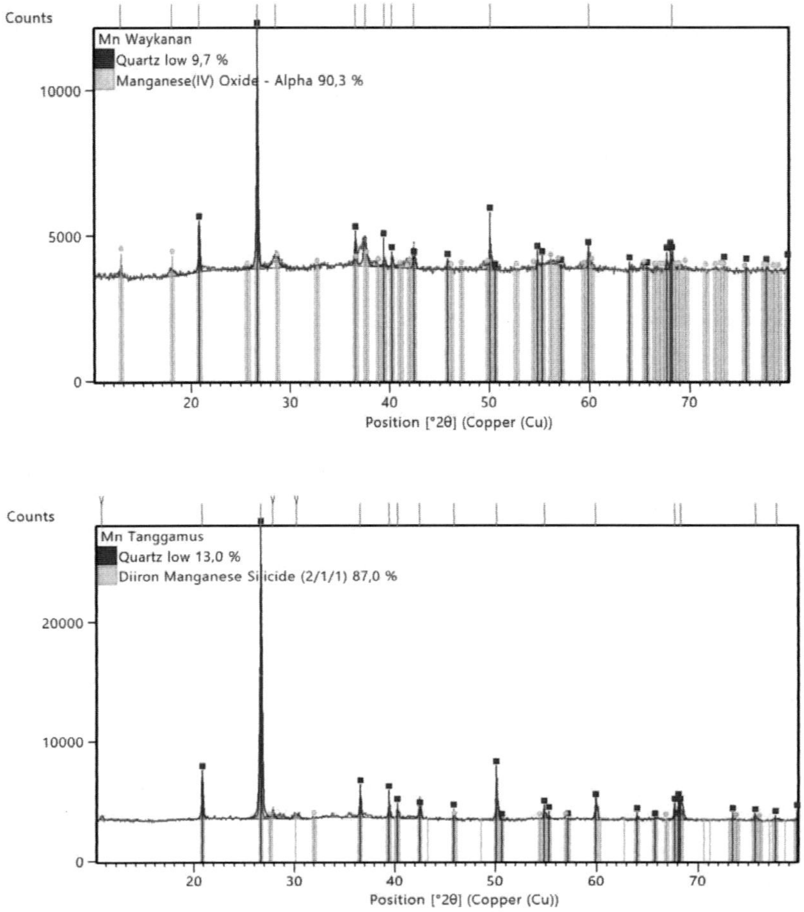

Figure 1. XRD pattern of manganese ores (a) WK ore; (b) TG ore

IOP Conf. Series: Materials Science and Engineering **478** (2019) 012014 doi:10.1088/1757-899X/478/1/012014

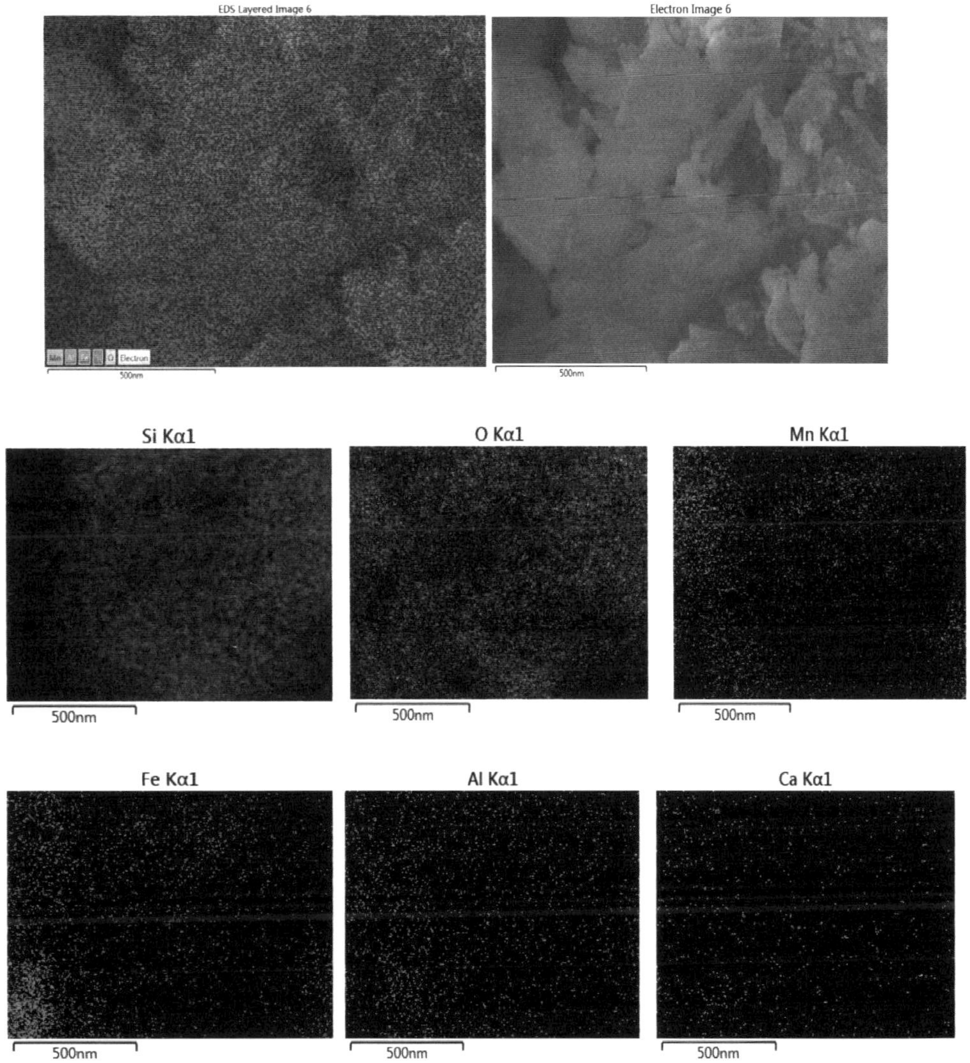

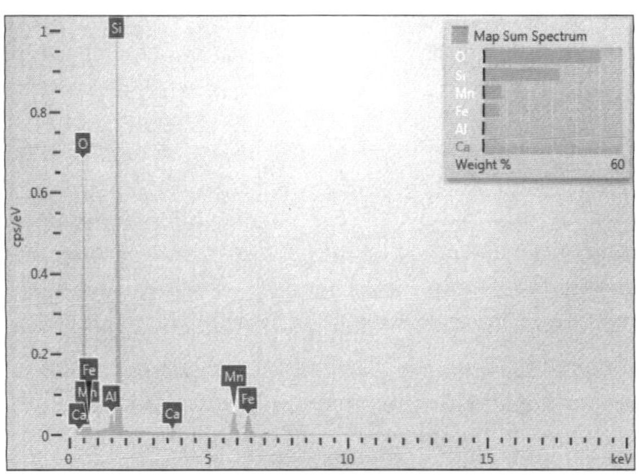

Element	Wt%
O	50.45
Al	1.24
Si	32.96
Ca	0.28
Mn	7.99
Fe	7.07
Total:	100.00

Figure 2. SEM/EDS metal mapping of TG ore

IOP Conf. Series: Materials Science and Engineering **478** (2019) 012014 doi:10.1088/1757-899X/478/1/012014

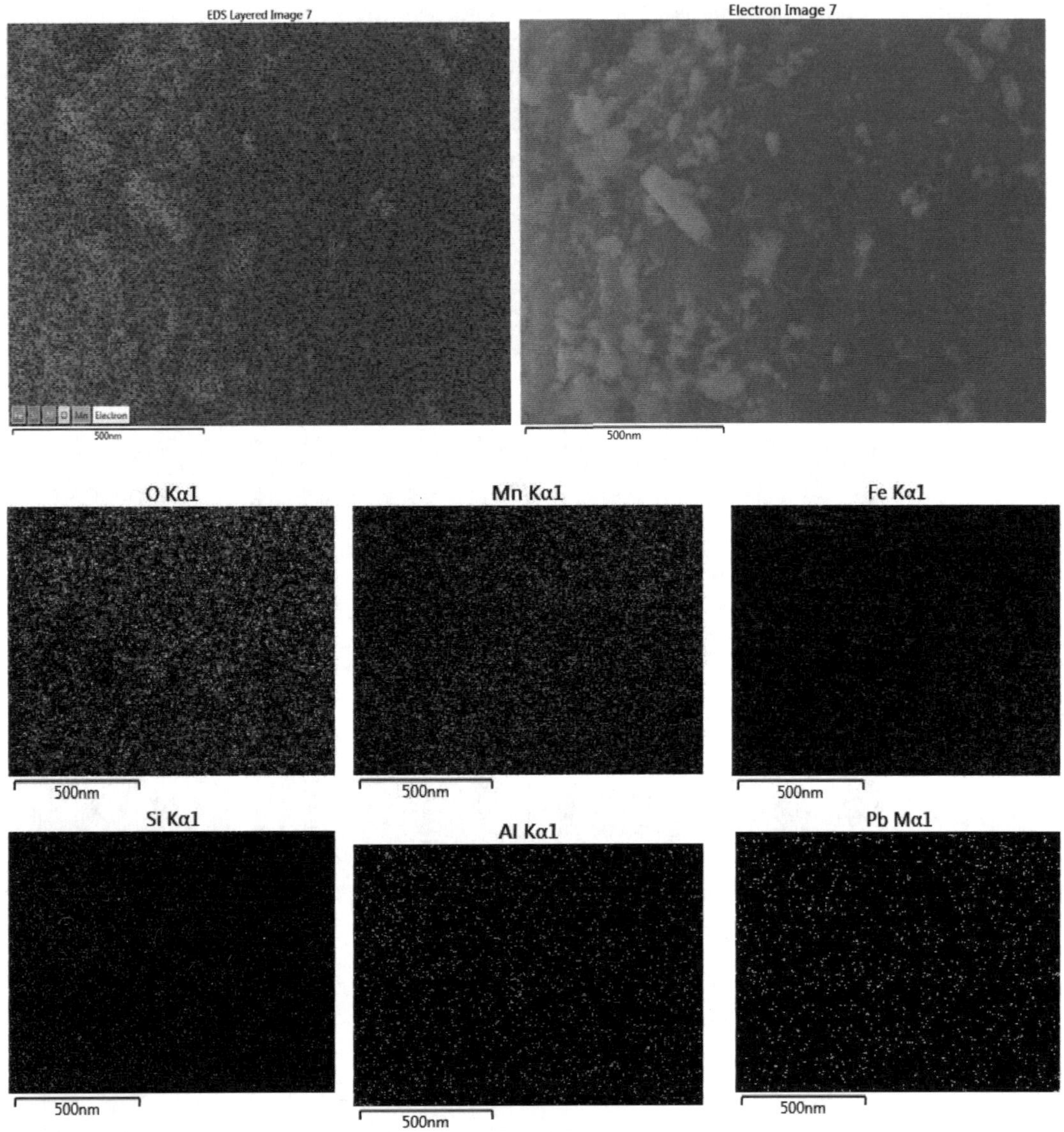

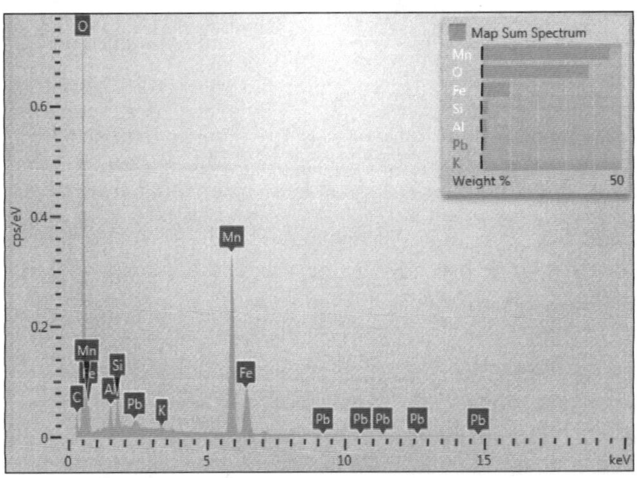

Element	Wt%
O	38.13
Al	2.06
Si	2.31
K	0.25
Mn	45.76
Fe	9.81
Pb	1.68
Total:	100.00

Figure 3. SEM/EDS metal mapping of WK ore

3.2. Effect of reagents on the non-reducing leaching

Leaching reagent is one of the most important parameters on the effectiveness and behavior of leaching. The effects of leaching reagents on the manganese recovery of this study were studied for concentration each acid of 1 M at 30°C, 150 rpm shaker speed, 8 hours of leaching period, and atmospheric pressure using <75 μm orc particle size and 5% (w/v) pulp density. Two manganese ores (WK ore and TG ore) were used to investigate the comparison of manganese leaching behavior with different leaching reagents. Figure 4 shows that without reducing agents, both manganese ores were very difficult to be leached by all leaching reagents.

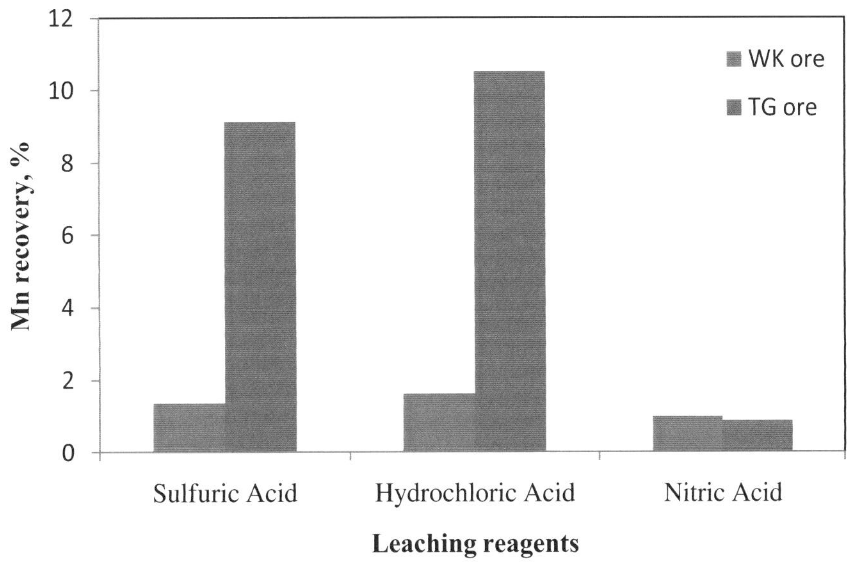

Figure 4. Manganese recoveries on the leaching of manganese ores (WK ore and TG ore) by various leaching reagents (acid concentration: 1M, pulp density: 5%, shaker speed: 150 rpm, temperature: 30°C, leaching period: 8 hours)

3.3. Effect of temperature on the non-reducing leaching

From the previous experiment, it can be seen that leaching reagents did not affect to the manganese leaching recoveries for both samples although the types of major manganese minerals contained in them were different. In order to increase the manganese recovery, a higher leaching temperature was used. The effects of leaching temperatures on the manganese recovery of this study were studied for sulfuric acid concentration of 1 M, 150 rpm shaker speed, 8 hours of leaching period, and atmospheric pressure using <75 μm ore particle size and 5% (w/v) pulp density with three various leaching temperatures (30, 50, 70°C). The results (Fig. 5) show that for both samples, manganese leaching efficiencies were not affected by leaching temperature. There were no significant changes when the leaching temperature were increased. Whereas usually the temperature greatly affects the leaching results and increases the recovery value, but in this study, the increasing temperature from 30 °C to 50 °C and 70 °C actually shows a decrease in manganese recovery value. It can concluded that the increase of temperature in the non-reducing leaching of manganese ore does not affected to the increase of manganese recovery.

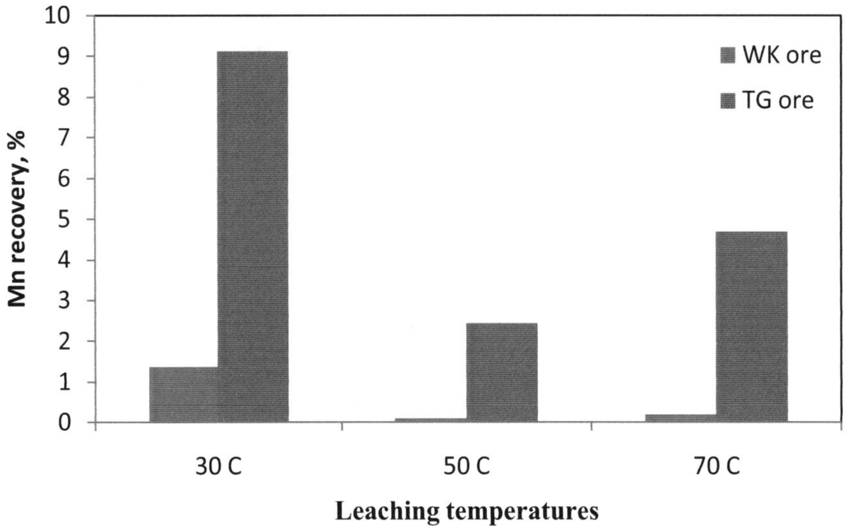

Figure 5. Manganese recoveries on the leaching of manganese ores (WK ore and TG ore) by various leaching temperatures (acid concentration: 1M, pulp density: 5%, shaker speed: 150 rpm, temperature: 30°C, leaching period: 8 hours)

3.4. Effect of various reducing agents on the sulfuric acid leaching

In some references, it was found that reducing agent was required in the leaching of manganese from manganese ore due to the stability of manganese in the manganese ore. The effect of various reducing agents on the sulfuric acid leaching of pyrolusite ore from WK ore and TG ore was conducted using 1 M sulfuric acid concentration at 30°C, 150 rpm shaker speed, 8 hours of leaching period, and atmospheric pressure using <75 μm ore particle size and 5% (w/v) pulp density. Sixteen reducing agents were used in this study that were molasses, glucosse, fructose, cane sugar, sodium sulfit, tannic acid, oxalic acid, citric acid, black tea, green tea, cellulose, sawdust of candlenut shell, sawdust of Acacia wood, sawdust of coconut shell, sawdust of Albizia chinensis wood, and hydrogen peroxyde. The results presented in Figure 6 show that the manganese leaching are strongly dependent on the reducing agents. The best reducing agents that produce the greatest manganese recoveries in both samples were tannic acid and oxalic acid. From this figure, it can also be found that manganese recovery from WK ore was higher than it from TG ore when tannic acid,oxalic acid, sodium sulfit, and cane sugar were used as reducing agents. However, manganese recovery from TG ore was greater than it from WK ore when the other reducing agents were used. In addition, figure 7 shows that iron was also leached simultaneously during leaching. This iron should be separated after leaching process if the manganese product is needed in the pure form.

IOP Conf. Series: Materials Science and Engineering **478** (2019) 012014 doi:10.1088/1757-899X/478/1/012014

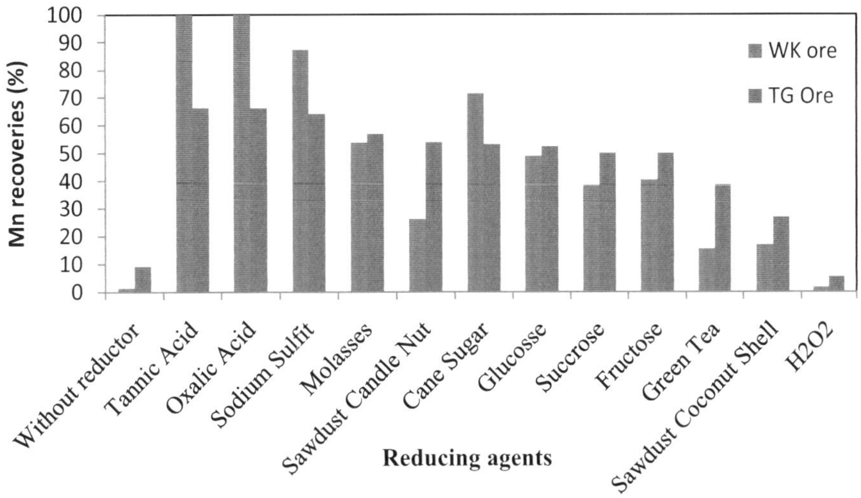

Figure 6. Manganese recoveries on the leaching of manganese ores (WK ore and TG ore) by various reducing agents (acid concentration: 1M, pulp density: 5%, shaker speed: 150 rpm, temperature: 30°C, leaching period: 8 hours).

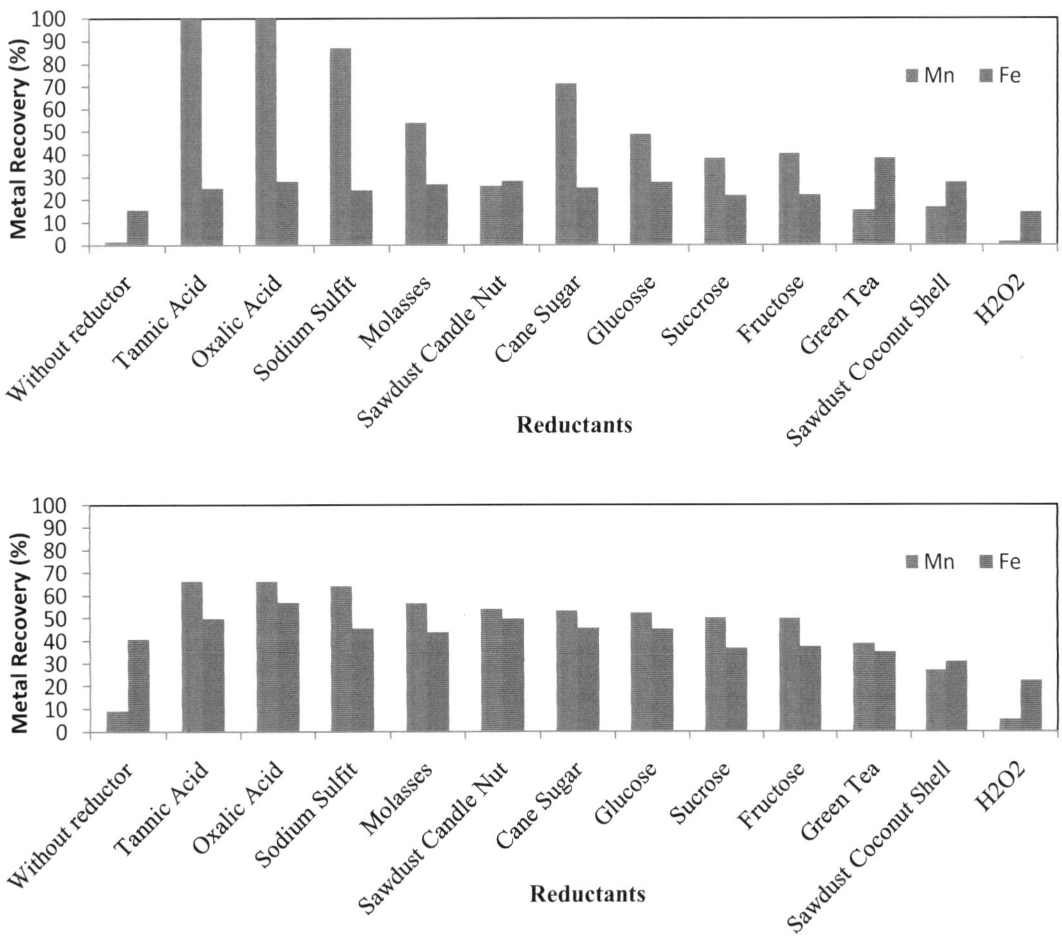

Figure 7. Comparison manganese and iron recoveries on the leaching of manganese ores (WK ore and TG ore) by various reducing agents (acid concentration: 1M, pulp density: 5%, shaker speed: 150 rpm, temperature: 30°C, leaching period: 8 hours) (a) WK ore (b) TG ore.

4. Conclusion

The chemical contents of the ores indicated that both samples were typical low-grade manganese ores. Tthe mineral composition of both samples differs. The WK ore contains quartz and pyrolusite as the main minerals, whereas the major phases in TG ore are quartz and diiron manganese silicide . The maximum leaching efficiency of manganese was achieved from WK ore (100%) and TG ore (66%) under the same leaching conditions, i.e., sulfuric acid concentration 1 M, leaching temperature 30 °C, leaching time 8 hours, ore particle size <75 μm, shaking speed 150 rpm, and pulp density 5% (w/v) using tannic acid as well as oxalic acid (2.5% w/v) as reducing agents. Manganese leaching rate from WK ore is always higher than that from TG ore under the same leaching conditions. From this study, it can be concluded that ore mineralogy is significantly affected to the leaching behaviour and leaching efficiency of manganese ore.

Acknowledgement

The authors gratefully acknowledge Pusat Pengembangan Teknologi Tepat Guna – LIPI and Balai Penelitian Teknologi Mineral – LIPI for the funding support of research and research facility.

References

[1] Su H, Wen Y, Wang F, Sun Y, and Tong Z 2008 *Hydrometallurgy* **93** 136–139
[2] Chao X, Longjun X, Tiefeng P, Kun C, and Jing Z 2013 *Chin.J.Geochem* **32** 222–226
[3] Das A P, Swain S, Panda S, Pradhan N, and Sukla L B 2012 *Geomaterials* **2** 70-72
[4] Hariprasada D, Mohapatra M, and Ananda S 2013 *J. Min. Metall. Sect. B-Metall.* **49(1)B** 97 – 106
[5] Ismail A A, Ali E A, Ibrahim, Ibrahim A, and Ahmed M S 2004 *The Canadian Journal of Chemical Engineering* **82**
[6] Li C, Zhong H, Wang S, Xue J, Wu F, Zhang Z 2015 *Trans. Nonferrous Met. Soc. China* **25** 1677−1684
[7] Nagib S and Abdelhameed R S 2014 *Int. Journal of Engineering Research and Applications* **4(1)** 281-286
[8] Naik P K, Nathsarma K C, Das S C, and Misra V N 2003 *Mineral Processing and Extractive Metallurgy (Trans. Inst. Min. Metall. C)* **112** C131
[9] Sahoo R N, Naik P K, and Das S C 2001 *Hydrometallurgy* **62** 157– 163
[10] Sun Y, Fu G, and Jiang L, 2017 *Minerals* **7** 83
[11] Tian X, Wen X, Yang C, Liang Y, Pi Z, and Wang Y 2010 *Hydrometallurgy* **100** 157–160
[12] Zhang Y, You Z, Li G, and Jiang T 2013 *Hydrometallurgy* **133** 126–132
[13] Zhanga C, Wanga S, Caoa Z, and Zhong H 2018 *Hydrometallurgy* **175** 240–249
[14] Aripin H, Joni I M, Busaeri N, Usrah I, Sudiana I N, and Sabchevski S 2017 *International Conference on Chemistry, Chemical Process and Engineering (IC3PE) AIP Conf. Proc.* **1823**
[15] Senanayake G, Childs J, Akerstrom B D, Pugaev D 2011 *Hydrometallurgy* **110** 13–32

Study: leaching of zinc dust from electric arc furnace waste using oxalic acid

R Kusumaningrum[1], A Fitroturokhmah[1], G S T Sinaga[1], A S Wismogroho[1,2], W B Widayatno[1,2], L J Diguna[3] and M I Amal[4]

[1]Nano Center Indonesia, PUSPIPTEK South Tangerang, Banten 15314 Indonesia
[2]Research Center for Physics, Indonesian Institute of Sciences, Bld. 440/442, PUSPIPTEK Serpong, Tangerang Selatan 15314, Indonesia
[3]Department of Renewable Energy Engineering, Prasetiya Mulya University, Kavling Edutown I.1, Jl. BSD Raya Utama, BSD City, Tangerang 15339, Indonesia
[4]Research Center for Metallurgy and Materials, Indonesian Institute of Sciences, PUSPIPTEK Tangerang Selatan, Banten 15314, Indonesia

Email: muha137@lipi.go.id

Abstract. A study has been conducted on the process of leaching of zinc dust using oxalic acid by hydrometallurgical method. The sample preparation process starts from the dust zinc washing. The leaching process was started by mixing the sample with solution of 5 M oxalic acid. There were three variations; the ratio of mole, temperature and leaching time to know the optimum condition. From the variation of leaching process that has been done, an optimum condition was obtained where the ratio of zinc mole dust to 5 M oxalic acid solution is 1:10 with temperature 90°C and leaching time for 150 minutes. The leaching filtrate was precipitated to obtain zinc oxalate with a solution of sodium hydroxide up to pH 7. The FTIR and XRD analysis showed that the leaching products confirmed as hydrozincite ($Zn_5(CO_3)_2(OH)_6$) and sodium bicarbonate ($NaHCO_3$). The carbonate compound may be caused by decarboxylation during the leaching process. The calcination of leaching products to convert hydrozincite turn out zinc oxide was done at 600 °C. The results of the XRD analysis show that there were zinc oxide and still found zinc carbonate and sodium carbonate.

1. Introduction

Zinc is very useful in many aspect of human life. For instance, in human body zinc have a role in growth and as an integral part of more than 200 enzymes [1,2]. Besides from natural resources, the utilization of zinc dust from electric arc furnace (EAF) can be used as an alternative source to obtain zinc. This dust contains a huge amount of zinc that proportional to 20-30 wt% zinc oxides [3].

Some studies to leach zinc from its dust have been reported. Two common methods for zinc dust recovery are pyrometallurgy and hydrometallurgy. The pyrometallurgy method requires expensive cost and high energy. Poorly, this method generated heavy and toxic air pollution that difficult to handling. The hydrometallurgy method has been known as an easy and simple technique. This method is feasible for scaling up to build a pilot plan. The important process in hydrometallurgy is leaching which means an extraction of value metal from its source using acid, base or water [4]. Mostly, leaching process is brought out in acidic solution such as sulfuric acid, nitrate acid and chloric acid. Yet, this waste treatment is harmful and needs high treatment cost.

Oxalic acid is a strong organic acid, environmental friendly and usually used as a leaching agent. Oxalic acid is suitable as selective precipitating agent for zinc and iron separation based on the solubility product constant (Ksp) values which 1.4×10^{-9} for zinc oxalate and $2 \times 10^{-7}$for iron oxalate [5]. The previous study gives information that zinc oxalate can be precipitated at pH 6.2-6.4 [6]. The leaching study of zinc dust using oxalic acid has not been reported intensively. In this study, oxalic acid was used as a leaching agent and sodium hydroxide as a precipitating agent of zinc oxalate.

2. Experimental Procedures

Zinc dust that used in this study was a waste of steel processing using electric arc furnace. Zinc dust was analyzed by X-Ray Diffraction (XRD) to confirm the containing phase in this dust. To remove the polar impurity, zinc dust was washed by demineralization water at 60 °C for one hour. The residues were dried at 100 °C.

Leaching process of zinc dust was driven by oxalic acid 5 M with mole ratio solid/liquid (s/l) 1/5 and 1/10. The leaching time and temperature was studied by calculation of yield product percentage. Zinc oxalate was precipitated by NaOH addition till reach pH 6-12. The white powder then was calcined at 600 °C to form zinc oxide. The leaching product and as-calcined product was analysis by X-Ray Diffraction and FTIR.

3. Result and Discussion

Zinc dust of EAF process was analyzed using XRD for analyzing the phase elements content. The percentages of elements content were analysis using X-Ray Fluorescence (XRF). The XRD pattern of zinc dust (Figure 1) showed that all phase was detected refers to ZnO. The content of elements in its dust was confirmed 96.3% Zn and 3.6 % Fe. Hence, the purification of zinc dust has to be examined.

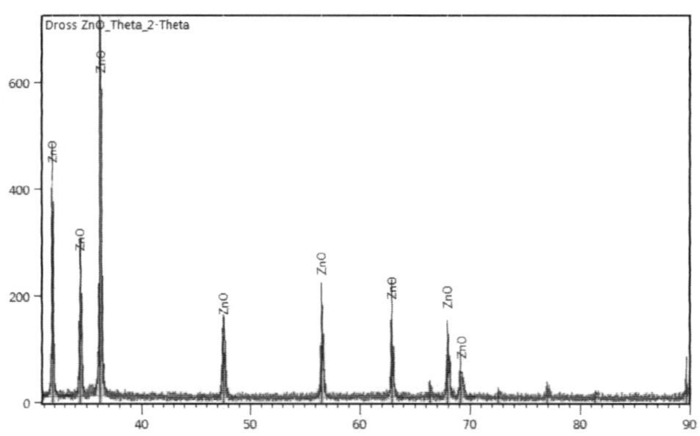

Figure 1. Diffractogram of zinc dust waste

Ratio mole s/l of zinc dust and oxalic acid solution were studies at 1:5 and 1:10 to investigate the optimum leaching condition. The reaction between oxalic acid and zinc dust (ZnO) were controlled at 60

°C for 120 minutes to form zinc oxalate. To precipitate the zinc oxalate, the sodium hydroxide was added until the pH of solution reach into 7 at room temperature. The calculation of yield percentage gave that mole ratio of s/l for 1:5 and 1:10 was 46.03% and 69.48%, respectively. The most effective leaching was given by the mole ratio of 1:10. It may be due to a great amount of oxalate ion that sufficient to extract more zinc from its dust. The possible reaction can be written as follows:

1. Oxalic acid running through dissociation in water:

$H_2C_2O_{4\,(aq)} \leftrightarrow H^+ + HC_2O_4^-$ (1)

$HC_2O_4^-{}_{(aq)} \leftrightarrow H^+ + C_2O_4^-$ (2)

Over all reaction could be written as

$H_2C_2O_{4\,(aq)} \leftrightarrow 2H^+ + C_2O_4^{2-}$ (3)

2. Extraction process to form zinc oxalate:

$ZnO_{(s)} + 2H^+ + C_2O_4^{2-} \rightarrow ZnC_2O_{4\,(aq)} + H_2O_{(l)}$ (4)

$FeO_{(aq)} + 2H^+ + C_2O_4^{2-}{}_{(aq)} \rightarrow FeC_2O_{4\,(aq)} + H_2O_{(l)}$ (5)

$Fe_2O_{3\,(aq)} + 2H^+ + C_2O_4^{2-}{}_{(aq)} \rightarrow Fe_2(C_2O_4)_{3\,(aq)} + H_2O_{(l)}$ (6)

The effect of pH in zinc oxalate precipitation was also studied. In pH 6 and 7, it was shown that the white precipitate was formed. Meanwhile, the addition of sodium hydroxide to obtain a base condition (pH 8-12) generated a brown precipitate that probably the iron elements. The addition of sodium hydroxide till pH 7 only resulted a zinc oxalate precipitation. The difference of Ksp value between zinc oxalate and iron oxalate which are 1,4 x 10^{-9} and 2 x 10^{-7}, respectively, may have caused the zinc oxalate to precipitate earlier. The precipitation was occurred at room temperature spontaneously.

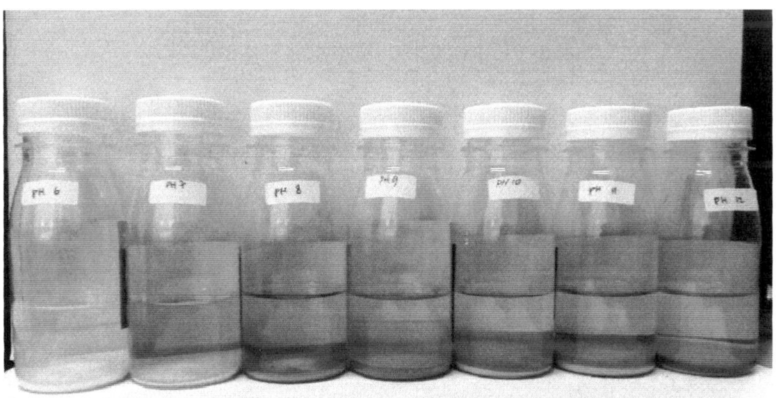

pH 6 7 8 9 10 11 12

Figure 2. Zinc oxalate precipitation by pH 6-11

The effect of temperature and time in leaching process were studied twice for every procedure. The yield percentage is shown in Figure 3. The increasing temperature in leaching process caused an increasing of zinc oxalate precipitation. It may be caused by the increasing of collision among molecules when more zinc was extracted. The optimum temperature was 90 °C with a yield percentage of 57.32%. For reaction time effect study, the leaching was done at 90 °C. Such as reaction time effect, the yield percentage increased significantly till reaction time of 150 minutes (65.78%). Characterization of zinc oxalate and zinc oxide synthesis products were taken at optimum condition based on yield percentage calculation which was temperature of 90 °C and reaction time of 150 minutes.

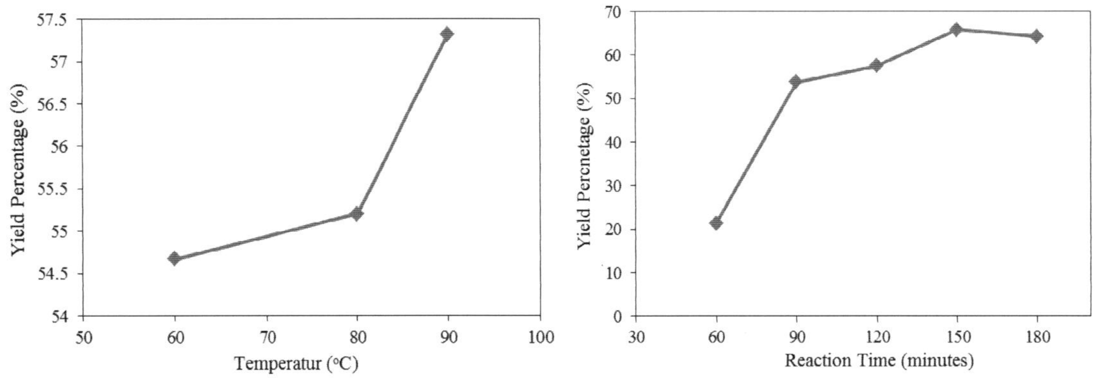

Figure 3. Yield percentage of leaching products by temperature and reaction times

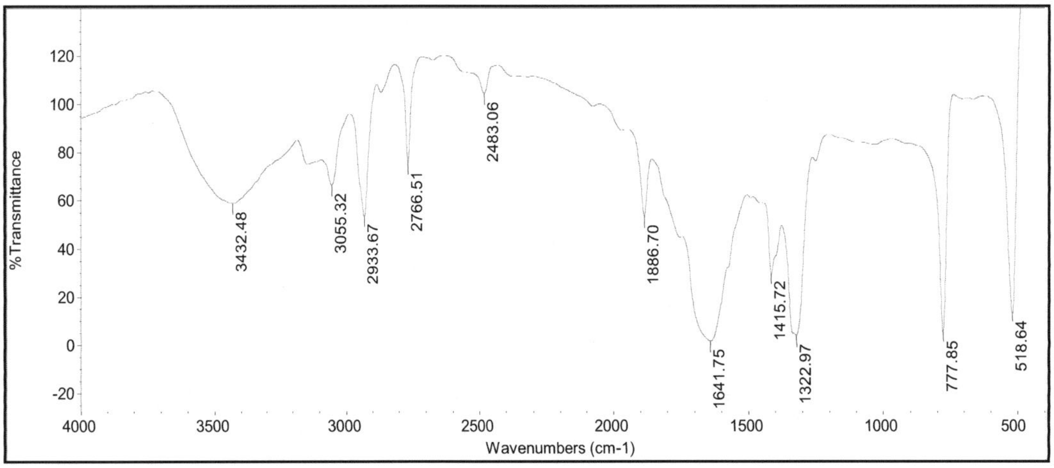

Figure 4. Spectra of leaching products

Zinc oxalate was characterized by FTIR is shown in Figure 4. Wavelengths number at 518 cm^{-1} referred to vibration of metal oxygen bonding. Wavelengths number give strong peak on 759 cm^{-1} ($\nu_{O=C=O}$ symmetric), 1349 cm^{-1} ($\nu_{C\text{-}O}$), 1415 cm^{-1} ($\nu_{C\text{-}C}$), 1655 cm^{-1} ($\nu_{C\text{-}O}$ asymmetric), and 2933 cm^{-1} ($\nu_{C\text{-}H}$). The weak peak supported $\nu_{O=C=O}$ at 1886 cm^{-1} and 2483 cm^{-1}. The wavelengths number at 3432, 3055 and 2766 cm^{-1} corresponding to water molecule that trapped crystal lattice. Confirmation of phase leaching product was analyzed by XRD. The pattern of XRD (Figure 6) showed that leaching product were in dominant by sodium bicarbonate, partly peaks as zinc oxide and the rest as hydrozincite ($Zn_5(CO_3)_2(OH)_6$). Carbonate could be emerged by decarboxylation during the leaching process. Probably after dissociation process of oxalic acid, the oxalate ions went through carbonate by carboxyl group release. This phenomenon caused the leaching products were in carbonate compound. Combine with FTIR analysis, sodium bicarbonate supported by appearance of C-H vibration at 2933 cm^{-1}. In line with the calcination product at 600 °C, XRD analysis confirmed into sodium carbonate, zinc carbonate and zinc oxide. Zinc carbonate occurred from dehydroxylation of hydrozincite. The other assume is zinc carbonate forming occur due to the existence of CO_2 in a large amount in the system and enable to drag the equilibrium of reaction into zinc carbonate formation. The other studies reported that carbon balance at atmosphere take a rule in the hydration-carbonation or hydration and carbonation reaction path in the $ZnO\text{-}CO_2\text{-}H_2O$ system [7].

1. Decarboxylation turn into carbonate acid

$$H_2C_2O_{4\,(aq)} \leftrightarrow CO_{(g)} + H_2CO_{3\,(aq)} \quad (7)$$

2. Hydrozincite Forming Reaction

$$5ZnO_{(s)} + 3\,H_2CO_{3\,(aq)} + NaOH_{(aq)} \rightarrow Zn_5(CO_3)_2(OH)_{6\,(s)} + NaHCO_{3\,(s)} \quad (8)$$

3. Zinc oxide Forming at 600 °C

$$Zn_5(CO_3)_2(OH)_{6\,(s)} \rightarrow 2\,ZnCO_{3\,(s)} + 3\,Zn(OH)_{2\,(s)} \quad (9)$$

$$Zn(OH)_{2\,(s)} \leftrightarrow ZnO_{(s)} + H_2O_{(g)} \quad (10)$$

$$ZnCO_{3(s)} \leftrightarrow 2\,ZnO_{(s)} + CO_{2\,(g)} \quad (11)$$

$$2NaHCO_{3\,(aq)} \rightarrow Na_2CO_{3\,(s)} + H_2O_{(g)} + CO_{2\,(g)} \quad (12)$$

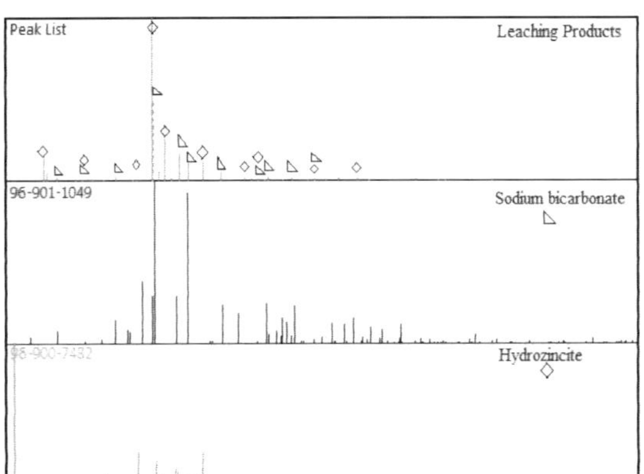

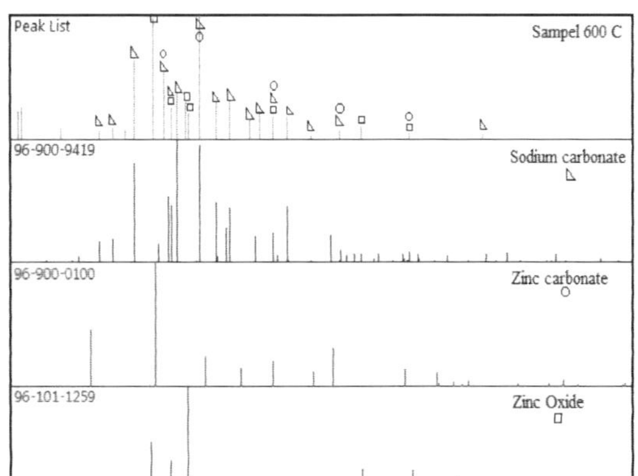

Figure 5. Difractogram XRD leaching product (left) and calcinated product (right)

4. Conclusion

Leaching zinc dust from the electric arc furnace waste by using oxalic acid at 90 °C for 150 minutes resulted to the hydrozincite and sodium bicarbonate formation. This perhaps caused by decarboxylation phenomenon during leaching process which encouraged the transformation of oxalate into carbonate. The existence of CO_2 in atmosphere caused the calcination process at 600 °C that resulted zinc carbonate, zinc oxide and sodium carbonate.

References

[1] Roohani N, Hurrell R, Kelishadi R, and Schulin R 2013 *J Res Med Sci.* **18(2)** 144-157
[2] Hsu C 2017 *Nanomed Nanotech.* **2(1)** 000114
[3] Shawabkeh RA 2010, *Hydrometallurgy* **104** 61-65
[4] Rahmani S A, Meidianto A, Amal M I, Wismogroho A S, Widayatno W B 2018 *J Phys Conf Ser* **985 (1)** 012033
[5] Pathak A, Roy A, and Manna M 2016 *Hydrometallurgy* **163** 161-166
[6] Marta L and Horouhz M 2003 *Leonardo Journal of Science* 72-78
[7] Vagvolgyi V, Hales M, Martens W, Kristof J, Horvath E, and Frost R L 2008 *J. Therm. Anal. Calorim.* **92** 911-916

Nickel production from laterites using electro metal electrowinning (EMEW) process

Sudibyo[1], A Junaedi[1], A S Handoko[1], M Amin[1], S Sumardi[1], F R Mufakhir[1], F Nurjaman[1], A H Tandoko[2], and Y I Supriyatna[1]

[1]Research Unit for Mineral Technology, Indonesian Institute of Sciences (LIPI), South Lampung, 35361, Indonesia
[2]Department of Physics, Engineering Faculty, Lampung University, Bandar Lampung, 3514, Indonesia

Email: sudibyo@lipi.go.id

Abstract. Laterite with low-grade nickel (< 0.5%) was very difficult to process using pyrometallurgy technique. One method which able to process this laterite is solvent extraction process. This process also able to separate cobalt from nickel solutions. The aqueous phase of solvent extraction process which has a high nickel content able to collect using electrowinning methods. Electrometal-electrowinning (EMEW) method is one of the electrowinning methods which has better performance to collect metal from solution. In this work, laterite was leached using H_2SO_4 then was extracted using Cyanex 272 in toluene as an organic solvent. In this research, the effect of a voltage, a duration time of operation, a concentration of boric acid and a temperature on EMEW process were studied and were optimized using Taguchi method.

1. Introduction

Indonesia has nickel laterite ore reserve of 1.391 billion tones which constitute 4.9% of nickel laterite reserves in the world and occupies the world's eighth largest position [1]. However, there are many laterites which have nickel concentration of 0.5%. This laterite is very difficult to be processed. However, it is able to process using solvent which able to process low-grade laterite to produce nickel and cobalt.

Nickel ore processing begins with a leaching process first, which aims to separate the desired components and undesired components [2]. The obtained leaching results still contain nickel, cobalt, and multi-component materials such as Cu, Co, Pb, V, and Fe. To separate the nickel with other multi components it will be extraction process. The separation of nickel from the multicomponent mixture of the material can be done well using cyanex 272 which has been diluted with toluene [3][4][5]

In this work, the nickel metal was processed from solvent extraction able to process using the electrowinning process [6]. Electrometal-electrowinning cell (EMEW) process is one of electrowinning process using the cylindrical electrode to catch the metal ions from the extracted leaching solution. This process is better than traditional electrowinning since it more has low operational cost, friendly for the environment, easy to install , design is compatible, it can be operated in large-scale process, acid electrolyte can be recycle, it can produce high purity of metal and it can be operated in low concentration of metal solution [7].

The parameters which have been studied and optimized such as temperature, time operation, voltage, boric acid concentration. The design of experiment (DOE) of Taguchi was used to optimize these parameters. Taguchi methods have many advantages such as it able to reduce economically the variability of the response variable and able to find out the optimum process conditions during laboratory experiments [8][9].

2. Experiments

In this experiment using raw materials of laterite nickel ore obtained from Pomala - Southeast Sulawesi (Integra mining Co. Ltd.). The composition of laterite nickel ore can see in Table 1. When the extraction process is used 90% of cyanex solution material from Anhui Jin'ao Chemical Co., Ltd. and 30% of toluene solution from Anugrah visi cemerlang Co. Ltd. The toluene is used as a cyanex solution diluent. The laterite was grinded using a ball mill to get the size less than 177 microns. The laterite was then leached using H_2SO_4 4 M for 8 hours [5]. Then the leaching solution was extracted using cyanex 10% v/v in Toluene. The solvent extraction was carried out in glass separating funnels with capacity 100 ml for 10 minutes. The aqueous phase was separated from the organic phase, the aqueous phase which rich with nickel was then processed using EMEW.

Aluminum cathode and nickel anodes were used in this EMEW process as shown in Figure 1. The design of experiments of Taguchi was used in this work as listed in Table 2. X-ray Fluorescence (Delta Premium XRF – Olympus) was used to analyze the material composition in electrodeposits. The results were analyzed using a signal-to-noise (S/N) ratio of "larger the better" was calculated for each factor level combination. The formula of the larger-is-better S/N ratio was shown in Equation 1.

$$S/N = -10*\log(\Sigma(1/Y2)/n) \quad (1)$$

where Y = responses for the given factor level combination and n = number of responses in the factor level combination.

Table 1. XRF analysis of laterite

El	wt (%)	El	wt (%)
Light Elements (LE)	79.2882	MnO	0.216
Fe	10.998	Co	0.046
Si	3.88	S	0.046
K	2.079	Sb	0.02
Al	0.701	Cd	0.013
Ni	0.488	Sn	0.014
Ca	0.65	Ti	0.022
Cl	1.253	Zn	0.0076
Cr	0.284		

Table 2. EMEW experiment design using Taguchi method

Run	Design of Experiment				Results
	Temp. (°C)	Time (hour)	Voltage (V)	Boric acid concentration (M)	Nickel electrodeposits (mg)
1	50	2	3	0.0	0.04
2	50	4	4	0.5	0.28
3	50	6	5	1.0	60.48
4	60	2	4	1.0	2.64

5	60	4	5	0.0	75.06
6	60	6	3	0.5	0.84
7	70	2	5	0.5	3.5
8	70	4	3	1.0	0.72
9	70	6	4	0.0	31.68

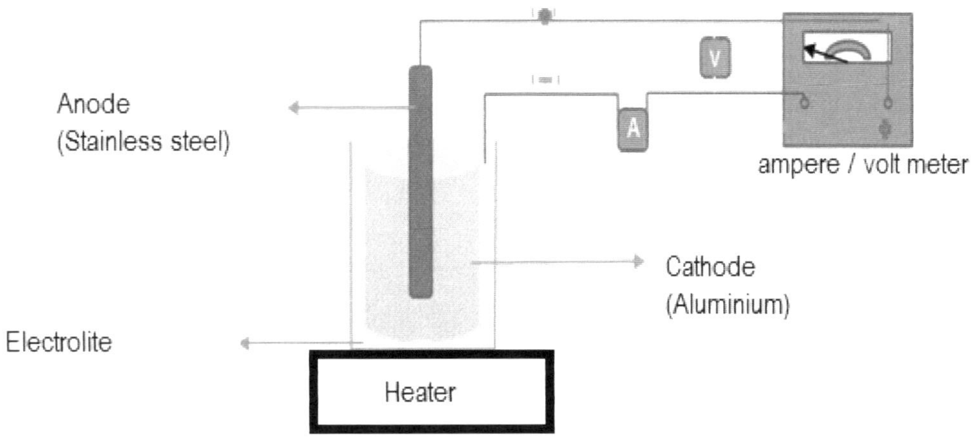

Figure 1. Experimental set up of electro metal-electrowinning (EMEW)

3. Results and discussion

3.1. Taguchi analysis for electrometal-electrowinning (EMEW)

The experiments have been conducted following Taguchi orthogonal array as shown in Table 2. A signal to noise (SN) ratio of larger the better of Taguchi was used to analyze using the results of this work as listed in Table 3. From the table shows that highest value of delta is the parameter of Voltage of DC power supply, hence this parameter is the first rank of the parameter which affects the amount of nickel electrodeposits. The second, third and fourth rank are the time, temperature and boric acid, respectively.

Table 3. Response for signal to noise ratios larger is better

	SN Larger is better			
Factor Level	**Temperature**	**Time**	**Voltage**	**Borid Acid Concentration**
1	-1.1278	-2.8818	-10.7755	13.1884
2	14.8086	7.866	9.1303	-0.5633
3	12.6812	21.3778	28.0073	13.737
DELTA	15.9364	24.2596	38.7828	14.3003
RANK	3	2	1	4

3.2. Effect of voltage on nickel products

To study the effect of Voltage towards the amount of nickel in electrodeposits, the voltages were varied from 3 to 5 volt as shown in Fig. 2. The result shows that the increment of voltage between the two electrodes will increase the weight of nickel in electrodeposits. This due to, the ncrease of this voltage increased the mass transport in an electrochemical cell. In addition, the electric field between the two electrodes is getting bigger. In a large electric field, the ions or electrons will move faster. The formation of layers on the cathode will be faster. Hence, the optimum of voltage for this process was 5 V.

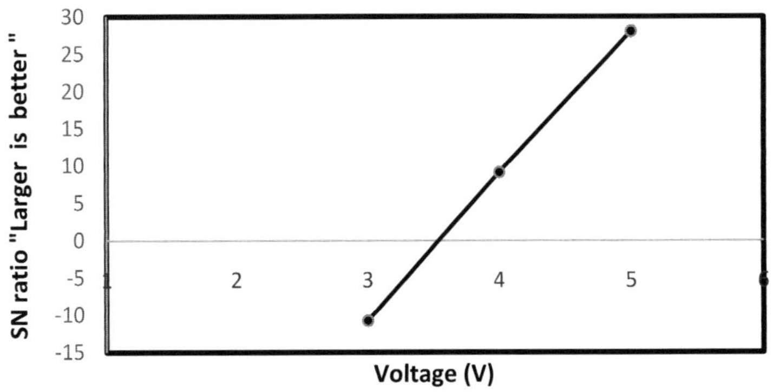

Figure 2. Effect of voltage on nickel products

3.3. Effect of the operation duration on nickel products

To study the effect of the operation duration towards the amount of nickel electrodeposits, the operation duration was varied from 2 to 6 hours as shown in Fig. 3. The figure shows that the amount of nickel electrodeposits were increase as the increase of the duration of the operation. Since the enhancing the operation duration will increase total mass transport during the electrochemical reaction. Hence the total amounts of nickel electrodeposits were increased. The Taguchi analysis shows that the 6 hours operation is the optimum of duration operation. At this condition, large number nickel ions in the solutions were deposited at the cathode.

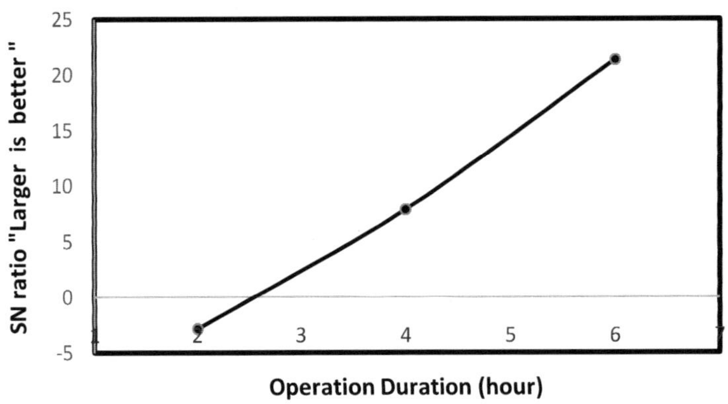

Figure 3. Effect of time on nickel products

3.4. Effect of boric acid on nickel products

The effect of boric acid as electrolyte additive towards the amount of nickel electrodeposits was studied by varied boric acid concentration from 0.5 to 2 M as shown in Fig.4. The results show that the increase of boric acid concentration caused the decrease of nickel electrodeposits since it caused the increment of electrolyte viscosity as shown in 1 M of boric acid. However, at the certain concentration of boric acid, it gives the better result of the amount of electrodeposits as shown in 2 M of boric acid. According to Zhang, et al [6] boric acid may increase the polarization of the cathode. In electrodeposition process, boric acid is absorbed on the surface of the cathode results in increasing the Ni^{2+} potential deposition. Boric acid and Ni^{2+} can form boric nickel which can increase hydrogen over potential and decrease the evolution of H_2 gas.

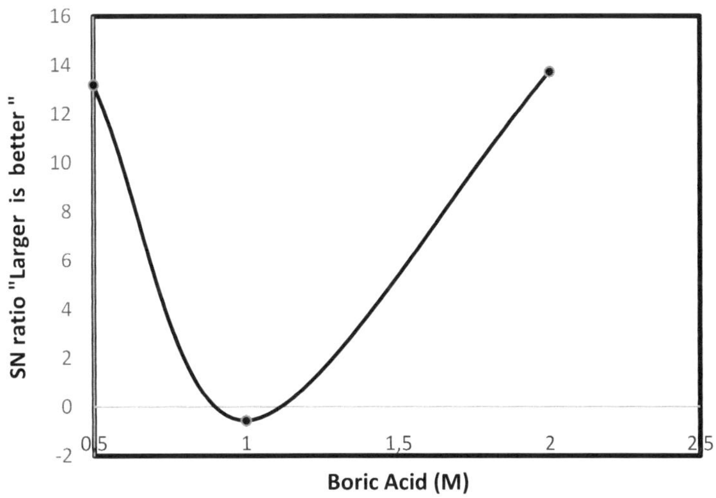

Figure 4. Effect of boric acid on nickel products

3.5. Effect of temperature on nickel products
The effect of temperature on nickel product has been studied by varied temperature from 50°C to 70°C as shown in Fig. 5. The result shows that the optimum temperature is at 60°C. The increment of temperature will increase the rate of electrochemical reactions [10]. The increasing of temperature also enhances the electrolyte conductivity and mass transfer [11]. However, the increment of temperature in the acidic electrolyte will enhance the hydrogen evolution which decreases the current efficiency. The increasing temperature also reduces the hydrogen over potential which able to reduce, hydrogen evolution problem [11]. The optimum temperature is important to solve the hydrogen evolution problem. At the temperature of 70°C, the mass of nickel electrodeposits decrease as shown in Fig. 3. The increasing temperature over the optimum condition caused the increasing the hydrogen evolution effect. The increment of temperature over the optimum condition will increase the overall mass transfer include the impurities mass transfer [10][11]. Hence it caused the decreasing of the mass of nickel electrodeposits.

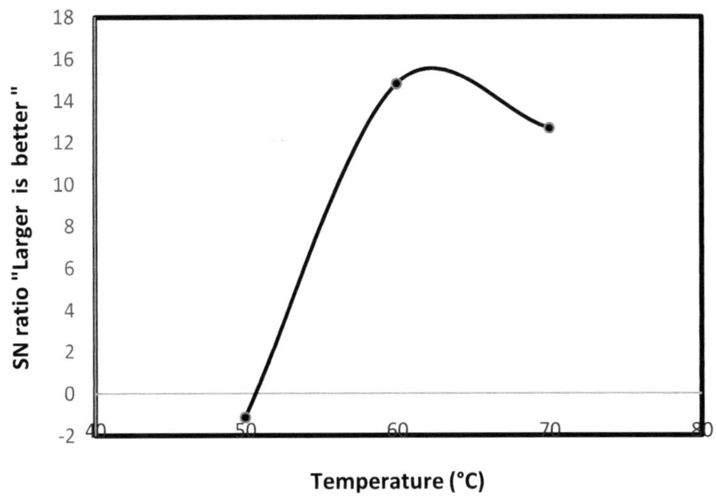

Figure 5. Effect of temperature on nickel products

4. Conclusions

Electrometal-electrowinning (EMEW) process to produce nickel from low-grade laterite was conducted in laboratory scale. Amount of nickel electrodeposits was affected by without boric acid additions, voltage, temperature, and operation duration. These results have been analyzed using Taguchi analysis of "Larger is better". The result shows that the optimum condition was at 60°C of the electrochemical reactor, 2 M of Boric acid concentration, 6 hours of operation duration and 6 V of DC power supply. From Taguchi analysis also shows that the voltage of DC power supply is the first rank of the parameter which affects the amount of nickel electrodeposits. Meanwhile, the second, third and fourth ranks are the time, temperature and boric acid, respectively.

Acknowledgments

This work was supported by Ministry of Research, Technology and Higher Education, Republic of Indonesia through INSINAS research grant no. RT-2016-0227 and Indonesian Institute of sciences through DIPA LIPI research grant are greatly acknowledged.

References

[1] Nurhakim N, Dwiatmoko U, NH R, and M A 2011 *Info-Teknik* **12(2)** 48–53
[2] Astuti W, Hirajima T, Sasaki K, and Okibe N 2016 *Miner. Eng.* **85** 1–16
[3] Iliev P, Stefanova V, Lucheva B, and Tzonevski A 2012 *Metal* **5** 23–25
[4] Kursunoglu S, Ichlas Z T, and Kaya M 2017 *Hydrometallurgy* **169** 135–141
[5] Guimarães A S, Da Silva P S, and Mansur M B 2014 *Hydrometallurgy* **150** 173–177
[6] Lu J, Yang Q H, and Zhang Z 2010 *Trans. Nonferrous Met. Soc. China* **20(1)** s97–s101
[7] Roux E, Roux E, Gnoinski J, Ewart I, Dreisinger D, and Vancouver N 2010 *Fourth South. African Conf. Base Met.* 27–44
[8] Rosa J L, Robin A, Silva M B, Baldan C A, and Peres M P 2009 *J. Mater. Process. Technol.* 209(3) 1181–1188
[9] Zainal N, Shukor S, and Razak K 2015 *J. Eng. Sci.* **11** 9–16
[10] Zhang W 2010 *Performance of lead anodes used for zinc electrowinning and their effects*

on energy consumption and cathode impurities Ph.D. Thesis Fac. DES Sci. GENIE Univ. LAVAL QUÉBEC

[11] Wei Q, Ren X, Du J,Wei S, and Hu S 2010 *Miner. Eng.* **23(7)** 578–586

Effect of temperature and acid concentration on lanthanum extraction from spent catalyst using organic acid

H T B M Petrus[1], A Wijaya[1] , A D P Putra[1], Y Iskandar[2], D Bratakusuma[2], F R Mufakhir[3], W Astuti[3], and Wiratni[1]

[1]Department of Chemical Engineering (Sustainable Mineral Processing Research Group), Universitas Gadjah Mada, Jl. Grafika 2, Yogyakarta 55281, Indonesia
[2]PT. Pertamina (Persero) RU VI, West Java, Indonesia
[3]Research Unit for Mineral Technology, Indonesian Institute of Sciences (LIPI), Jl. Ir. Sutami Km. 15, Tanjung Bintang, Lampung Selatan, Indonesia

Email: bayupetrus@ugm.ac.id

Abstract. Rare earth elements (REE) are nowadays being used widely in many industries from electronics to petroleum industries as catalysts. However, their disposal caused serious problems to the environment. With the sharp growth in its usage, there is a better way to use and utilize REE from secondary sources such as their disposal rather than using new raw materials. The aim of this work is to study the potential of citric acid as a leaching agent to extract one of REE (i.e. lanthanum) in various acid concentrations and leaching temperatures. The raw material used in this work is spent catalyst from Pertamina Refinery Unit VI, Balongan, Indonesia. The spent catalyst is decarbonized with a heat treatment at 725°C for 10 minutes before the leaching process. The leaching process used 0.1; 1; and 2 M of citric acid with a varied temperature of 30, 60, and 80°C. The lanthanum recovery was calculated by comparing the mass percentage of lanthanum before leaching process and after leaching process using Energy Dispersive X-Ray Spectroscopy (EDX). The results were analyzed by response surface methodology (RSM) and are proved to be a reliable method to depict and analyze the leaching characteristics. The independent variables used in the research show significant effect on the response, especially the acid's concentration. A second order polynomial equation was used to correlate response and independent variables. The coefficient of determination (R^2) shows a satisfactory result of 86.53% and the normal probability plot ensure the model's adequacy.

1. Introduction

The sharp growth in information technology such as tablet PCs, smart phones, and other gadgets in the last years leads to a huge demand for REE (Rare Earth Elements). On the other hand, due to their special characteristics, rare earth materials are also nowadays important part of the catalysts used in petroleum chemical industries[1]. However, aligned with its frequent use, the waste is also rising by a significant amount. Disposal by land filling is the oldest and was the most widely used method years ago. Due to increased environmental consciousness and regulations, land filling is, fortunately, less and less used. The issue is that liability still belongs to the waste producer as long as this waste has not been destroyed, and thus lasts on long term[2]. Today the wastes are becoming a serious global threat[3]. This is because spent catalysts are composed of heavy metals and REE such as iron, chromium, nickel, palladium, manganese, and lanthanum. Instead of using new raw materials, there is a wiser choice that is extracting REE from secondary raw materials such as their disposal.

The special characteristics of REE make them considered as highly valuable metals. Nowadays, their price and presence have become a serious concern to many countries which also led to the rapid studies and researches for their extraction from secondary sources[4, 5, 6]. One of the most widely used REE which is still lacking studies of its extraction from secondary sources is lanthanum. Lanthanum is a ductile and soft metal which easily tarnishes when exposed to air.

Lanthanum extraction from industrial wastes becomes important because any disposal may affect the environment. Previous research by Jinxia et al. (2018), stated that Lanthanum may affect the soil ecosystems at a concentration slightly above natural condition level (6.6-50.0 mg La/kg dry soil)[7].A study by Abhilash et al. (2014), found that the maximum recovery of lanthanum (99,9%) from Indian red mud in sulfuric acid extraction can be achieved in 3 M of H_2SO_4 and 35°C temperature[8]. Another study by Kuang et al. (2017), stated that among inorganic acids such as H_2SO_4, HNO_3, HCl, HCl was the most effective stripping solvent[9]. In addition, a work by Wei et al. (2011), stated that lanthanum can be easily stripped–reaching 95% recovery from chloride medium with organic acetic acids when the acidity is more than 0.03 M[10]. Astuti et al. (2017) also found that approximately 90% of lanthanum can be leached from hydro-processing spent catalyst by inorganic acid (HNO_3, HCl, H_2SO_4) with 4M acid concentration at 90°C temperature[5].

However, there is still lack of study in terms of REE extraction, especially lanthanum, from spent catalyst using organic acid. Because so, this work was done to study the effect of temperature and acid concentration on lanthanum extraction from spent catalyst using citric acid.

2. Experimental

2.1. Raw material characterization

The spent catalyst was obtained from PT. Pertamina Refinery Unit VI, Balongan, Indramayu, Indonesia. The spent catalyst was covered with carbon and it has to be decarbonized using a muffle furnace. The de-carbonization process was carried out at 725°C for 10 minutes. The spent catalyst was then analyzed using Energy Dispersive X-ray spectroscopy (EDX).The result is shown in Table 1.

Table 1. EDX result of de-carbonized spent catalyst

Constituents	%Mass
Al	50.66
Si	40.17
La	**3.56**
Ni	**2.66**
Pr	0.77
Fe	0.59
Ca	0.49
S	0.47
P	0.33
K	0.20
Co	0.09
Zr	0.01
Sr	0.01

2.2. Leaching process

The de-carbonized spent catalyst was leached by citric acid with a concentration of 0.1; 1; and 2 M. As much as 10 grams of de-carbonized spent catalyst were put in a 50 ml of acid in an Erlenmeyer flask.

The process was carried out for 4 hours in a laboratory water bath (Fig. 1) at a temperature of 30, 60, and 80°C.

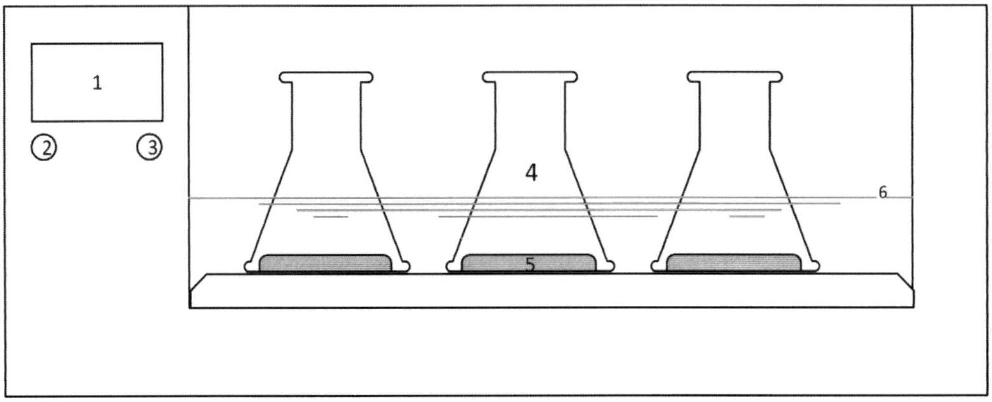

Figure 1. Laboratory water bath as a reactor for leaching process (1.Temperature display; 2.On/off button; 3.Temperature controller; 4.Erlenmeyer flask; 5.Leached spent catalyst; 6.Water level)

2.3. Recovery analysis
After 4 hours of leaching, the leachate was separated from the spent catalyst using filter paper. The remaining solid was dried in an oven at 50°C for 5 hours to remove its moisture. The dried spent catalyst was then analyzed again using EDX to determine its metal contents. The overall process is shown in figure 2.

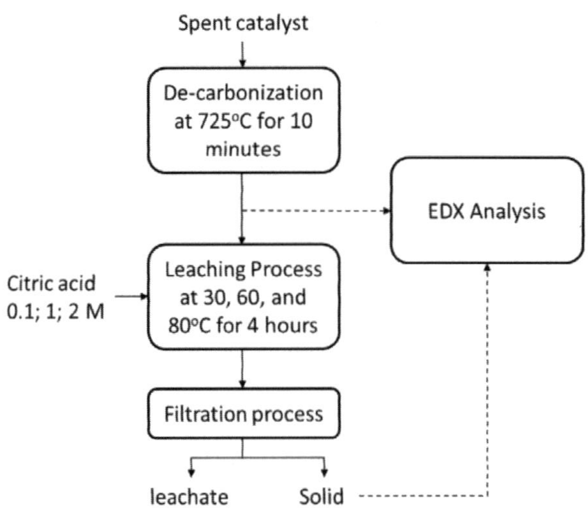

Figure 2. Overall experimental process

The spent catalyst's REE content after leaching process will be surely reduced. Therefore, the recovery calculation must be carried out with constant denominator instead of the total mass. Silica is considered to be one of the most inert constituents in this research. Previous studies show that silica can only be leached by a strong acid such as hydrochloric acid[11]. Furthermore, acid leaching methods are widely used as silica purification from heavy metals[12]. The formulated equation can be seen below:

$$\%R = \frac{W_0 - W_t}{W_0} x100 \quad (1)$$

Where, %R is the recovery percentage; W_0 and W_t are the weight of lanthanum before and after leaching process, in mg.

2.4. Design of experiment and statistical analysis

Response surface methodology (RSM) was chosen to analyze the result of this research statistically. A full three-level factorial design was used with two independent variables, acid concentration and temperature. The dependent variables, of course, are the recovery percentage of lanthanum. The design is shown in table 2.

Table 2. Experimental range and level of independent variables

Independent variables	Range and level		
	+1	0	-1
Acid Concentration, M (X_1)	0.1	1	2
T, °C (X_2)	30	60	80

A second-order polynomial equation was used to match the response (dependent variable) and the factors (independent variables). The polynomial equation has a general form which written as follows[13]:

$$Y = \beta_0 + \sum_{i-1}^{k} \beta_i x_i + \sum_{i-1}^{k} \beta_{ii} x^2{}_{ii} + \sum_{1 \le i \le j}^{k} \beta_{ij} x_i x_j + \varepsilon \quad (2)$$

Where Y is the response; x_i and x_j are the factors (i and j are the range from 1 to k); β_0 is the constant coefficient; β_i, β_{ii}, β_{ij} are the coefficients for the linear; quadratic and interaction effect; and ε is the error. The model's accuracy was analyzed using the coefficient of determination (R^2). The R^2 has a value, ranged from 0 to 1. If R^2 is close to 1, the model is highly accurate[14].

3. Results and Discussion

3.1. Lanthanum recovery percentage statistical analysis

The experiment was done based on a full three-level factorial design to analyze the factors' influence on the lanthanum recovery. In the result, it is clear that both acid concentration and temperature play a significant role in the recovery of the lanthanum. Lanthanum can reach 100% of recovery using 2 M of citric acid in all temperatures. However, in a 0.1 M of the citric acid solution, the maximum lanthanum recovery was 43.94% (at 80°C), making acid concentration as the most significant variable in the result.

On the other hand, there is a declining trend aligned with lower temperature. This phenomenon can be seen as an incomplete reaction as it is widely known that temperature affects the kinetics of chemical processes. But, it has to be proved in further study of leaching kinetic behavior of lanthanum in organic acids. The result is shown in Table 3 below.

Table 3. Lanthanum recovery result and the predicted value

Temperature (°C)	Acid Concentration (M)	%R	Predicted %R
30	0.1	40.54	29.89
	1	53.25	73.72
	2	100.00	90.69
60	0.1	41.07	47.29
	1	100.00	89.74
	2	100.00	105.18
80	0.1	43.94	49.39
	1	100.00	90.92
	2	100.00	105.34

Minitab software was used to process the data with the satisfactory result as it can be seen in Table 3 and Table 4. The empirical correlation between response *(%R)* and independent variables (acid concentration and temperature) are expressed in the equation 3 below.

$$\%R = -11.3 + 68.6X_1 + 1.44X_2 - 16.7X_1^2 - 0.0095X_2^2 - 0.051X_1 X_2 \quad (3)$$

The data satisfactoriness and the model fitness were assessed by regression and analysis of variance (ANOVA), a statistical technique that distributes the total variability into several component parts associated with particular sources of variation for the testing of model parameters[15]. In addition, F-test and P value are also assessed to determine the statistically significant variables. The results can be seen in Table 4.

As it can be seen from the table, the P-value for all linear independent variables was practically smaller than the square or 2-way interaction mode. For a value lower than the level of significant, that is 0.05, the variable is considered as statistically significant for the proposed model[16] and in this case, X_1 (acid's concentration) is the most significant variable.

Table 4. ANOVA Analysis Results

Source	DF	Adj. SS	Adj. MS	F-Value	P-Value
Model	5	5,970.93	1,194.19	3.86	0.148
X_2	1	416.84	416.84	1.35	0.330
X_1	1	5,067.17	5,067.17	16.36	0.027
X_2^2	1	64.80	64.80	0.21	0.678
X_1^2	1	450.90	450.90	1.46	0.314
$X_1.X_2$	1	6.01	6.01	0.02	0.898
Error	3	928.99	309.66		
Total	8	6,899.92			

3 2. The validity of the model

To verify the proposed model, it is important to ensure the adequacy of predicted values with actual (experimental) values. In this case, the experimental and predicted values are fitted as it can be seen in Figure 3. The data points on this plot are considered close to the straight lines. The results suggest that the model used in this study is compatible to depict the conditions (with an R^2 value of above 0.86) with considerably high accuracies.

Figure 4 shows whether the data are normally distributed, other variables are affecting the responses, or outliers exist in the data. In addition, residual values explain the variation between predicted and experimental values. A highly accurate model is characterized by its small residual value[17]. The results show that the data are normally distributed, as it visually visible because the data points are positioned close to the reference line and the outlier data are minimal.

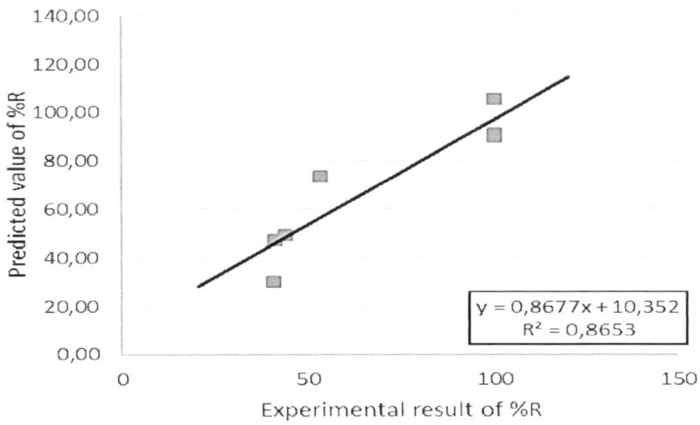

Figure 3. Comparison between experimental and predicted value of %R

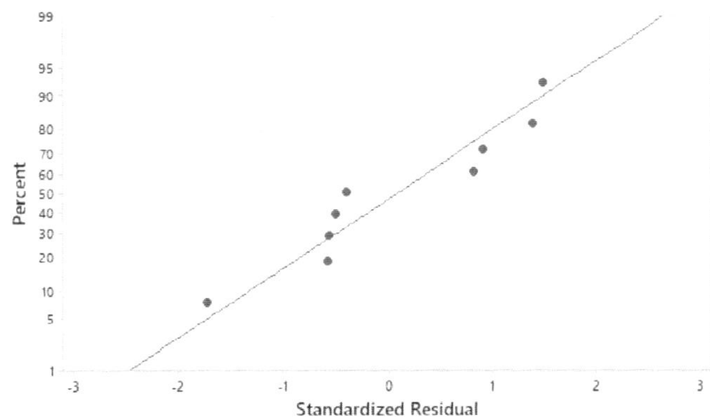

Figure 4. Normal probability plot of *%R*

4. Conclusion

In this study, a full three-level response surface design is proved to be a reliable method to depict and analyze the characteristics of lanthanum leaching from spent catalyst using citric acid. The independent variables used in the research show a significant effect on the response, especially the acid concentration. A second order polynomial equation was used to correlate response and independent variables. The coefficient of determination (R^2) shows a satisfactory result of 86.53% and the normal probability plot ensure the model's adequacy.

References

[1] Wangcheng Z, Yun G, Xueqing G, Yanglong G, Yanqing W, Guanzhong L 2014 *Chinese J Catal* **35** 1238–1250
[2] Dufresne P 2007 *Appl Catal* **322** 67–75
[3] Zhao Z, Qiu Z, Yang J, Lu S, Cao L, Zhang W, Xu Y 2017 *Hydrometallurgy* **167** 183–188
[4] Thompson R L, Bank T, Roth E, Granite E 2016 *Fuel* **185** 94–101
[5] Astuti W, Prilitasari N M, Iskandar Y, Bratakusuma D, Petrus H T B M 2018 *IOP conferece series Materials Science and Engineering* **285(1)** 012007

[6] Sari D P, Tawfiequrrahman Y, Petrus H T B M, Mufakir F R, and Astuti W 2018 *Seminar Nasional Teknik Kimia Kejuangan 4*
[7] Jinxia L, Verweij R A, and Van G C A M 2018 *Chemosphere* **193** 412–420
[8] Abhilash, Sinha S, Sinha M K, and Pandey B D 2014 *Int J Miner Process* **127** 70–73
[9] Kuang S, Zhang Z, Li Y, Wei H, and Liao W 2017 *J Rare Earths. doi: 10.1016/j.jre.2017.09.007*
[10] Wang W, Yang H, Cui H, Zhang D, Liu Y, and Chen J 2011 *Ind Eng Chem Res* **50** 7534–7541
[11] Santos I C, Alves A P G O N, Santos C S, Almeida M and Afonso M H 1990 *Hydrometallurgy* **23** 237–246
[12] Faizul C P, Abdullah C, Fazlul B and Noorina H J 2014 *Key Eng Mater* **595** 329–333
[13] Bezerra M A, Santelli R E, Oliveira E P, Villar L S and Escaleira L A 2008 *Talanta* **76** 965–977
[14] Vedaraman N, Sandhya K V, Charukesh N R B, Venkatakrishnan B, Haribabu K, Sridharan M R, and Nagarajan R 2017 *Chem Eng Process Process Intensif* **114** 46–54
[15] Prakash M J and Manikandan S 2012 *Dye Pigment* **95** 465–472
[16] Sinha K, Saha P D, and Datta S 2012 *Ind Crops Prod* **37** 408–414
[17] Agarry S E 2012 *J Environ Prot (Irvine, Calif)* **3** 748–759

Characterization of gold ore from Tanggamus mine Lampung Province

F R Mufakhir, W Astuti, E Purwaningsih, E Prasetyo, S Sumardi, Sudibyo, A S Handoko, A Junaedi, N M Prilitasari, and Y I Supriyatna

Research Unit for Mineral Technology, Indonesian Institute of Sciences (LIPI), Lampung 35361, Indonesia

Email: fika001@lipi.go.id

Abstract. The present paper reports thecharacterization of gold ore from Tanggamus mine Lampung Province bychemical and geological analysis.The presence high content of Fe and S indicated that the ore issulfides which is form as pyrite crystal confirmed by XRD and SEM analysis. Auas trace element detected by fire assay and wet analysis comprises of 4 ppm in an amount.It is not yet clear that the existence of Au with minor of amount, merelythe possibility Au associated with pyrite or quartz matricesas micro inclusion.The gold ore is classified as refractories ore.

1. Introduction

The Tanggamus gold ore deposit is located on the southern tip of the island of Sumatra direct in west from Bandar Lampung, which consists mainly of the volcano-sedimentary rock unit. Hulu simpang formation rocks show that the mineralogical composition of rocks varies from acid to base while alteration processes are characterized by crystallization of chlorite, carbonate (epidemic), epidote, silica, and clay mineral groups and associated with metal mineral deposits such as pyrite, chalcopyrite, sphalerite, magnetite, galena, tetrahidrite and gold [1,2]. The reserves of Tanggamus ore are estimated about 605.927 Mt [3].

In general, gold ores can be classified as free gold, free milling and refractory, depending on the response to cyanide leaching [4,5]. The illustrative-drawing can be seen in Figure 1. The types of free gold and free milling are also classified as "visible gold" that can be seen directly or through the milling process ahead. Gold refractory is also classified as "invisible gold, but this definition includes a native gold type found in micro-inclusions in sulfide or oxide crystals that are not detected by conventional microscopy techniques [6]. Lack of knowledge about the characteristics of gold ores will be causing optimum recovery not fulfilled.

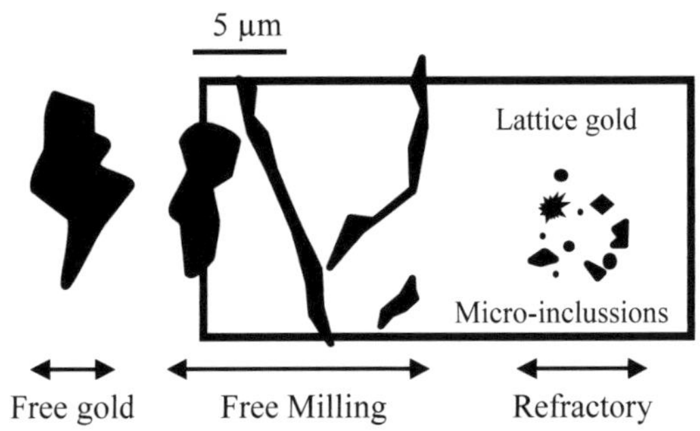

Figure 1. The illustrative drawing of the difference types of Au associated with oxides or sulfides

Nowadays, most gold are recovering by cyanidation at industrial scale and amalgamation at artisanal/small miner. Other processes proposed including halogenation e.g. bromination and iodination; chlorination; thio-ligan complexation [7] etc. could be an alternative, although the economic value still needs to be studied. When found in electrum (AuAg, conc. Ag>18%) or native form (AuAg; conc. Ag 8-13%) is usually easy after sufficient ore comminution. The presence of gold in other forms such as Au is contained within sulfides crystal, commonly pyrite (FeS_2), arsenopyrite (FeAsS) and galena (PbS) often describe as refractories ore.

Efforts to characterize gold ores are influential not only for geology but also for economics. Analysis methods for "trace metals" that contribute data on concentration, homogeneousness, mineralogy and distribution between crystals will be very interesting to study as a fundament for assessing the use of beneficence concentration and extraction methods. The aim of this subject was to approach, cautiously and comprehensive of gold ore in chemical and mineralogical characterizations pairing several analytical techniques in order to guidance to an enrich proficiency of their properties. The preliminary research will be proposed to next gold processing process method.

2. Materials and Methods

2.1. Materials
50 kg of gold ore was carried from mine located in Lengkukai, West Kelumbayan, Tanggamus Regency, Lampung Province. Gold ores were crushed and ground by jaw crusher and ball mill. The chemical analysis was carried out using high-performance spectrometry X-ray Fluorescence (XRF, PANanalytical Epsilon 3 XLE with Rh anode material and Si-drif detector (>135eV@5.9 keV)), wet chemical analysis using Atomic Absorption Spectrometer (AAS, Shimadzhu type AA-7000 Japan), and Induced Couple Plasma – Optical Emission Spectrometer (ICP-OES, Agilent 5100) for Au, Pd, Pt elements after fire assay and digestion in aqua regia. The X-ray Diffraction (XRD) was performed on PANalytical X'Pert3 Powder, with following conditions of operation: radiation Cu Kα (20 mA, 40 kV); goniometer speed of 0.05^{o} 2θ per step with 29.8450 seconds per step of counting time and collected 10 – 79.9^{o} 2θ. The interpretation using X'pert High Score Plus (HSP) with ICDD PDF2 and PAN-ICSD database. The studies for finding of the chemical composition, grain size, mineralogical association and liberation assortment was performed with inverted optical microscope Nikon type Eclipe MA 100 and Scanning Electron Microscope (SEM) Phenom World type Phenom ProX with CeB6 cathode.

2.2. Methods
The collected ores were crushed to reduce particle size below 4000 μm. After crushing, the ore was divided using riffle splitter into two part samples. One part of the sample was ground in order to allow better study with respect to liberation. After grinding, the sample was divided using the coning quartering method into 2 small parts. The sampling method was carried out in Duplo method to ensure

the homogeneous sample. Further, one of the small parts of the sample was dried at 120° C for 24 hours in order to divest the water on the ore surface. The dried sample was classified into the following particle sizes: (+149) μm; (-149+105) μm; (-105+74) μm; (-74+53) μm; (-53) μm.

3. Results and discussions

3.1. Particle Size

Figure 2 showed the results of particle size distribution of the gold ore before and after crushing and milling. It is ascertained that after using the jaw crusher there was a considerable difference in particle size distribution between coarse (+149 μm) and fine (-53 μm) fractions. Efforts to reduce particle size of ore are needed to ensure better evaluation of mineral liberation ranges. A ball mill is used to overcome this deficiency. Furthermore, degree of liberation and fines ranges are subject to consider, since the material has more than 20 % in fine fraction (< 53μm).

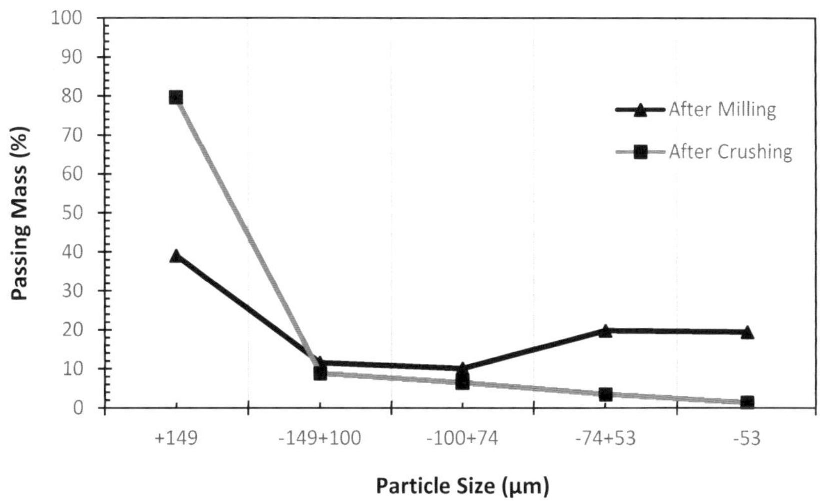

Figure 2. Particle size distribution results of gold ore sample before and after milling.

3.2. Chemical composition

The results of the chemical analysis by XRF of the gold ore for each fraction showed in Table 1. The results signalize the presence of Si as the main element followed by Fe and S. It was also noted that the Ag and Ti presence, which is component of great interest due to its economic value. The high presence of S indicated that the ore is sulfides. Moreover, Au was not detected and the presence of As will interfere with the gold dissolution process and environmental issue [8,9,10].

Table. 1. The chemical composition of the particle size distribution of the gold-bearing ore using XRF

Element*	Unit	Fraction (μm)				
		+149	-149+100	-100+74	-74+53	-53
Mg	%	0.521	0.344	0.357	0.373	0.397
Al	%	8.685	8.399	8.407	8.508	9.812
Si	%	34.344	34.843	34.457	34.561	34.394
P	%	0.507	0.464	0.443	0.430	0.466
S	%	16.086	17.937	17.451	16.661	13.564

Element*	Unit	Fraction (µm)				
		+149	**-149+100**	**-100+74**	**-74+53**	**-53**
K	%	3.141	2.884	2.841	2.827	3.650
Ca	%	1.081	1.048	1.104	1.111	1.076
Ti	%	0.354	0.315	0.324	0.331	0.404
Mn	%	0.785	0.659	0.587	0.530	0.481
Fe	%	32.300	31.264	32.024	32.639	33.378
Zn	%	0.174	0.096	0.123	0.107	0.141
As	%	0.494	0.450	0.467	0.472	0.505
Ag	%	0.322	0.265	0.256	0.240	0.233
Sn	%	0.115	0.174	0.204	0.223	0.170
Eu	%	0.274	0.248	0.032	0.223	0.226
Pb	%	0.664	0.487	0.558	0.604	0.932

*Element with the concentration in ppm not showed

Table 2 shows the results of wet chemical composition analysis by AAS and ICP OES. The result points out the presence of Au as the minor elements, that Pd and Pt were not detected. It is also prominent the Fe, as mayor element followed As, Zn and Ag confirmed the XRF result. The presence of Cu depends on detection limits of the instrument that is confirmed using XRF.

Table 2. The chemical composition of the gold ore using AAS and ICP OES

Elements	Au	Pd	Pt	Ag	Cu	Zn	Fe*	As
Concentration (ppm)	4.03	<0.01	<0.02	90	144	607	26.41	1692

* concentration in %

3.3. X-Ray Diffraction

Figure 3 shows the result of XRD analysis of the bulk material at fine fraction < 53µm in size. In this figure is viable to track of the presence of minerals as quartz, pyrite, and hematite. The micrograph results confirmed the presence of quartz and pyrite showed Figure 4.The presence of that mineral's have also confirmed through the geological approach of Hulu simpang formation rocks. The metallic Au frequently co-exists in the same host mineral, but distribution within the pyrite and arsenopyrite[6], quartz, hematite is predominantly not uniform [6].

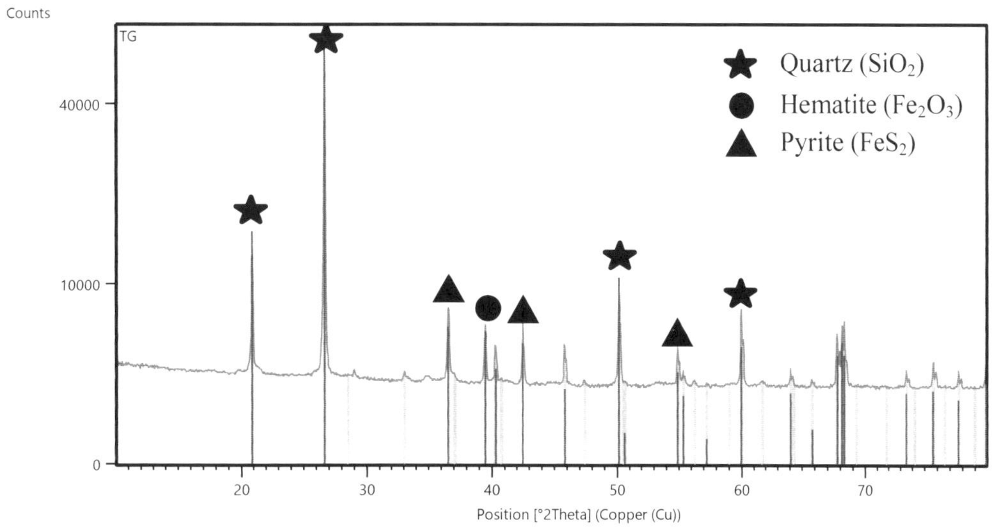

Figure 3. XRD pattern of TG gold ore. Radiation Cu Kα (20 mA, 40 kV)

3.4. Optical and Scanning Electron Microscope

Complementary optical microscope and SEM-BSE-EDS point analyses were carried out to determine the nature of the micro-inclusions contained in the heterogeneous sulfides phase. Figure 4 shows the mineralogical analysis of the ore was pyrite and quartz. Au does not seem to be visible, we assume Au is associated on the pyrite phase. To be more aware of the existence of gold, enlargement is needed using SEM BSE and EDS which is easy to see the difference in atomic weight between Au and another element. As the show's in figure 5 (a) pyrite matrix is confirmed which is consist of Fe as the main element followed S, O, and Si. Figure 5 (b) shows the point of very light shape, that Galena (PbS) micro inclusions were detected (Fig. 5).

Figure 4. Presence of quartz and pyrite.

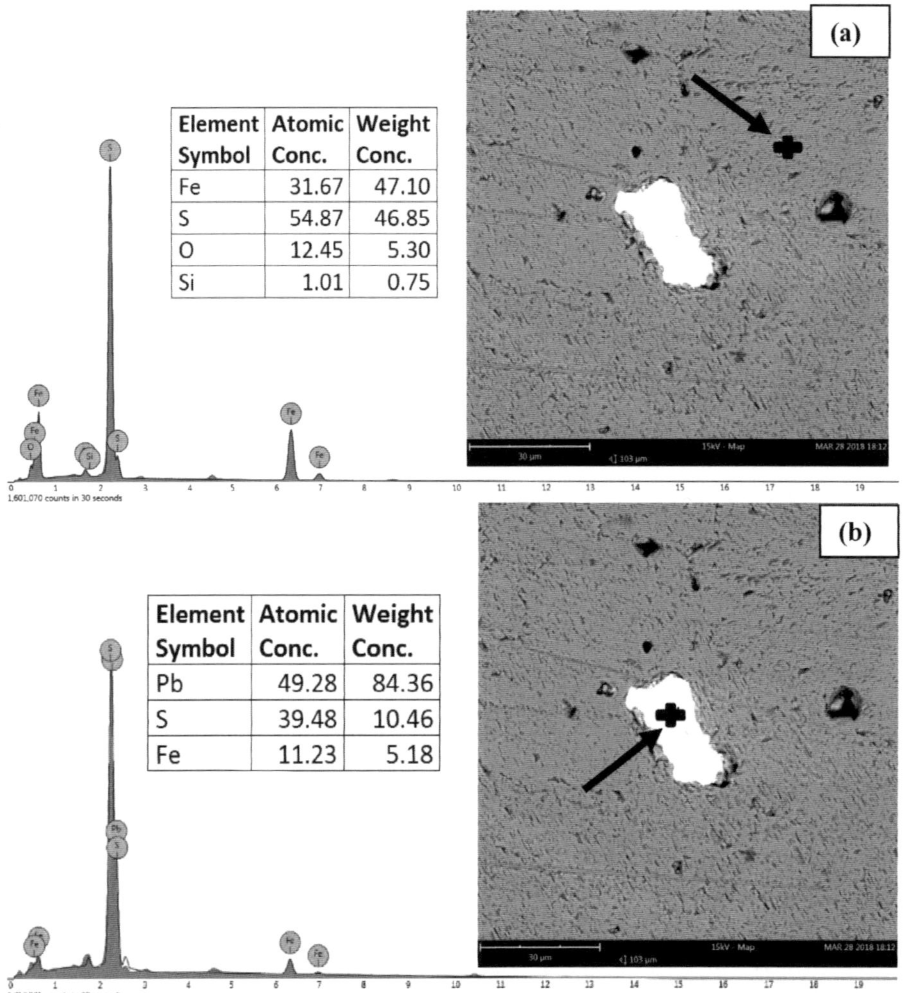

Element Symbol	Atomic Conc.	Weight Conc.
Fe	31.67	47.10
S	54.87	46.85
O	12.45	5.30
Si	1.01	0.75

Element Symbol	Atomic Conc.	Weight Conc.
Pb	49.28	84.36
S	39.48	10.46
Fe	11.23	5.18

Figure 5. BSE and EDS

Au is very difficult to be detected even using SEM analysis, although by ICP analysis Au was detected with minor concentration. This study concludes that Tanggamus gold ore classification was refractory ore with high sulfides mineral. The SEM-EDS analysis is still necessary on the other side of the surface sample specimen. The possibilities presence of Au is in pyrite, arsenic, galena or quartz crystal as micro inclusion. In addition, a more appropriate method of analysis is diagnostic leaching, where this method will provide information on the presence of Au in certain crystals precisely, even with more complicated-stages. Direct cyanide moreover amalgamation was not economically method that is Au difficult to be dissolved. It is proposed that the ore can be extracted with the complete processes including comminution, flotation, oxidation, and leaching extraction.

4. Conclusions

This paper shows that each analytical technique has evidence-information that has been confirmed mutually. The Au presence was only detected after analyzing by fusion process followed by digest and ICP. It needs to the fully establish the nature of Au in sulfides ores by the multi-analytical approach.

Direct observation when Au is well visible using back scattered electron image and EDS Mapping or scan line or point can reveal micro-inclusion on the host of the crystal.

Acknowledgment
The authors would like to thank Indonesian Institute of Sciences through Penelitian Mandiri scheme for funding the research and the research.

References

[1] Setiawan I, Zulkarnaen I, Indarto, and Sudarsono 2005 *Riset Geologi dan Pertambangan* **15(1)** 13-26
[2] Indarto S, Setiawan I, Zulkarnaen I, Sudarsono, Fiqih F M, and Fauzi A, 2007 *Prosiding Seminar Geoteknologi Kontribusi Ilmu Kebumian dalam Pembangunan Berkelanjutan* 165 – 173
[3] Taryono 2014 *Website : http://lampung.tribunnews.com/2014/10/03/lampung-punya-17-juta-ton-bijih-emas* accessed on 16 August 2018.
[4] Celep O, Alp I, Deveci H, and Vicil M 2009 *Trans. Nonferrous Met. Soc. China* **19** 707 -713
[5] La Brooy S R, Linge H G, and Walker G S 1994 *Journal of Minerals Engineering* **7(10)** 1213-1241
[6] Benzaazoua M, Marioun P, Robaut F, and Pinto A 2007 *Mineralogical Magazine* **71(2)** 123-142
[7] Gökelma M, Birich A, Stopic S, and Friedrich B 2016 *Journal of Materials Science and Chemical Engineering* **4** 8-17
[8] Curreli L, Garbarino C, Ghiani M, and Orru G 2009 *Hydrometallurgy* **96** 258 - 263
[9] Paktunc D, Foster A, Heald S, and Laflamme G 2004 *Geochimica et Cosmochimica Acta* **86(5)** 969-983
[10] Robins R G and Jayaweera L D 1992 *Mineral Processing and Extractive Metallurgy Review* **9(1-4)** 255-271

IOP Conf. Series: Materials Science and Engineering **478** (2019) 012019 doi:10.1088/1757-899X/478/1/012019

Effect of calcination temperature on the characteristics of TiO_2 synthesized from ilmenite and its applications for photocatalysis

Y Aristanti[1], Y I Supriyatna[2], N P Masduki[1], and S Soepriyanto[3]

[1]Department of Metallurgical and Material Engineering, Bandung Institute of Technology and Science, Bekasi 17530, Indonesia
[2]Reasearch Unit for Mineral Technology, Indonesia Institute of Sciences, Lampung 35361, Indonesia
[3]Department of Metallurgical Engineering, Bandung Institute of Technology, Bandung 40132, Indonesia

Abstract. Titanium dioxide (TiO_2) contained in the Ilmenite Bangka mineral was synthesized by varying the calcination temperature to obtain characteristics that can improve the photocatalic ability.It can be a solution to handling of waste in the textile industry.Synthesis starts from a caustic fusion process using NaOH to decompose of ilmenite into the titanate compound were more soluble in acid, so that can reduce acid consumption in the leaching process. The leaching process uses 8 M H_2SO_4 in 5% solid to produce $TiOSO_4$ as a filtrate. Then the filtrate was hydrolyzed to obtain the precipitate TiO_2-Fe_2O_3. Calcination with variation of temperature was came after leached the precipitate with the water until neutral pH is obtained. The temperature used is: 350 °C (TiO_2 T_1), 450 °C (TiO_2 T_2), 550 °C (TiO_2 T_3), 650 °C (TiO_2 T_4), and 750 °C (TiO_2 T_5). The results of photocatalytic activity test using a solution of Rhodamin B (RhB) and Methylene Blue (MB) revealing that the highest photocatalyst efficiency was achieved by TiO_2 which calcined at a temperature of 650 °C, with 69.33% (RhB) and 83.41% (MB).Where in this temperature, rutile phase began to form about 15.38%, the band gap energy on 2.4 eV, and the growth of crystalite size become 53.41 nm which affected the increase in photocatalytic activity.

1. Introduction

Titanium dioxide (TiO_2) is one of semi-conductor material which is one of the products from ilmenite oxide minerals ($FeTiO_3$) [11]. In Indonesia Ilmenite reserves are very abundant as a result of tin ore processing by-products on the island of Bangka [1]. This material has been widely used in applications in various fields such as paint pigments, cosmetics, solar cells, sensors, photocatalysts and others [4].

The synthesis process of TiO_2 uses large amounts of acid reagent with relatively high production prices. As Aristanti (2016) stated, that to make 1 ton of TiO_2 will produce 8 tons of dilute sulfate waste, so to reduce the negative impact, a new alternative method used namely caustic fusion [18]. The caustic fusion aims to decompose a mineral or sample using reactive alkali compounds such as NaOH and KOH [4]. This will make it easier for the TiO_2 compound to dissolve in acidic solutions [2, 3, 5].

In this research, we want to know about the photocatalytic ability of TiO_2 for being catalyst to decompose the waste of textile industry. Photocatalysis itself is a chemical reaction that utilizes the absorption of ultraviolet (UV) light energy which is accelerated by the presence of a catalyst [2, 3, 4]. Photon energy absorption will activate the catalysis process and produce hydroxyl radicals which will degrade an organic pollutant and produce an environmentally friendly end product [16]. Photocatalytic reactions begin with the formation of electron-hole pairs in semiconductor particles. If a semiconductor absorbs energy equal to or greater than the energy of the band gap, then electrons (e^-) in the valence band (valence band, VB) will be excitation to the conduction band (CB), leaving a positive hole (hole, h^+) Thermodynamically, the energy level of a conduction band (E_{CB}) is a measure of electrons' reduction strength in semiconductors, whereas the energy level of a valence band is a measure of hole oxidation power [16].

Many factors will affect to thephotocatalitic activity of TiO_2 in the synthesis process. One of them is the calcination temperature. Calcination aims to eliminate the water content which trapped in TiO_2 [2, 3]. Besides that calcination will change the structure of TiO_2 which is still amorphous to crystalline. The change of crystalite structure is accompanied by a phase transformation from anatase to rutile phase. This transformation in crystalite structure is closely related to energy balance as a function of crystallite size [2]. The anatase phase tends to be unstable at high temperatures, so that the higher calcination temperature of the anatase particles stick together to form the larger particles and the interface of the anatase particles then becomes the rutile phase spinning point [6]. During the transformation process an electron-hole pair will be formed on the crystal structure that will activate the photocatalyst process [7].

In addition to TiO_2, Fe_2O_3 also dissolves and is extracted into impurities. However, according to Rias and Das (2004), this will have a good effect on photocatalyst applications because it can slow down phase transformations from anatase to rutile and also reduce the band gap energy of the TiO_2 itself. In addition, Fe_2O_3 can also prevent recombination of electrons from the conduction band to the valence band so that photocatalyst activity increases [8].

2. Experimental Procedures

2.1 Synthesis of TiO_2

200 mesh of high grade ilmenite concentrate was mixed with Merck 99% NaOH with a mol ratio of 1: 4 [5],then done the caustic fusion at a temperature of 650 °C for 2 hours until got the frit as the product. To remove the salt compounds from the caustic fusion product, frit was leached by water with a ratio of 1: 5 (w / v) [18]. After that, leached the frit used the sulfate method to dissolve TiO_2 compound sand produce $TiOSO_4$ as a precursor. Then the precursors was hydrolyzed with the sol-gel method [15] by adding NH_4OH 10% to get a pH of 1.00 with a temperature of 90 °C for 2 hours to form $TiO(OH)_2$ precipitates. Finally calcination process was conducted with temperature variations 350 °C (TiO_2 T_1), 450 °C (TiO_2 T_2), 550 °C (TiO_2 T_3), 650 °C (TiO_2 T_4), and 750 °C (TiO_2 T_5) with 1 hour holding time.

2.2 Photocatalytic Activity TiO_2

To find out the efficiency of photocatalyst activity, TiO_2 was applied to Rhodamin B and Metilen Blue solutions. The concentration used in the solution is 50 ppm with 0.1 gram of TiO_2. The solution that has been mixed with TiO_2 then left under solar radiation for 1.5 hours.

2.3 Instrumentations

In this research, we use 4 instruments to characterize the samples. To find out the compounds contained in ilmenite concentrate XRF was carried out. To determine the Ti and Fe concentrations found in $TiOSO_4$ precursors, AAS analysis was carried out. From the concentration of Ti and Fe, recovery can be obtained from the extraction results. Then on synthetic TiO_2 samples XRD tests were performed in order to identify phases and calculate the size of crystallites formed at each temperature change using the Scherrer equation, $L = k\lambda/\beta \cos\theta$[14]. And the last to find out the efficiency of photocatalysis, UV-Vis analysis was performed on RhB and MB solution resulting from photocatalysis test. In addition to UV-Vis on RhB and MB solutions, UV-Vis was also carried out for TiO_2 synthesis powder to determine band gap energy using the Kubelka Munk method.

3. Results and Discussion

3.1 Analysis of Tio2 Synthesis Results

XRF results conducted by Mulyono (2017) stated that the TiO_2 content of high grade Bangka ilmenite had the highest percentage compared to other compounds that are 45.13% while Fe_2O_3 content is 32.15% [3]. Based on the visuals of titanium dioxide from calcination temperature variations, it can be seen that the higher of temperature, then more brown the color produced as shown in Figure 1. This is due to the thermal expansion of the lattice which occurs reversible. Besides that brownish color was obtained because of the many compounds of Fe_2O_3 which are oxidized at high temperatures.

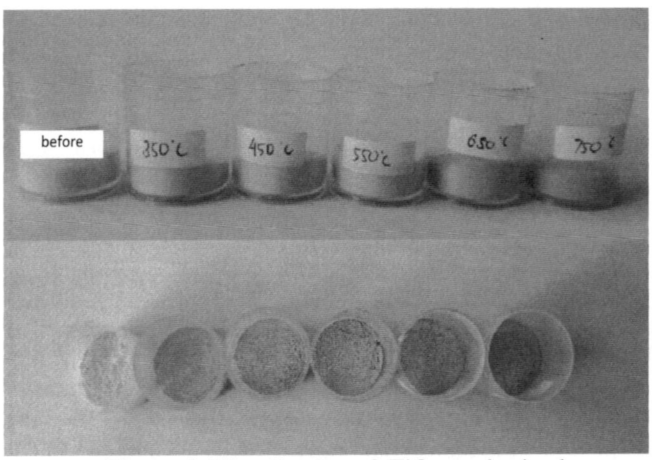

Figure 1. Visualization of TiO_2 synthesized

3.2 Effect of Calcination Temperature on TiO_2 Characteristics

The effect of calcination temperature on phase formation can be seen from the XRD test results which shown in Figure 2. In the literature, it was explained that TiO_2 compound with the crystal structure of anatase marked by the maximum intensity appears at an angle of 2θ about 25° [19]. If the maximum intensity changes it is an indication that the rutile phase appears.

Figure 2 explained that the transformation from anatase to rutile was slow, because rutile appeared in T4 (Temperature of calcination at 650 °C) with a percentage of 15.38%. But, in T5 (Temperature of calcination at 750 °C) percentage of rutile rises significantly up to 89.57%. It caused the maximum intensity has changed to an angle of 2θ about 27°. The slowness of phase transformation occurs due to the presence of Fe_2O_3 impurities in the synthesis product [2, 8]. Furthermore, Fe_2O_3 can reducing the band gap energy [8] and prevent recombination of electrons from conduction band to valance band [10].

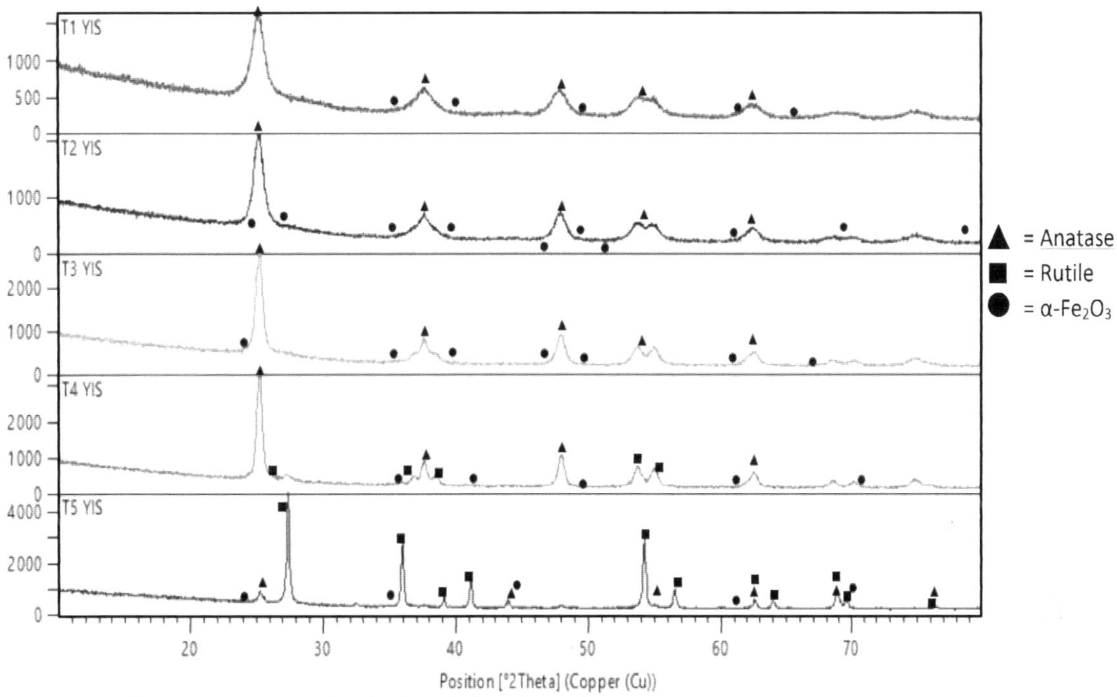

Figure 2. XRD pattern of TiO_2 synthesized

Besides the phase transformations, from the diffraction pattern can determine the size of crystallites using the Scherrer equation. The results of Scherrer's calculations can be seen in table 1. The higher calcination temperature will produce larger crystallite sizes. As Zang (2006) stated, the instability of the anatase phase at high temperatures causes anatase particles stick together to form larger particles and the interface of the anatase particles will become the rutile phase nucleation [6].

Table 1. The size of crystallite TiO_2 at each increase in calcination temperature

Sampel	Size(nm)
T1	5,66
T2	8,8
T3	11,31
T4	53,41
T5	104,89

The temperature of calcination indirectly affected to the band gap energy value. The curve in Figure 3 obtained from UV-Vis powder testing using the Kubelka Munk method. More higher of calcination temperature make the band gap energy will decrease. It caused by presence of Fe_2O_3 as impurity which have a theoritical band gab energy around 2.2 - 2.3 eV [9]. Furthermore, appearance of Fe as impurities also changes the electron configuration of the TiO_2 system, so the band gap energy to decrease [17].

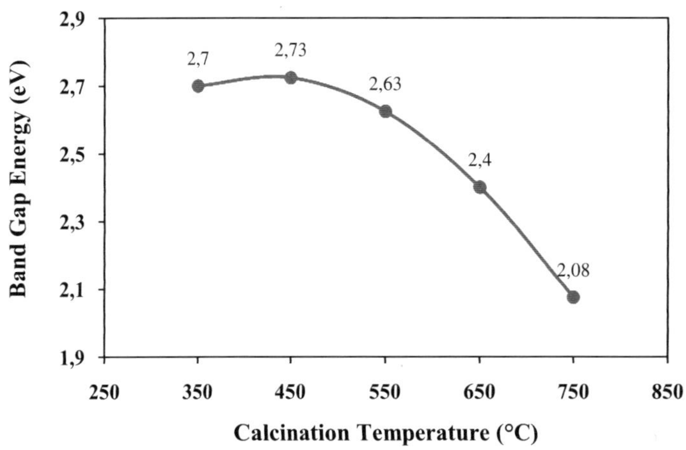

Figure 3. Effect of calcination temperature on TiO_2 band gap energy

The decreasing of band gap energy value affected to the photocatalytic ability for TiO_2. Photocatalysis mechanism start from the excitation of electrons (e^-) from valence band to conduction band and leaves a positive hole (h^+) in the valence band. The positive hole (h^+) will react with H_2O to form OH^- free radicals as oxidizing agents. On the other hand, the electrons (e^-) in conduction band will react with O_2 to form O^{2-} free radicals as reducing agents. That pairs agent can reduce the amount of metal ion and organic compound in solution and produce the friendly environment solution.

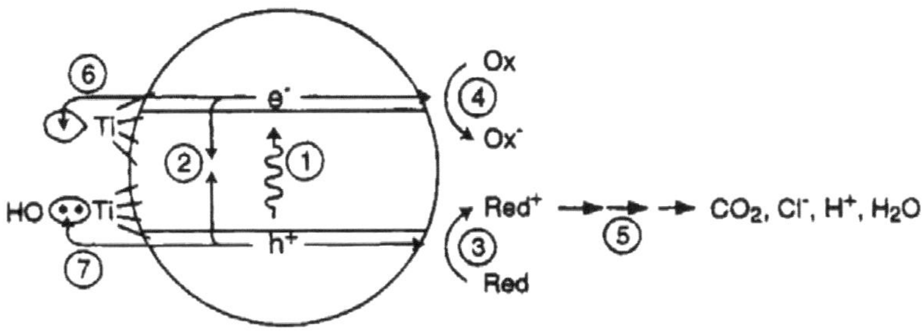

Explanation:
1. Excitation
2. Recombination
3. Oxidation
4. Reduction
5. Hydrolysis
6. Trapping with (O^{-2})
7. Trapping with (OH^{-})

Figure 4. Photocatalysis Mecanism [16]

Low band gap energy causing the process runs faster and make the forming of reducing and oxidizing agent runs fast too. Because we just need a smaller energy to make the electrons (e^{-}) excited from valence band to conduction band rather than the larger band gap energy. Consequently, the degradation process will reaching the maximum level and photocatalytic activity of TiO_2 increases

3.3 Photocatalytic Activity of TiO_2

The discussion above is very affected for photocatalyst activities. The phase, the crystal size, and the band gap energy of TiO_2 will affect the decomposition level of RhB and MB solutions. In a previous study conducted by Fujisima and Honda (1972) explained that the anatase phase produces optimal photocatalysts. However, after further investigation and research, TiO_2 with the anatase phase system and rutile mixture showed higher photocatalyst activity than each phase individually [13]. This provides a synergistic effect that can improve photocatalyst efficiency more significantly.

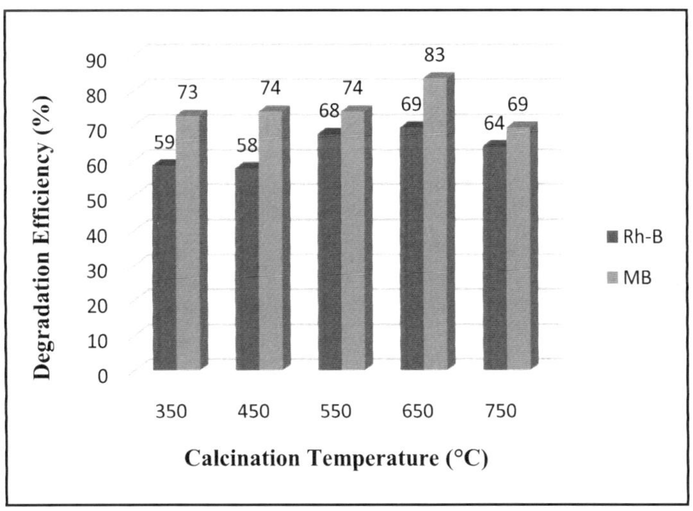

Figure 5.The degradation efficiency of TiO_2 with the calcination temperature variable

Evidently, the results of photocatalyst testing using the Rh-B and MB solution reached the maximum level at 650 °C of calcination temperature. When the rutile phase begin to form about 15.38%, the crystal size growth to 53.41 nm and the band gap energy decreases to 2.4 eV. As shown in Figure 5 the degradations efficiency a chief. However, at 750 °C temperature degradation efficiency dropping off. It caused by the size of the crystallite is too large so that the specific surface area decreases. Then, at this temperature the percentage of rutile rises significantly up to 89.57% makes the band gap energy too small and causing the recombination of electrons from the conduction band to the valence band. Recombination makes the formation of electron pairs and holes to bind pollutants unstable.

4. Conclusion

The synthesis of TiO_2 was not pure, there is Fe_2O_3 compound as impurities. Because of Fe_2O_3, rutile phase growth occurs at a temperature of 650 °C. Calcination temperature variations affect the TiO_2 crystal structure. The more higher temperature of calcination, the rutile phase will increase than make the crystallite size will increase too. Because in high temperature will trigger crystallite growth and encourage the occurrence of rutile phase nucleation. The presence Fe_2O_3 as impurity in semiconductor TiO_2 will decrease the amount of band gap energy from 3,2 eV to 2,7 eV. In addition, presence of rutile is causing the band gap energy become more lower. When the percentage of rutile is 89.57%, the band gap energy will decrease until 2.08 eV. It causes poor photocatalytic properties for TiO_2 because of the recombination. So, the maximum photocatalytic activity of TiO_2 was obtained at 650 °C of calcination temperature, when the rutile phase begins to form about 15.38%, the band gap energy is 2,4 eV and the crystallite size was 53.41 nm, causing the highest degradation efficiency reach 69% in Rh-b solution and 83% in MB solution.

Acknowledge

The author would like to thanks to PT. Timah Tbk., Laboratory of Solid Oxide System - ITB, ITSB, and LIPI Lampung which has provided facilities during the research.

References

[1] Wahyuningsih S 2016 *Pembentukan TiO_2 Mesopori dari $TiOSO_4$ Hasil Pelarutan Ilmenit*.Penelitian Dasar Ilmu Pengetahuan dan Teknologi Nuklir. Bandung.
[2] Aristanti Y 2018 *IOP Conf. Series: Materials Science and Engineering* **285**
[3] Mulyono J E and Soepriyanto S 2017 AIP Conference Proceedings **1805** 030010
[4] Gaetzel M and McEvoy J A 2004 *Asian J. Energy Environ* 197-210
[5] Mostavan A and Soepriyanto S 2015 *Experience on Caustic-Fusion Ilmenite to Produce TiO2 for DSSC.* in Proceedings of 31st European Photovoltaic Solar Energy Conference, (Fraunhover, Hamburg) 223–225
[6] Zhang J, Li M Feng Z, Chen J and Li C 2006 *J. Phys. B* 927-935
[7] Cook G, Blillman L, and Adcock R 2013 *Photovoltaic Fundamentals Scientific Report*, Solar Energy Reasearch Institute, Washington DC
[8] Riyas S and Das P N 2004 *British Ceramic Transitions* 23-28
[9] Gilbert B, Frandsen C, Maxey R E, Sherman M D 2009 *Phys. Rev. B* **79** 035108
[10] Nowotny J 2012 *Oxide Semiconductors for Solar Energy Conversion Titanium Dioxide*. CRC Press.
[11] Chatterjee K K 2007 *Uses of Metals and Metalic Minerals.* New Delhi: New Age International Publishers
[12] Fujishima A and Honda K 1972 *Nature* **238** 37-38

[13] Bickley R I, Gonzales-Varreno T, Lees J S, Palmmisano L, and Tilley R J 1991 *Journal of Solid State Chemistry* 178-190
[14] Abdullah M 2009 *Pengantar Nanosains* Bandung Penerbit ITB
[15] Lalasari L H, Yuwono A H, Rochman N T, Harjanto S, and Suharno B 2013 *Trans Tech Publications* 34-40
[16] Hoffmann M R, Martin S T, Choi W, and Bahnemann D W 1995 *American Chemical Society* **96(1)**
[17] Lopez R and Gomez R 2012 *J Sol-gel Sci. Technology* **6** 1-7
[18] Soepriyanto S and Aristanti Y 2016 *Journal of The Australian Ceramic Society* **52** 131-137
[19] Theivasanthi T and Algar M 2013*Titanium Dioxide (TiO_2) Nanoparticles -XRD Analyses- an Insight* Centre for Research and Post Graduate Department of Physics, Ayya Nadar Janaki Ammal College, Sivaksi – 626124, Tamilnadu, India

Biooxidation pretreatment of low grade refractory gold tailings using a sulfur-oxidizing mixotrophic bacterium

I Purnomo[1], S K Chaerun[1,2], and M Z Mubarok[1]

[1]Department of Metallurgical Engineering, Faculty of Mining and Petroleum Engineering, Institut Teknologi Bandung, Ganesha 10, Bandung 40132, Indonesia
[2]Geomicrobiology-Biomining & Biocorrosion Laboratory, Microbial Culture Collection Laboratory, Biosciences and Biotechnology Research Center (BBRC), Institut Teknologi Bandung, Ganesha 10, Bandung 40132, Indonesia

E-mail: zaki@mining.itb.ac.id

Abstract. An inefficient gold leaching in the cyanidation process from refractory gold tailings was caused by gold particle locked in specific minerals. The aim of this study was to harness the ability of a sulfur-oxidizing bacterium to oxidize iron and sulfur and its behaviour on ferrous sulfate heptahydrate ($FeSO_4.7H_2O$) and pyrite (FeS_2) as the medium in the biooxidation process for gold and silver extraction. The results showed that the gold extracted in ferrous sulfate medium was 20.48%, while that in pyrite medium tended to reduce from direct cyanidation to 14%. The silver extraction increased from 80.96% to 99.26% in ferrous sulfate medium and 91.30% in pyrite medium. This result indicated that ferrous sulfate heptahydrate was more suitable than pyrite, which resulted in the higher extraction of gold.

1. Introduction

The increase of the gold demand leads to searching for more developed gold resources [1]. Most of the world gold (±2/3) are obtained from the processing of non-refractory gold ores because the ore proves to be easier to process using conventional gravity or cyanidation concentration methods. Generally, the method that is commercially undertaken is through cyanidation process. The processing of gold ore is carried out by cyanidation process followed by recovery of gold through cementation using zinc powder, adsorption with activated carbon or ion exchange resin. The problem in this process is gold losses caused by the loss of Au-Ag in fine carbon fraction, which is destroyed and carried into the tailings. In addition, the gold leaching process does not undergo optimally if gold ore that is processed is refractory. Gold minerals are called refractory when gold metal particles are very small (50 μm in diameter) and are included in carbonaceous minerals, sulfides, tellurides, and cyanicides. With the presence of these minerals, the cyanidation process becomes ineffective and tends to reduce gold recovery to below 80% even though ore particles are crushed [2]. Sulfide minerals also restrict access to leachate reagents to liberate the gold ore that is included. Hence, this refractory gold ore in its processing requires pretreatment methods before cyanidation.

Some conventional pretreatment methods such as the pre-aeration method by chlorination, roasting, pressure oxidation, ultrafine grinding, and flotation-intensive leaching are proven to be less suitable to process refractory-type gold ores because such methods have several disadvantages such as expensive investment costs, low gold recovery and environmental risks [3]. Recently, biooxidation pretreatment that was first time introduced in 1986, has been used for about 5% of total world gold production [4]. Industries and researchers have been concerned with this method because of its lower investment costs, easier operation and more environmentally friendly way. Pretreatment of gold ores in the industry through biooxidation generally uses acidophilic bacteria, which only live at a low pH <2. These acidophilic bacteria cannot grow and move well at pH >3 and are not resistant to organic compounds. On the other hand, the mineralogy of the gold ore was found more complex, for example, gold was found to be associated with acid-consuming minerals, such as carbonate minerals and some organic carbon compounds in the ore. A previous study has reported that the gold extraction can be increased by harnessing the ability of the mixotrophic bacterium capable of oxidizing sulfur (i.e., *Citrobacter youngae*) to oxidize iron and sulfur and increase gold recovery which reached the recovery value of above 90% from high sulfides ore [5,6,7]. The bacterium has the ability to live in a wider pH range from pH 4 to 8 than chemolithotrophic bacteria. *C.youngae* uses organic and inorganic carbon as its carbon sources and can oxidize sulfur to sulfuric acid and also oxidize iron. Basically, the mechanism of iron and sulfur oxidation by bacteria occurs in two mechanisms, direct and indirect mechanism by oxidizing Fe^{2+} to Fe^{3+} according to the reactions in equations 1 and 2 [8].

$$2Fe^{2+} + 0.5O_2 + 2H^+ \xrightarrow{\text{bacteria}} 2Fe^{3+} + H_2O \quad (1)$$

$$MS + 2Fe^{3+} \xrightarrow{\text{bacteria}} M^{2+} + S^o + 2Fe^{2+} \quad (2)$$

Based on these mechanisms, the iron sources could vary from any compounds such as $FeSO_4$ and pyrite. In this study, the effects of $FeSO_4$ and pyrite addition in biooxidation system on gold and silver recovery were investigated by taking into account several aspects of biooxidation and leaching processes.

2. Materials and Methods

2.1 Sample characterization

In this study, a total amount of 50 kg samples were received from a gold processing plant tailing in West Java. As received samples were mostly in fine particles of about P75-74 μm, where the coarser size was ground to get the desired size. The samples were then prepared by quartering-coning methods and subsequently sampled with riffle splitter and then prepared for mineralogical analysis. The semi-quantitative X-ray diffraction (XRD) analysis and chemical composition were showed in Table 1 and Table 2, respectively. A diagnostic leaching was conducted to determine the gold distribution in the tailing more clearly [9,10,11]. Diagnostic leaching steps were conducted using a variety of acids aimed to destroy specific minerals, followed by residual cyanidation from each stage (data not shown). At the end of the cyanidation process, dissolved gold was analyzed by using atomic absorbtion spectroscopy (AAS) [12,13]. The result indicated that the more than 30% particle of gold was distributed throughout pyrite mineral.

Table 1. Semi-quantitative XRD analysis of the sample

Minerals	Weight percent
Quartz	42.1
Calcite	23.5
Sanidine	26.5
Albite	6.3
Pyrite	1.7

2.2 Biooxidation pretreatment

In the present study, the bacterium used was a sulfur-oxidizing bacterium (i.e., *Citrobacter youngae*) as one of our bacterial collections. This bacterium was cultured in the LB-broth medium that was sterilized by autoclave at 121°C for 20 min. For the initial condition, the pH medium was set for optimal grwoth of the bacterium. The pH setting was perfomed to control the dominant species in the biooxidation system. Also, it could maximize the oxidation rate and prevent the formation of jarosite during biooxidation [14,15]. LB-broth medium consisted of 5 g/L $Na_2S_2O_3.5H_2O$, 7 g/L FeSO4.$7H_2O$, 3 g/L molasses, 5 g/L yeast extract, and 10 g/L NaCl.

Table 2. Chemical composition of the sample

Element	Weight percent	Element	Weight percent
Si	60.06	S	0.53
Ca	13.53	Mn	0.54
Al	7.95	Ti	0.33
K	7.16	Cr	0.07
Fe	5.49	Cu	0.06
Mg	3.34	Zn	0.06
C	1.92	Ni	0.03
Au (g/t)	0.5	Ag (g/t)	18

Before being adapted with the tailings, the bacterium was cultured until the growth reached 75% of the exponential phase (for 3 days) under aerobic condition. Bacterial adaptation included biooxidation and cyanidation processes of biooxidation residue. Biooxidation experiments were conducted in duplicate in 500 mL Schott bottles with a working volume of 350 mL. This working volume consisted of 15% (v/v) bacterial inoculum, 5% solids of the sample and medium. The composition of the medium at this stage was the same as that at the inoculation stage. The pH and redox potential of the suspensions were measured daily at the interval of 24 hours for 5 days. To evaluate the changes in the concentration of soluble Fe, the suspension (5 mL) was collected daily and centrifuged at 5000 rpm for 10 min. In centrifugation step, the supernatant was collected and prepared for AAS analysis, while the pellet was returned to biooxidation suspension by top-up solution. After 5 days, the residue of biooxidation was filtered and prepared for cyanidation step.

2.3 Leaching procedure

In the leaching step, all residues from biooxidation were leached using cyanide. A series of experiment were carried out in duplicate in the beaker glass for 24 hours and all parameters in this process were kept constant. Leaching parameters under study included NaCN concentration of 1000 ppm, pH 10.5-11, 30% solid, dissolved oxygen concentration adjusted to 7-10 ppm by H_2O_2 addition and pH adjustment by hydrated lime addition. At the end of the leaching, the slurry was filtrated and repulped by NaCN wash solution. The filter cake was then dried to determine the mass loss before a subsequent step. The analysis of Au and Ag in the residue was performed by digestion using aqua regia. In this procedure, 5 g of dried residues were dissolved in 100 ml of a mixture of concentrated HCl and HNO_3 and then stirred at 350 rpm at 90 °C for 8 hours. The filtrate from this process was separated from silicate residue and submitted for analysis of Au and Ag by AAS. Direct cyanidation was also conducted in the same parameters to compare the result from biooxidation pretreatment.

3. Results and Discussion

3.1 Biooxidation pretreatment

Daily monitoring data of biooxidation system are presented in Figs. 1-3. Figure 1 showed the solution pH of the biooxidation system supplemented with $FeSO_4$ or pyrite. Both biooxidation systems showed the increase of pH values although the pH was adjusted to ~5 at the onset of the experiment. This behaviour might be caused by dissolution of alkaline minerals [16] as well as a large number of calcite mineral in the tailings (~23.50 wt.%). Calcite mineral reacted with H^+ ions in the solution might form calcium hydroxide and carbon dioxide [17]. Therefore, the pH increase in the biooxidation system containing pyrite tended to be lower than that containing $FeSO_4$. This indicated that the bacterium was more capable of providing proton ions by oxidizing sulfide mineral (herein pyrite) than $FeSO_4$.

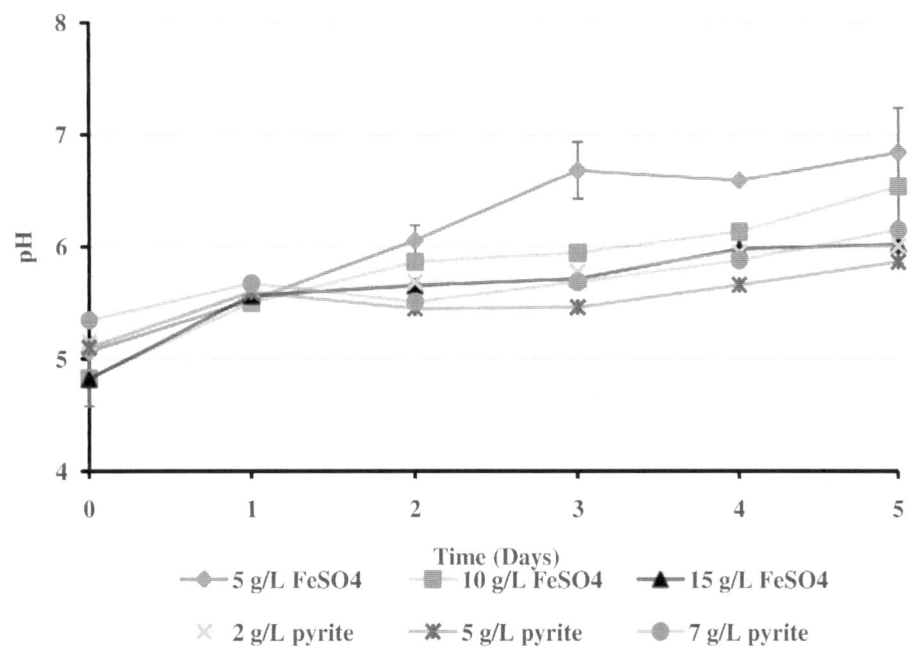

Figure 1. Solution pH values of the biooxidation system supplemented with $FeSO_4.7H_2O$ (5, 10, 15 g/L) or pyrite (2, 5, 7 g/L) using a sulfur-oxidizing mixotrophic bacterium over a period of 5 days

Figure 2 revealed the solution Eh values of the biooxidation system supplemented with $FeSO_4$ or pyrite. The Eh values tended to decrease caused by physicochemical or bacterial activity. The values indicated the efficiency and bacterial activity during the biooxidation process. It is reported that non-adaptive bacteria reach solution potential under 650 mV (vs. SHE) [7]. The performance of the bacterium used in this study might be influenced by the complexity of the tailing characteristics.

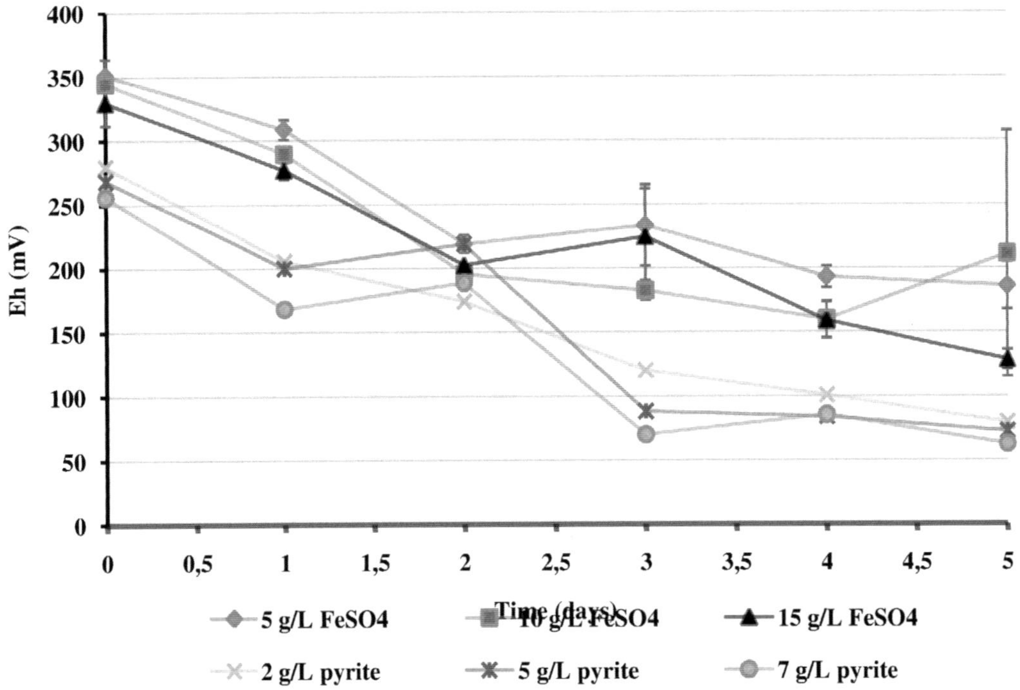

Figure 2. Solution Eh (mV vs. SHE) values of the biooxidation system supplemented with $FeSO_4.7H_2O$ (5, 10, 15 g/L) or pyrite (2, 5, 7 g/L) using a sulfur-oxidizing mixotrophic bacterium over a period of 5 days

Figure 3 represented soluble iron (Fe) in the biooxidation system supplemented with $FeSO_4$ or pyrite. The concentration of soluble Fe in the biooxidation system supplemented with $FeSO_4$ was higher than that with pyrite. This behaviour occurred because $FeSO_4$ dissolved immediately at the first day of biooxidation, whereas pyrite (in solid form) required the role of the bacterium to dissolve. The decreased concentration of soluble iron in the biooxidation system supplemented with $FeSO_4$ might be due to an increase in pH during the biooxidation pretreatment. It is reported that high pH values leads to an unstable condition for Fe ions, thus causing precipitation [5, 17].

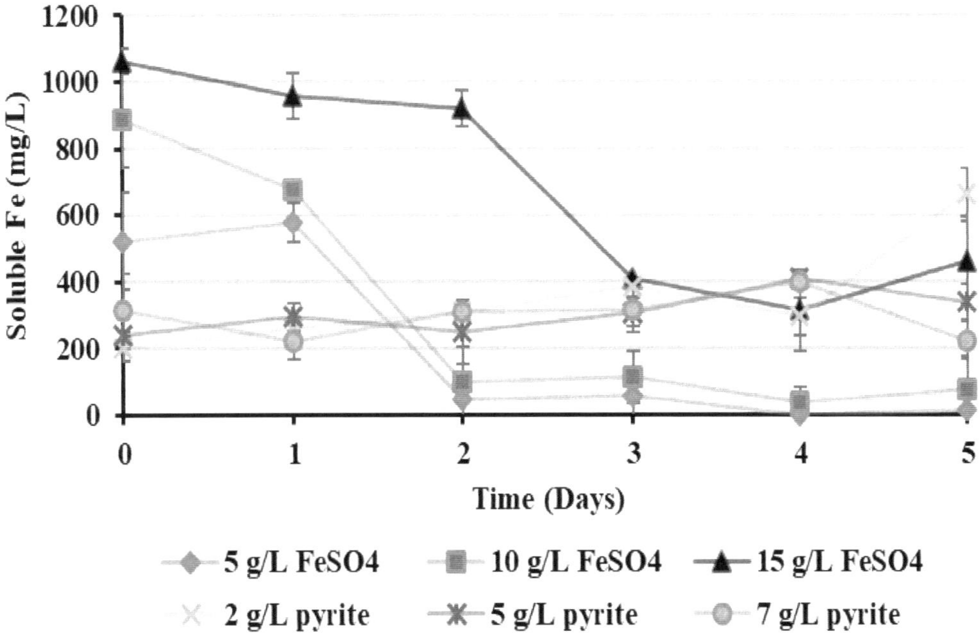

Figure 3. Soluble iron (mg/L) in the biooxidation system supplemented with $FeSO_4.7H_2O$ (5, 10, 15 g/L) or pyrite (2, 5, 7 g/L) using a sulfur-oxidizing mixotrophic bacterium over a period of 5 days

3.2 Cyanidation leaching

The gold and silver recovery in the biooxidation system supplemented with $FeSO_4$ or pyrite was presented in Table 3. It was demonstrated that direct leaching without biooxidation pretreatment reached the recovery of 17.85% and 80.96% for gold and silver, respectively. The extraction levels of gold and silver in the biooxidation system supplemented with $FeSO_4$ was greater than those in direct leaching, which reached an optimum recovery at 7 g/L $FeSO_4$ addition for gold and 5 g/L for silver. It appeared that an increase in $FeSO_4$ concentration (15 g/L) resulted in the decreased extraction values of gold and silver. The addition of $FeSO_4$ in high concentration increased the iron and sulfur ions soluble in bulk solution. Consequently, this condition decreased the probability of the bacterium to oxidize the associated minerals of gold because the bacterium preferred to oxidize the iron and sulfur ions. The gold recovery in the biooxidation system supplemented with pyrite tended to decrease with increasing pyrite concentration and was lower than direct leaching (without biooxidation pretreatment). This behaviour might be caused by the formation of silver sulfides between sulfide ions, which were not oxidized by the bacterium, and silver in the tailings. This silver sulfide leads to passivation layer, which can limit the access of cyanide solution and cause the lower gold and silver recovery [18].

Table 3. Gold and silver extraction from biooxidation residue

Medium	Concentration (g/L)	%Extraction	
		Au	Ag
$FeSO_4.7H_2O$	5	18.97	99.26
	7	20.48	67.08
	10	17.12	96.90
	15	17.17	93.94

Pyrite	2	14.00	91.30
	5	13.28	87.20
	7	12.88	88.55
Direct leaching		17.85	80.96

4. Conclusions

In the present study, *Citrobacter youngae* was able to increase the gold recovery from 17.85% to 23.95% under biooxidation parameters of 5% solids, $FeSO_4.7H_2O$ concentration of 7 g/L and particle size under 200 mesh. Based on these result, *C. youngae* was able to recover gold trapped in pyrite by 20.33%, while pyrite addition tended to reduces the recovery of gold caused by passivation layer forming in the cyanidation process.

Acknowledgements

The authors thank all the members and students of the Chaerun Laboratory for their assistance and cooperation. IP would also like to thank the Indonesia Endowment Fund for Education (LPDP-RI) for the scholarship provided during his study at Institut Teknologi Bandung.

References

[1] World Gold Council. Gold Demand Trends Q2 2018. 2018 [cited on 1 August 2018] Available from https://www.gold.org/
[2] Afidenyo J 2008 *Microbial pre-treatment of double refractory gold ores* [Dissertation]
[3] Winarko R 2016 *Study of the Effects of Several Variables in Biooxidation of a Sulfidic Gold Refractory Concentrate with Mixotrophic Bacteria on Gold Extraction Percentage Improvement* [thesis]. Institut Teknologi Bandung
[4] Johnson D B 2014 *Curr. Opin. Biotechnol* **30** 24–31
[5] Mubarok M Z, Winarko R, Chaerun S K, Rizki I N and Ichlas Z T 2017 *Hydrometallurgy* **168** 69–75
[6] Winarko R 2015 *Biooxidation of sulfidic gold concentrate with Aciditiobacillus ferooxidans, Alicyclobacillus ferooxydans and Citrobacter youngae at neutral pH* [thesis]. Institut Teknologi Bandung
[7] Marsden J O and House C L 2009 T*he Chemistry of Gold Extraction. Society for Mining, Metallurgy, and Exploration*, Inc. USA. Second Edition
[8] Celep O, Alp I, Deveci H and Vicil M 2009 *Trans. Nonferrous Met. Soc. China* **19** 707–713
[9] Saba M, Mohammad Y A, Rashchi F and Moghaddam J 2011 *Miner. Eng.* **24** 1703–1709
[10] Henley K J, Clarke N C and Sauter P 2001 *Minerals Engineering* **14(1)** 1-12
[11] Lorenzen L 1995*Minerals Engineering* **8(3)** 247-256
[12] Lorenzen L and Van Deventer J S J T 1992 *Minerals Engineering* **5(10-12)** 1377-1387
[13] Pooley F D, Shrestha G N, Errington M T and Gibbs H E 1987 S*eparation Processes in Hydrometallurgy* 58-67
[14] Southwood M J 1986 *Mineralogical Aspects of the Bacterial Leaching of Auriferous Sulphide Concentrates, and a Mathematical Model for the Release of Gold*
[15] Marchevsky N, Quiroga M B, Giaveno A, and Donati E 2017 *Transactions of Nonferrous Metals Society of China* **27(5)** 1143-1149
[16] Ubaldini S, Veglio F, Beolchini F, Toro L, and Abbruzzese C 2000 *International Journal of Mineral Processing* **60(3-4)** 247-262
[17] Adam K, Stefanakis M, and Kontopoulos A 1989 *Proc. Biotechnology in Minerals and Metal Processing*, SME, New York

[18] Soefihara M D U and Bakhtiar A 2017 *Study of the Effect of Biooxidation of Sulfidic Gold Concentrate on Gold Recovery Through Intensive Leaching* [Thesis]. Institut Teknologi Bandung

Screening of bacteria for coal beneficiation

S R Nurhawaisyah[1], E Sanwani[1], S K Chaerun[1,2], and M A Rasyid[1]

[1]Department of Metallurgical Engineering, Faculty of Mining and Petroleum Engineering, Institut Teknologi Bandung, Ganesha 10, Bandung 40132, West Java, Indonesia
[2]Geomicrobiology-Biomining & Biocorrosion Laboratory, Microbial Culture Collection Laboratory, Biosciences and Biotechnology Research Center (BBRC), Institut Teknologi Bandung, Indonesia

Email: skchaerun@metallurgy.itb.ac.id

Abstract. Biobeneficiation is considered to be a clean coal technology as an alternative method to reduce sulfur and ash content present in coal by using microbes, in particular bacteria. Recent studies have demonstrated that bacteria can oxidize iron and reduced sulfur compounds present in coal because they promote the oxidative conversion of the reduced forms of sulfur to soluble, easily washed-out compounds. Moreover, the use of bacteria in coal beneficiation has the following advantages, such as simple operation, extensive installation, performed under the mild condition without harmful product, low energy consumption and eco-friendly method. Hence, this study dealt with the investigation of the bacteria capable of oxidizing iron and sulfur compounds and also reducing ash content present in coal. Nine bacteria isolated from mercury-contaminated gold mine sites in Bandung, West Java Province, Indonesia were screened for their ability to oxidize iron and sulfur compounds and reduce ash content. Of the nine bacteria studied, two bacteria (i.e., *Pseudomonas plecoglossicida* and *Pseudomonas hibiscicola*) were able to oxidize iron and sulfur compounds and reduce ash content at pH of ~4. The findings of this study provide evidence that both bacteria could be employed as an oxidizing agent that will be applicable for the beneficiation of coal.

1. Introduction

Coal is one of the most widely used fossil fuel energy sources to meet the needs of various sectors, including electricity generation, iron and steel production, cement plants and as liquid fuels. Therefore, coal production is estimated to increase every year. The high demand for coal is due to the cheapest fossil fuel. Meanwhile, the latest data from the Geological Agency of the Ministry of Energy and Mineral Resources (ESDM), the Republic of Indonesia, presently coal reserves of Indonesia are 26.2 million tons in which 56 years later, coal reserves will exhaust if it is assumed that there are no new coal reserves.

However, not all of the coal reserves in Indonesia have the same characteristics. Coal from different location contains variable amounts and types of sulfur, depending on the geological location of coal. Low sulfur of coal has high industrial prospects, while the presence of high sulfur restricts its utilization and is often regulated for use. Coals containing high sulfur cause many environmental problems. When coal is burnt, its sulfur content combines with oxygen to produce sulfur oxides which contribute to both pollution and acid rain. The best possible way to prevent our environment from sulfur oxides is to reduce the amount of sulfur in coal before combustion. Sulfur compounds contained in coal are generally divided into two major groups on the basis of the chemistry of sulfur, i.e., inorganic sulfur and organic sulfur. The inorganic sulfur in coal usually consists of sulfide and sulfate minerals. Sulfide minerals include pyrite (FeS_2), sphalerite (ZnS), galena (PbS), and others [1]. The pyrite is generally the preponderant inorganic sulfur in coal [2]. Particles of pyrite are randomly distributed as crystals throughout the coal but are not bound to it [2]. The sulfate minerals include barite ($BaSO_4$), gypsum ($CaSO_4.2H_2O$) and a number of iron sulfates and others [2]. The organic sulfur in coal is covalently bound into its large complex structure and is difficult to remove physically and chemically [2].

Recently, biobeneficiation offers a clean alternative method to reduce sulfur content present in coals. It is well known, compared to the physical and chemical methods, the biobeneficiation of coal has the following advantages, such as simple operation, extensive installations, and low energy consumption [3][4]. Moreover, biobeneficiation process, promoting the oxidative conversion of the reduced forms of sulfur to soluble and easily washed-out compounds, is performed under mild conditions without harmful products [5][6].

Keeping in view of the importance of coal biobeneficiation, in this study, nine bacteria were screened to find out whether these bacteria are able to oxidize sulfur and iron, which can be developed for the process of coal biobeneficiation. The characteristics of coal in pre-treatment and post-treatment using bacterial process were compared in terms of proximate analysis, total sulfur and FTIR analysis.

2. Materials and Methods

2.1. Coal samples

The coal biobeneficiated in this study was obtained from Kalimantan island on the coal mine, Indonesia. The coal samples were crushed in roll crusher, further ground in tumbling mill and separated into various particle size fractions by sieving machine to obtain the grain size of -200 +325 mesh (-74+44 μm).

2.2. Bacterium and growth medium

The nine bacteria used in this study are mixotrophic bacteria that have the ability to oxidize sulfur and iron. They were isolated from a mercury-contaminated gold mine site in Bandung, West Java Province, Indonesia. The growth medium used was SKC broth medium (molasses and seawater).

2.3. Experimental procedure

Screening of bacteria was conducted in two stages. The first stage was to make the bacterial growth curve to know bacterial growth phases by a spectrophotometric method. It is a method for calculating the number of cells using a spectrophotometer to see optical density through the absorbance value produced. The bacterial growth medium used was SKC broth (molasses and seawater). Furthermore, bacterial culture was incubated at room temperature and agitated at a speed of 180 rpm. Turbidity level and pH of the bacterial cultures were monitored periodically, every 24 h for 10 days. From the growth curve, the time length of incubation of the bacteria in the medium was able to be determined so that the bacteria were ready to be used in coal biobeneficiation experiments.

In the second stage, screening of bacteria was carried out by adapting coal with bacteria in duplicate in a sterile 500 ml Erlenmeyer flask containing 250 ml of growth medium supplemented with 10% (v/v) bacterial inoculum and 25% (w/v) coal. Cultures were incubated with agitation (180 rpm) at room temperature for 10 days under aerobic condition. The pH and Eh of the suspensions were observed periodically (every 24 h) and subsequently the coal was separated, dried, weighed and prepared for further experiments and for analysis.

2.4. Analytical measurements

Measurement of turbidity level was carried out using a UV-VIS spectrophotometer. The pH and Eh of the suspensions were observed using Lutron pH-207. To determine the chemical properties of coal before and after treatment, a proximate analysis consisting of moisture, ash, volatile materials and carbon content was carried out according to the standard method of ASTM D 7582-12. In addition, the sulfur content in coal was also determined according to the standard methods of ASTM D 2492-02 and D 3177-02. FTIR analysis of coal were carried out before and after biobeneficiation to determine changes in chemical bondings in the wavenumber range of 400 to 4000 cm^{-1}. The dried coal sample was mixed and ground with 200 mg KBr for FTIR measurement.

3. Results and Discussion

3.1. Bacterial growth curve

Screening of bacteria was firstly performed by making a growth curve. The growth curve provides an overview of the overall bacterial growth cycle, which includes the lag phase, exponential phase, stationary phase, and death phase [7]. Bacterial growth was measured spectrophotometrically on the basis of turbidity level for 10 days (Figure 1) while monitoring the pH of the suspensions (Figure 2). From Figure 2, it can be seen that all the bacteria grew well in the pH range of 3 to 6, indicating that by using SKC broth medium the bacteria were able to produce organic acids without the addition of sulfuric acid or hydrochloric acid. Organic acids produced by bacteria were expected to be able to dissolve sulfur contained in coal. Therefore, based on the bacterial growth curve and pH of the suspensions, the two best bacteria were obtained, namely *Pseudomonas plecoglossicida* (designated M9) and *Pseudomonas hibiscicola* (designated M10).

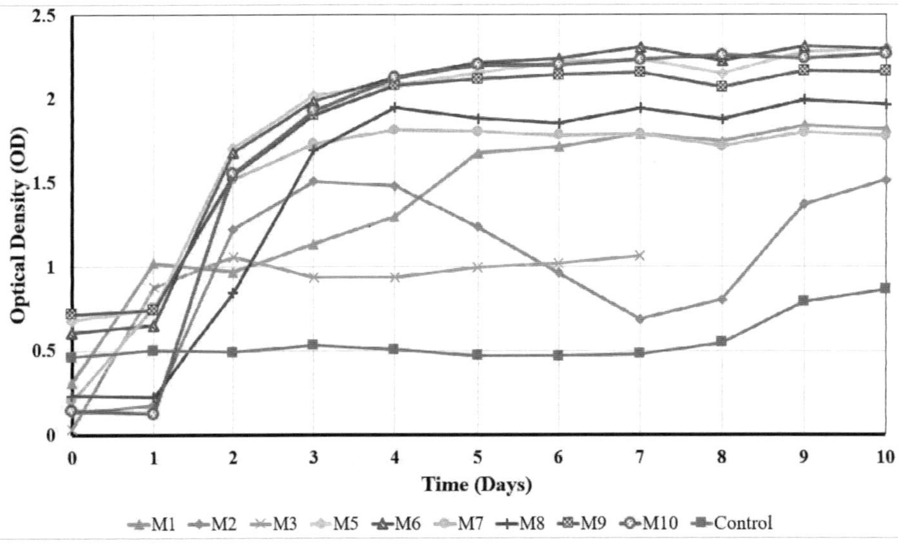

Figure 1. Growth curve of the nine bacteria (designated M1, M2, M3, M5, M6, M7, M8, M9) based on optical density measurement. Control: turbidity level of the solution in the experimental system without bacteria

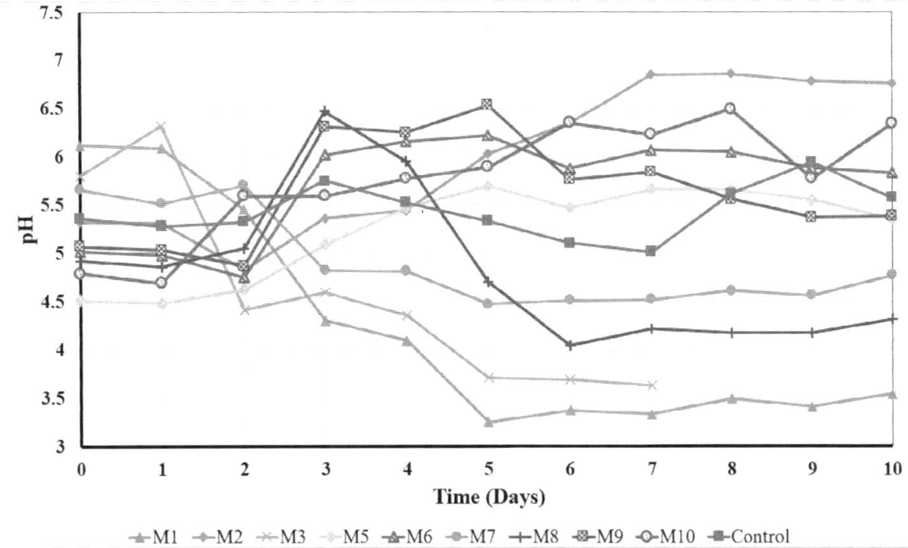

Figure 2. Changes in the pH of the suspensions of the experimental systems containing coal and the bacterium (designated M1, M2, M3, M5, M6, M7, M8, M9) during bacterial growth over 10 days of the experiment. Control: Suspension pH in the experimental system without bacteria

3.2. Coal biobeneficiation

Figure 3 demonstrated changes in pH and Eh of the experimental system for coal biobeneficiation with the bacterium *Pseudomonas plecoglossicida* (designated M9) or *Pseudomonas hibiscicola* (designated M10) over the course of 10 days. It can be seen that the pH of the suspension in the experimental system decreased over time. This was clearly influenced by bacterial activity in which the bacteria accelerated pyrite oxidation in coal by oxidizing iron and sulfur which generated H_2SO_4. This interpretation was also supported by the measurement of the redox potential (Eh) of the suspension. With reference to the Pourbaix diagram [8], sulfur can dissolve into H_2SO_4 in the pH range of 3 to 7 with Eh value in the range of 100 to 800 mV. Therefore, it was suggested that the bacterium *Pseudomonas plecoglossicida* (designated M9) and *Pseudomonas hibiscicola* (designated M10) were capable of oxidizing sulfur in coal under this study.

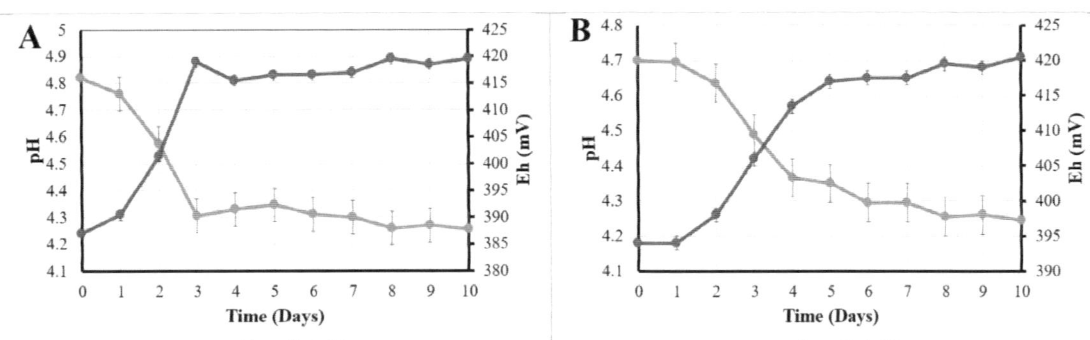

Figure 3. Changes in pH and Eh (mV vs. SHE) of the suspensions in experimental system containing coal and the bacterium *Pseudomonas plecoglossicida* (designated M9) or *Pseudomonas hibiscicola* (designated M10) over the 10-day experimental period

3.3. Effect of biobeneficiation process on sulfur and ash elimination in coal

Coal biobeneficiation was carried out to investigate the changes in the total sulfur and ash in coal after treatment with bacteria (Table 1). It was found that the total sulfur content of the initial coal sample was 2.82%, and under optimal condition, 15.25% and 16.31% of the total sulfur were removed from the coal after treatment with *Pseudomonas plecoglossicida* and *Pseudomonas hibiscicola*, respectively. The changes in the fixed carbon content, volatile matter and ash content of coal were also shown (Table 1). Fixed carbon and volatile matter content in the coal after biobeneficiation increased slightly. This increase might be due to the fact that lower volatile substances were converted into highly volatile substances in the beneficiation process. Furthermore, there was a decrease in ash content of coal sample after treatment with bacteria from 8.82% to 7.175% (for *Pseudomonas plecoglossicida*) and 6.77% (for *Pseudomonas hibiscicola*), indicating that there was a relationship between the extent of coal biobeneficiation and ash elimination. However, some researchers have reported that the reduction in ash content may be mainly attributed to dissolution of minerals [9]. The decrease of ash content in coal has a positive impact on the environment.

Table 1. Composition of untreated and microbially treated coal samples

Analysis (% adb)	Initial coal	Coal treated with the bacterium	
		M9	M10
Ash	8.82	7.175	6.77
Volatile matter	39.76	40.95	41.18
Fixed carbon	40.81	44.745	45.065
Total sulfur	2.82	2.39	2.36

3.4. Fourier transform infrared (FTIR) analysis

The FTIR spectra of the initial coal and the coal after biobeneficiation with the bacterium *Pseudomonas plecoglossicida* (designated M9) or *Pseudomonas hibiscicola* (designated M10) were shown in Figure 4. FTIR spectrum is widely used to recognize bonds of organic compounds in the coal structure [10][11]. However, an overlap of the organic and inorganic bands was common when the coal sample was analyzed [9][12]. Assignments of FTIR spectral bands were summarized in Table 2.

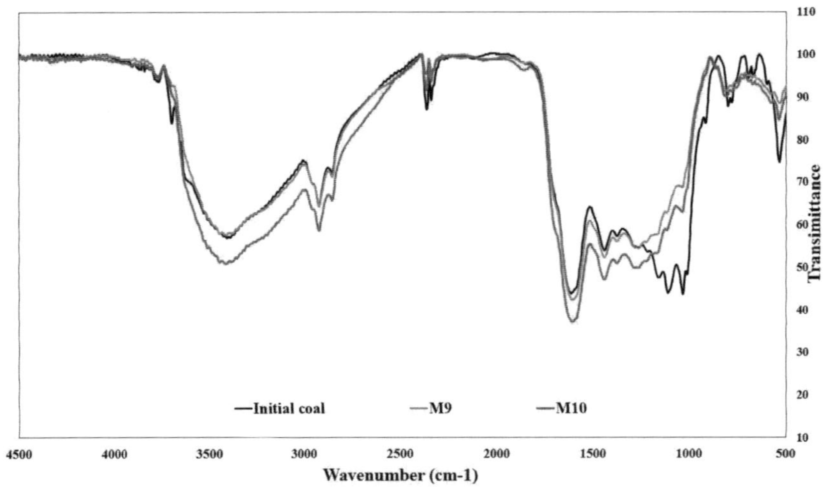

Figure 4. FTIR spectra of initial coal (or starting coal) and the coal after biobeneficiation with the bacterium *Pseudomonas plecoglossicida* (designated M9) or *Pseudomonas hibiscicola* (designated M10)

Table 2. FTIR spectra assignments of initial and biobeneficiated coal samples

Assignment	Initial coal	After treated with the bacterium	
		M9	M10
O-H stretching	3408.22	3413.04	3414.005
C−H stretching	2922.16	2922.16	2922.16
C−H stretching	2852.72	2851.755	2852.72
C≡C stretching	1612.49	1608.63	1609.595
$-CH_3$ bending	1442.75	1441.79	1442.75
$-CH_3$ bending	1377.17	1377.17	1377.17
−S=O stretching	1286.52	1274.945	1285.555
−S=O stretching	1159.22	1166.93	-
C−O stretching	1111.00	1112.93	-
S−O stretching	1033,85	1037.7	1036.735
S−O stretching	796.60	811.065	808.175
C−S stertching	692.44	692.44	-
S−S disulfide	534.28	559.355	-
C−C bending	470.63	472.56	536.21

4. Conclusions

Nine bacteria isolated from a mercury-contaminated gold mine site of West Java Province, Indonesia were screened for coal beneficiation from Kalimantan island of Indonesia. Of the nine bacteria screened, two bacteria have the capability for reducing total sulfur and ash content in the coal sample after treatment. The two bacteria were identified as *Pseudomonas plecoglossicida* (designated M9) and *Pseudomonas hibiscicola* (designated M10). *P. plecoglossicida* and *P. hibiscicola* were capable of removing total sulfur (15.25% and 16.31%, respectively) and the ash content (18.65% and 23.24%, respectively) from the coal. The biobeneficiation process takes place at pH of around 4.

Acknowledgments

This study was supported by Research Program, Community Service, and Innovation (P3MI) of ITB, Geomicrobiology-Biomining & Biocorrosion Laboratory of ITB and Beasiswa Unggulan Kemendikbud.

References

[1] Calkins W H 1994 The chemical forms of sulfur in coal, a review *Fuel* **73 (4)** 475-484
[2] Prayuenyong P 2002 Coal biodesulfurization processes *Songklanakarin Journal of Science and Technology* **24(3)** 493-507
[3] Meruane G, Salhe C, Wiertz J, and Vargas T 2002 Novel electrochemical–enzymatic model which quantifies the effect of the solution Eh on the kinetics of ferrous iron oxidation with *Acidithiobacillus ferrooxidans Biotechnology and Bioengineering* **80 (3)** 280-288
[4] Machnikowska H, Pawelec K, and Podgórska A 2002 Microbial degradation of low rank coals *Fuel Processing Technology* **77** 17-23
[5] Bressler D C and Fedorak P M 2001 Identification of disulfides from the biodegradation of dibenzothiophene *Applied and Environmental Microbiology* **67(11)** 5084-5093
[6] Soleimani M, Bassi A, and Margaritis A 2007 Biodesulfurization of refractory organic sulfur compounds in fossil fuels *Biotechnology Advances* **25(6)** 570-596
[7] Madigan M T, Martinko J M, and Parker J 1997 *Brock Biology of Microorganisms* vol 11 Upper Saddle River NJ: Prentice Hall
[8] Pourbaix M 1974 *Atlas of Electrochemical Equilibria in Aqueous Solutions* NACEmTX
[9] Cardona I C, Márquez M A 2009 Biodesulfurization of two Colombian coals with native microorganisms *Fuel Processing Technology* **90** 1099–1106

[10] Aller A, Martínez O, Linaje J De, Méndez R, Morán A 2001 Biodesulphurization of coal by microorganisms isolated from the coal itself *Fuel Processing Technology* **69** 45–57
[11] Tripathy S S, Kar R N, Mishra S K, Twardowska I, Sukla L B 1998 Effect of chemical pretreatment on bacterial desulphurization of Assam coal *Fuel* **77** 859–864
[12] Huggins F 2002 Overview of analytical methods for inorganic constituents in coal *International Journal of Coal Geology* **50** 169–214

On the determination of manganese in the presence of iron in pregnant leach solution using derivative spectrophotometry

E Prasetyo*, E Purwaningsih, and W astuti

Research Unit for Mineral Technology, Indonesian Institute of Sciences

Email: erik.prasetyo@lipi.go.id

Abstract. Mn determination in samples with high Fe content using derivative spectrophotometric method was proposed in this paper, which formaldoxime was used as chromogenic reagent. The derivative method is useful to manipulate spectra data in order to eliminate the noise spectra i.e. Fe and isolate the spectra of elements of interest (Mn). Formaldoxime was chosen since the reagent reacted both to Mn and Fe at the same pH condition. In this paper, the accuracy in Mn determination in the presence of Fe using proposed method would be assessed and discussed. Mn and Fe could be simultaneously determined Mn using zero crossing point of first derivative Fe-formaldoximate spectra at 530 nm for Mn determination, while Fe could be predicted using zero crossing point of second derivative Mn-formaldoximate spectra at 469 nm. Maximum concentration which obeyed the linearity of calibration curve is 2 and 20 mg/L for Mn and Fe, respectively. The error in Mn and Fe prediction depends on ratio of both elements.

1. Introduction

The increasing needs for manganese to satisfy the demand in many industrial applications such as steel-alloy, chemicals and electronics [1] have sparks the interest for manganese resource exploration including its processing even for the low grade ones. To support these activities, reliable analysis method is mandatory. Several standard Mn determination methods in laboratory include atomic absorption spectrophotometry (AAS), induced couple plasma-atomic emission spectrometry (ICP-AES) and UV Vis spectrophotometry or photometry, which had been applied in all stages of mineral industry from prospecting phase until product quality control. The Mn content in sample is generally considered the key factor in choosing the method. In the case of Mn content falls in the range 0.1-100 ppm, UV Vis spectrophotometry is the best method due to its versatility and low capital cost.

Numerous protocols to determine Mn using UV Vis spectrophotometry had been proposed, which depended on special chemical reagent (chromogenic reagent) to develop the colour, including [2], [3], [4], [5]. Generally, these chromogenic reagents are not selective for Mn determination and reacts with other elements especially iron (Fe), which is omnipresent with high content compared to Mn. Hence, preliminary separation step such as ion exchange [3,5] and addition of masking agent to make other elements unreacted toward chromogenic reagents are required in the protocols.

To address the problem in Mn determination in samples with high Fe content, in this research derivative spectrophotometric method would be proposed, which formaldoxime was used as chromogenic reagent. The derivative spectrophotometry method is useful tools to manipulate spectra data in order to eliminate the noise spectra i.e. Fe and isolate the spectra of elements of interest (Mn). Formaldoxime was chosen since the reagent reacted both to Mn and Fe at the same pH condition. In this paper, the accuracy in Mn determination in the presence of Fe using proposed method would be assessed and discussed.

2. Experimental

The chromogenic reagent, formaldoxime was prepared by mixing 7.9 ml formalin solution (38%) with 7 gr of hydroxylammonium hydrochloride (Merck, Darmstadt). The mixture was then diluted with deionized water to 100 ml in standard volumetric flask. Other chemicals such as potassium sodium tartrate, sodium hydroxide, Mn and Fe stock solution (1000 mg/L) for AAS were purchased from Merck, Darmstadt and used without further purification. Series of Mn and Fe standard solution to obtain calibration curve were made by mixing an aliquot of Mn or Fe stock solution with 0.5 ml of potassium sodium tartrate solution (20%), 1 ml formaldoxime solution, 0.25 sodium hydroxide solution (10 M) in standard volumetric flask and marked with deionized water. Binary mixtures of Mn and Fe with varied content and ratio, treated as samples were prepared as same manner.

The absorption spectra of Mn and Fe formaldoximate complex were collected using Perkin-Elmer Lambda 365 Spectrophotometer (USA), using 1 mm standard cuvette. Spectral bandwith and scan speed were 1 nm and 200 nm/min, respectively. Derivative spectra were generated using UV Express software, scan mode and smoothed using Savitsky-Golay algorithm for spectra smoothing.

3. Results and discussion

3.1. Mn direct determination

Mn in the form of MnO_2 in basic solution would react with formaldoxime to form brown colored complex Mn-formaldoximate. Other Mn species such as Mn^{2+} in basic solution was tended to be oxidized by oxygen in the solution to form MnO_2. Other species with higher oxidation state such as MnO_4^- would be reduced by formaldoxime to tetravalent Mn and then complexed by the formaldoxime [4]. Based the scanning result, the maximum absorption was attained at λ 450 nm (Figure 1). At this wavelength, Lambert-Beer law is obeyed for Mn content 0.1-2 mg/L (R^2 0.9948) and molar absorptivity value ε 1.04×10^4 L mol^{-1} cm^{-1}.

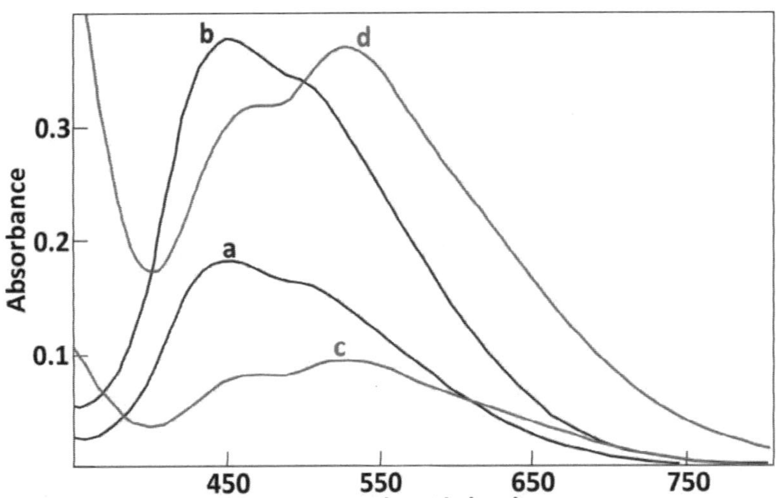

Figure 1. Mn-formaldoximate spectra (a) 0.91 mg/L and (b) 2.03 mg/L, Fe-formaldoximate-tartrate spectra (c) 2.55 mg/L and (d) 11.35 mg/L.

3.2 Fe direct determination

The hydrolysis of trivalent iron (Fe^{3+}) in basic solution was hindered by the presence of tartrate, and reacted with formaldoxime to form purple colored complex. Spectra scanning of Fe-formaldoximate-tartrate is shown in Figure 1, with maximum absorption occurred at λ 528 nm. At this wavelength, Lambert-Beer law is obeyed between 0.6 and 20 mg/L Fe, with molar absorptivity value ε 2.02 x 10^3 L mol^{-1} cm^{-1}.

3.3 Mn determination in the presence of Fe

As explained before, both Mn and Fe reacted with formaldoxime to form colored complex in basic condition. Derivative spectra of both complexes, reconstructed using UV Express software are shown in Figure 2. In the figure, at first derivative spectra, the derivative values of Fe-formaldoximate complex become zero regardless the Fe concentration at λ 398 and 530 nm. These wavelengths are the zero crossing point of Fe spectra and at these wavelengths Mn could be determined. The linear calibration curve to determine Mn at 398 nm is $A = (2.666C + 0.123)x10^{-3}$ (R^2 0.9985), and at 530 $A = (-1.061C - 0.016)x10^{-3}$ (R^2 0.9994), where A and C are derivative value (y axis) and Mn concentration, mg/L (x axis) respectively

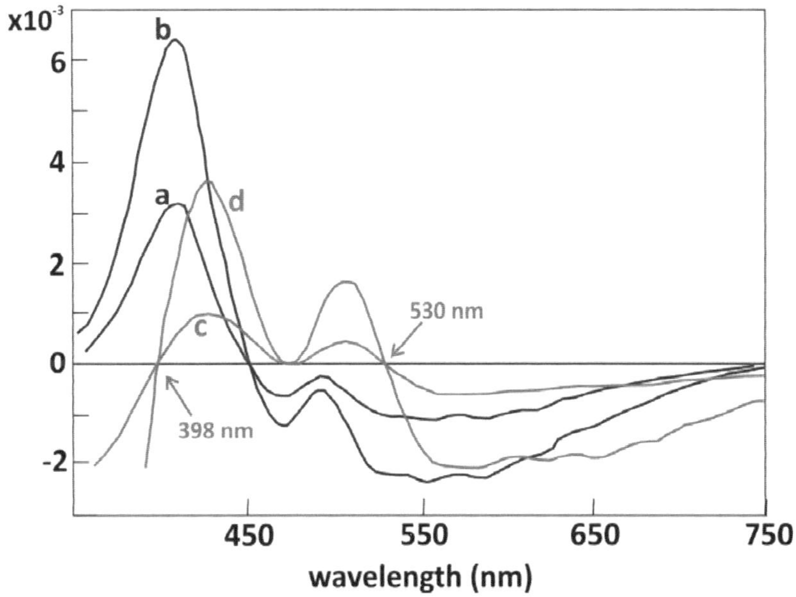

Figure 2. First derivative of Mn-formaldoximate spectra (a) 0.91 mg/L and (b) 2.03 mg/L. Fe-formaldoximate-tartrate spectra (c) 2.55 mg/L and (d) 11.35 mg/L. Zero crossing points of Fe-formaldoximate-tartrate spectra are indicated by arrows.

To determine the Fe content in samples, second derivation of spectra is used, shown in Figure 3. In second derivative spectra, zero crossing point for Mn-formaldoximate complex occurred at λ 469 nm. The linear calibration curve to determine Fe content using second derivative spectra at this wavelength is $A = (-0.0373C-0.0467)x10^{-4}$ (R^2 0.9989), where A and C are derivative value (y axis) and Mn concentration, mg/L (x axis) respectively.

Table 1 shows the results of Mn and Fe determination using first derivative at λ 398 and 530 nm (Mn) and second derivative at λ 469 (Fe) including the error. Based on the error value in Table 1, the calibration curve using zero crossing point of Fe-formaldoximate at 530 nm of first derivative spectra is better in estimating Mn content than of 398 nm. At 530 nm, Mn in six out of seven samples could be predicted with error value less than 7%. Higher error tended to occur in samples with high Fe content. In the case of Fe, using zero crossing point of Mn-formaldoximate spectra at 469 nm, out of seven samples, six could be predicted with error less than 7%. The relation between error and Mn/Fe mass ratio in samples is demonstrated in Figure 4. In Mn prediction, first derivative at 530 nm had better correlation than of 398 nm, but both agreed that higher ratio of Mn/Fe would result in better prediction. In the case of Fe prediction using second derivative spectra at 469 nm the figure shows that higher error occurred for samples with high Mn/Fe ratio (Fe content was smaller than Mn). Based on the figure it could inferred that if the maximum error is set at 5, then the minimum ratio of Mn/Fe for Mn determination is 0.125 or maximum 0.14 for Fe determination.

Table 1. Results of Mn and Fe determination in samples

Added (mg/L)		Mn Found (mg/L)	Error (%)	Mn Found (mg/L)	Error (%)	Fe Found (mg/L)	Error (%)
Mn	Fe	1st, 398		1st, 530		2nd, 469	
1.11	0.96	1.09	-1.78	1.11	-0.28	0.93	-2.24
1.41	19.36	1.26	-10.75	1.50	6.51	18.80	-2.93
1.04	14.79	1.04	-0.24	1.15	10.55	13.80	-6.68
0.90	4.75	1.12	24.42	0.92	2.61	5.05	6.21
1.99	19.04	1.81	-9.10	2.08	4.60	19.45	2.14
2.07	14.66	2.01	-2.92	2.19	5.76	14.09	-3.83
1.99	11.38	2.00	0.65	2.02	1.82	14.28	25.43

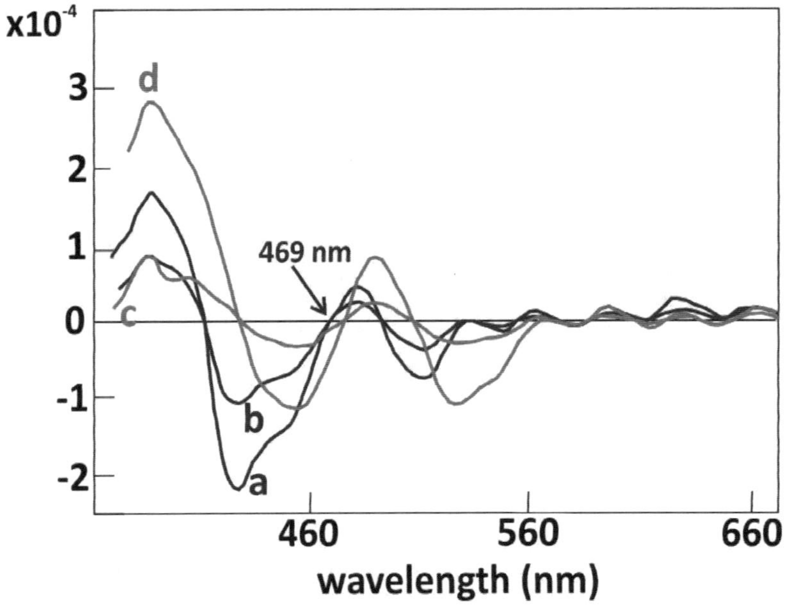

Figure 3. Second derivative of Mn-formaldoximate spectra (a) 0.91 mg/L and (b) 2.03 mg/L. Fe-formaldoximate-tartrate spectra (c) 2.55 mg/L and (d) 11.35 mg/L. Zero crossing points of Mn-formaldoximate spectra are indicated by arrows

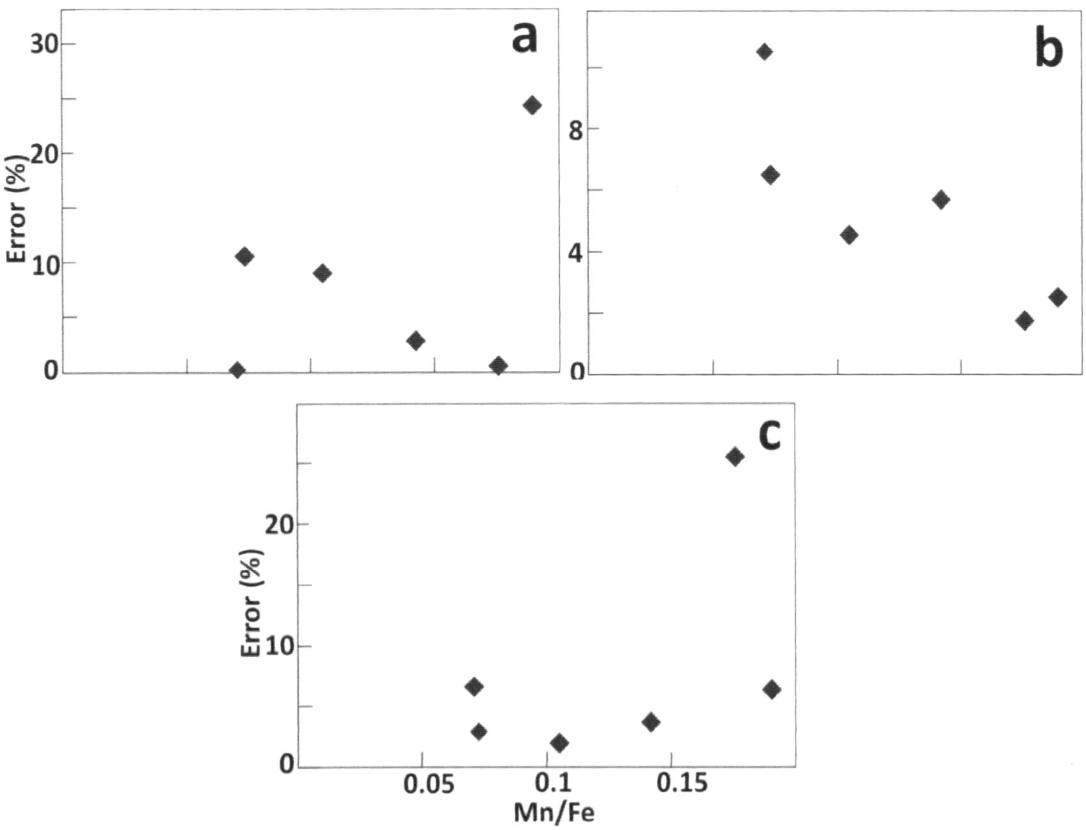

Figure 4. Correlation between Mn/Fe ratio and error (%) for Mn prediction using 1st derivative at 398 nm (a), at 530 nm (b), and Fe prediction using 2nd derivative, 469 nm (c)

4. Conclusions

Mn and Fe could be simultaneously determined using spectrophotometric method using formaldoxime as chromogenic reagent. Mn could be predicted using zero crossing point of first derivative Fe-formaldoximate spectra at 530 nm, while Fe could be predicted using zero crossing point of second derivative Mn-formaldoximate spectra at 469 nm. Maximum concentration which obeyed the linearity of calibration curve is 2 and 20 mg/L for Mn and Fe, respectively. The error in Mn and Fe prediction depends on ratio of both elements.

References

[1] Su H, Wen Y, Wang F, Sun Y, and Tong Z 2008 *Hydrometallurgy* **93** 136–139
[2] van Staden J F, Mulaudzi L V, and Stefan R I 2003 *Analytica Chimica Acta* **499(1–2)** 129-137
[3] Li Z, Yang G, Wang B, Jiang C, and Yin J 2002 *Journal of Chromatography A* **971(1–2)** 243-248
[4] Marczenko Z and Balcerzak M 2000 *Separation, Preconcentration and Spectrophotometry in Inorganic Analysis* Elsevier Amsterdam
[5] Melek Doğutan M, Hayati Filik H, and Apak R 2003 *Analytica Chimica Acta* **485(2)** 205-212

IOP Conf. Series: Materials Science and Engineering **478** (2019) 012023 doi:10.1088/1757-899X/478/1/012023

Clean coal technology using an iron- and sulfur-oxidizing mixotrophic bacterium

S R Nurhawaisyah[1], E Sanwani[1], S K Chaerun[1,2], and M A Rasyid[1]

[1]Department of Metallurgical Engineering, Faculty of Mining and Petroleum Engineering, Institut Teknologi Bandung, Ganesha 10, Bandung 40132, West Java, Indonesia
[2]Geomicrobiology-Biomining & Biocorrosion Laboratory, Microbial Culture Collection Laboratory, Biosciences and Biotechnology Research Center (BBRC), Institut Teknologi Bandung, Indonesia

Email: skchaerun@metallurgy.itb.ac.id

Abstract. Coal is one of the important energy sources in the world. It has been part of the important roles for centuries, not only for generating electricity but also the main fuel for steel and cement production as well as other industrial activities. However, the quality of the coal is generally too low for the practical, economical utilization. High sulfur and ash contents come up as one of the barriers in the productive usage of indigenous coal. Clean coal technology such as coal biobeneficiation can appear as a panacea for upgrading the coal reserves with high sulfur and ash contents. In the current study, the removal potential of the sulfur and ash content from coal of Kalimantan, Indonesia was investigated by using the bacterium *Pseudomonas moraviensis*. It was reported for the first time that the bacterium had the capability to remove about 17.37% of total sulfur and ash content from the coal. The results revealed that the bacterium *Pseudomonas moraviensis* used in this study could remove sulfur and ash content from the coal and could thus be used in the pre-combustion operation with appropriate arrangement.

1. Introduction

Coal is one of the important energy sources in the world. The usage of coal as energy sources is more useful than gas, oil, nuclear, water and substitute resources. Coal has played a very important role for centuries, not only generating electricity, but also the main fuel for steel and cement production, as well as other industrial activities. Coal energy consumption certainly has an impact on the environment which is the emergence of pollutants such as sulfur and nitrogen oxides (SO_x and NO_x). Sulfur oxides (SO_x) and nitrogen (NO_x) are emitted at various levels during the burning of fossil fuels. These gases give chemical reactions to water vapor and other substances in the atmosphere and form acids which then settle when it rains into the acid rain. Therefore, the use of low sulfur coal is the most economical way to control the emissions produced. However, the fact is that not all of coal has a low sulfur content. Hence, the utilization needs special processing to reduce the sulfur content in coal in which processing methods include physical, chemical and biological processes. In the physical coal processes, the coal is crushed, ground and washed. However, depending on the type of coal, a considerable amount of finely distributed pyrite as well as organic sulfur can remain in and attach to the coal particles [1]. The inability of physical methods in removing inorganic and organic sulfur causes the development of chemical methods. Chemical methods include carbonization, water oxidation, wet oxidation Meyers process, chlorination and extraction with sodium hydroxide, copper chloride and a solution of ethanol [2]. However, the chemical processes produce the detrimental product and require high costs and affect of the structural integrity of coal. Therefore, the methods of coal processing biologically have begun to be developed which are based on the degradation of sulfur compounds by microorganisms which provide many advantages over conventional physical and chemical processes. Microorganisms require sulfur for their growth and biological activity that takes place in mild condition without harmful reaction products or an environmentally friendly method. Therefore, this study dealt with the method of producing clean coal technology by using bacteria. The used bacteria are mixotrophic bacteria that are able to oxidize sulfur and iron which are expected to reduce the amount of sulfur contained in coal. The characteristics of coal pre- and post-microbial process were compared in terms of proximate and sulfur total analysis.

2. Materials and Methods

2.1. Coal

Coal used in this study was collected from Kalimantan, Indonesia. The coal was crushed in roll crusher, further ground in tumbling mill and separated into various particle size fractions by sieving machine to obtain the grain size of -200 +325 mesh (-74+44 μm).

2.2. Bacterium and growth medium

The bacterium *Pseudomonas moraviensis* used in this study was a mixotrophic bacterium which was isolated from a mercury-contaminated gold mine site of West Java Province, Indonesia. The growth medium used was SKC broth. The photomicrograph of bacterial cells of the bacterium and its colonies was shown in Figure 1.

A

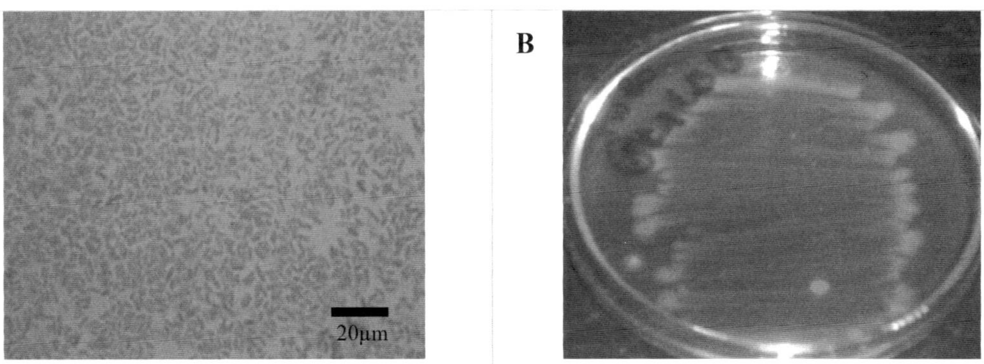

Figure 1. Photomicrograph of bacterial cells (A) and colonies (B) of *Pseudomonas moraviensis* grown on LB agar

2.3. Experimental procedure

Experiments were conducted in duplicate in 250 mL Erlenmeyer flasks containing 100 mL of growth medium with 10% (v/v) inoculum of *Pseudomonas moraviensis* (designated M1). The bacterial growth medium used SKC broth which was incubated at room temperature and agitated (180 rpm) under aerobic condition. Bacterial growth (as measured by optical density) and pH of solution were monitored daily for 10 days to determine the growth curve of the bacterium. From the growth curve, the time of incubation of the bacterium in the medium was then used for study of bacteria-coal interaction. Furthermore, the interaction of coal with the bacterium *Pseudomonas moraviensis* was conducted in duplicate in 500 mL Erlenmeyer flask containing 250 ml of SKC broth with 10% (v/v) inoculum of *Pseudomonas moraviensis* and 25% (w/v) coal. Cultures of bacteria-coal interaction were incubated with the agitation (180 rpm) at room temperature for 10 days under aerobic condition. The pH and redox potential (Eh) of the suspension were observed periodically (every 24 hours) and then the coal was separated, dried, weighed and prepared for further experiments and for analysis.

2.4. Analytical techniques
Optical density (OD) measurements were carried out using a UV-VIS spectrophotometer. The pH and Eh of the suspensions were observed using Lutron pH-207. To determine the chemical properties of coal before and after treatment, a proximate analysis consisting of moisture, ash, volatile materials and carbon content was carried out according to the standard method of ASTM D 7582-12. In addition, the sulfur content in coal was also determined according to the standard methods of ASTM D 2492-02 and D 3177-02.

3. Results and Discussion

3.1. Bacterial growth curve
Before interacting *Pseudomonas moraviensis* with coal, it is necessary to know the bacterial growth curve in SKC broth medium. The growth curves provide an overview of the overall bacterial growth cycle, which includes the lag phase, exponential phase, stationary phase, and death phase [3]. In addition, measurement of suspension pH was also carried out. Based on the growth curve of *Pseudomonas moraviensis* in Figure 1, it can be seen that *Pseudomonas moraviensis* grew well in SKC broth medium in the pH range of 3 to 5. This suggested that by using SKC broth medium, the bacterium was able to produce organic acids without the addition of chemical reagents such as sulfuric acid to control the pH of the solution, and the organic acids was expected to dissolve sulfur in coal.

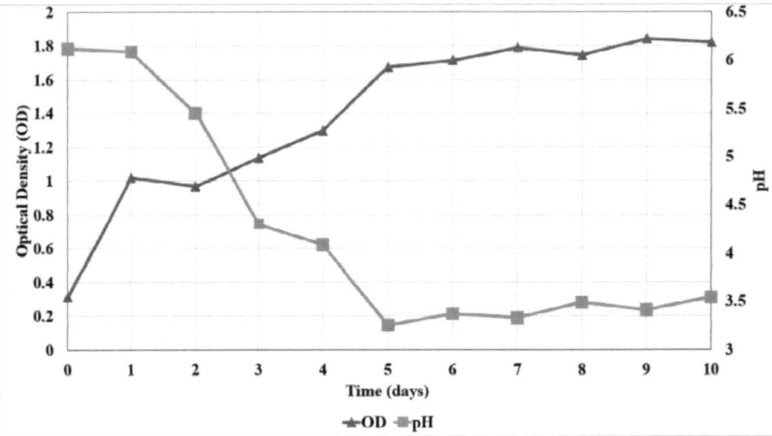

Figure 2. Growth curve of the bacterium *Pseudomonas moraviensis* (designated M1) based on optical density measurement

3.2. Clean coal technology using the bacterium Pseudomonas moraviensis based on pH and Eh analysis
Changes of pH and Eh in the experimental system of bacteria-coal interaction for 10 days were shown in Figure 3. It can be seen that the pH of the suspension decreased over time. Changes in suspension pH of experimental systems resulted in a speedy bacterial growth since the bacterium accelerated pyrite oxidation by oxidizing iron and sulfur which generated H_2SO_4, thus lowering the pH of the suspension [4]. This interpretation was also supported by the measurement of the redox potential (Eh) of the suspension according to the Pourbaix diagram [5]; sulfur can dissolve into H_2SO_4 in the range of

pH 3 to 7 with Eh range of 100 to 800 mV. Therefore, it can be inferred that the bacterium *Pseudomonas moraviensis* used in this study was able to oxidize sulfur in coal.

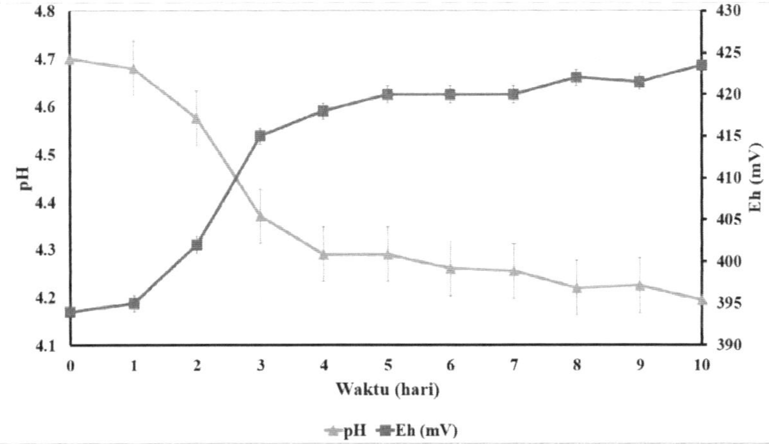

Figure 3. pH and Eh (mV vs. SHE) of the suspensions of the experimental systems containing coal and the bacterium *Pseudomonas moraviensis* (designated M1) over 10 days of the experiment

3.3. Effect of the bacterium Pseudomonas moraviensis on clean coal technology

Coal biobeneficiation was carried out to investigate the result of pre- and post-microbially treated coal (Table 1). Compared with the total sulfur content of the starting coal of 2.82%, the total sulfur content was reduced to 2.33% after the starting coal was treated with the bacterium *Pseudomonas moraviensis*. It was found that 17.37% of the total sulfur was removed from the coal after treatment with *Pseudomonas moraviensis* under optimal condition. Although the reduced total sulfur has not yet meet the coal quality standards including HBA (Harga Batubara Acuan = Indonesian coal price references) based on the index average of Indonesia Coal Index (ICI), Newcastle Export Index (NEX), Global Coal Newcastle Index (GCNC) and Platts-5900, the coal quality meets the maximum permissible total sulfur content of 0.8 wt.% (according to the Government Decree of the Ministry of Energy and Mineral Resources of the Republic of Indonesia No.1917 K/30/MEM/2018). Thus, further studies are needed to optimize the coal biobeneficiation parameters in order to minimize environmental impacts caused by sulfur content in coal, since the presence of sulfur in coal contributes to SO_2 generation during the process of combustion; the sulfur gets oxidised into various oxides of sulphur [6], which in gaseous forms are released primarily as SO_2 and in a much smaller quantity as SO_3, including gaseous sulphates. The sulphur oxides react with water molecules to form acid rain, which deteriorates the environment and infrastructure [7][8]. In addition, the quality of coal after biobeneficiation was also studied including ash content, fixed carbon, and volatile matter which were shown in Table 1. Ash content of coal after treatment with *Pseudomonas moraviensis* decreased from 8.82% to 7.27%, indicating the presence of interaction between the extent of coal biobeneficiation and the ash elimination. The reason for the reduction in ash content may be attributed to dissolution of mineral matters [9]. The reduction in ash content of coal has a positive impact on the environment, especially in relation to heavy metals in coal and fly ash after combustion of untreated coal. Fixed carbon content of coal after treatment with *Pseudomonas moraviensis* increased slightly, and volatile matter content increased by around 2%. This increased volatile matter content might be due to low volatile matters converted into the high volatile matters in the beneficiation process.

Table 1. Composition of untreated and microbially treated coal samples

Analysis (% adb)	Starting coal	After treated with the bacterium *Pseudomonas moraviensis*
Ash	8.82	7.27
Volatile matter	39.76	40.59
Fixed carbon	40.81	44.79
Total sulfur	2.82	2.33

4. Conclusions

The low rank coal from Kalimantan island in Indonesia was able to be beneficiated by using the bacterium *Pseudomonas moraviensis* on laboratory scale. It was found that 17.37% of the total sulfur was removed from the coal after treatment with *Pseudomonas moraviensis* under optimal condition. Chemical characterization revealed a reduction in ash content which indicated the presence of bacteria-coal interaction for coal biobeneficiation as well as ash elimination. Thus, the bacterium has the potential to be used in clean coal technology as a green process.

Acknowledgments

This study was supported by Research Program, Community Service, and Innovation (P3MI) of ITB, Geomicrobiology-Biomining & Biocorrosion Laboratory of ITB and Beasiswa Unggulan Kemendikbud.

References

[1] Klein J 1998 Technological and economic aspects of coal biodesulfurisation *Biodegradation* **9(3-4)** 293-300

[2] Yaman S, Ersoy-Mericboyu A, and Küçükbayrak S 1995 Chemical Coal Desulphurization Research In Turkey *Fuel Science & Technology International* **13(1)** 49-58

[3] Madigan M T, Martinko J M, and Parker J 1997 *Brock Biology of Microorganisms* vol 11 Upper Saddle River NJ: Prentice Hall

[4] Dwyer R, Bruckard W J, Rea S and Holmes R J 2012 Bioflotation and bioflocculation review: microorganisms relevant for mineral beneficiation *Mineral Processing and Extractive Metallurgy* **121(2)** 65–71

[5] Pourbaix M 1974 *Atlas of Electrochemical Equilibria in Aqueous Solutions* NACEmTX

[6] Bailey RA, Clark H M, Ferris JM, Krause S, Strong R L 2002 Chemistry of the Environment 2nd edition *US Academic Press* 698–99

[7] Ryan B, Ledda A 1998 A review on sulphur in coal: With specific reference to the Telkwa deposit, North-Western British Columbia *Geological Fieldwork* 129–50

[8] Fecko P, Sitavancova Z, Cvesper L, Cablik V 2006 Bacterial desulfurization of coal from mine CSA most *Journal of Mining and Metallurgy* **42B** 13–23

[9] Acharya C, Kar R and Sukla L 2001 Bacterial removal of sulfur from three different coals *Fuel* **80** 2207–2216

Effect of Leaching Temperature on Lithium Recovery fromLi-Montmorillonite (Bledug Kuwu's Mud)

L H Lalasari[1], M Rohmah[2], I Setiawan[1], N C Natasha[1], L Andriyah[1], T Arini[1], F Firdiyono[1], and J Wahyuadi[3]

[1]Research Center of Metallurgy and Materials, Indonesian Institute of Sciences, Kawasan PUSPIPTEK Gedung 470, Tangerang Selatan, 15314, Indonesia
[2]Kalimantan Institute of Technology, Kampus ITK, KM 15 Karang Joang, Balikpapan, Indonesia
[3]Metallurgy and Material Engineering Department, Universitas Indonesia, Kampus Baru UI, Depok, 16424, Indonesia

Email: ifa_sari@yahoo.com

Abstract. The availability of lithium resources in the world is important along with the development of electronic technology and electrical transportation, especially lithium batteries. In Indonesia, lithium is found in brine associated with hot spring, clay associated with a geothermal activity, and spodumene minerals. The Bledug Kuwu's Mud contained a Li-Montmorillonite phase with lithium content that was 0.0029%. For lithium recovery, Li-Montmorillonite was leached with distilled water on the S / L ratio was 1: 2 for 3 hours. In this experiment, the leaching temperature was varied as follows: 25°C, 30°C, 35°C, and 45°C. Experimental results showed that lithium recovery from Li-Montmorillonite was successfully carried out by water leaching at atmospheric pressure. This result was supported by XRD analysis on residues. It indicated that there was no Li-Montmorillonite phase at the residues. The highest percentage of lithium recoveries in this research obtained at leaching temperature of 25°C was 87.2% (15.78 ppm).

1. Introduction

Lithium is critical metal that can be used as the future energy in the world.According to physical and chemical properties. Lithium is a lightweight metal that provides higher cell voltage, high specific energy, high energy density, high durability, discharge characteristics excellent, long service life, and high environmental compatibility when compared with alloys of lead, copper, vanadium, and nickel-zinc, [3]. Therefore, lithium is very suitable to be applied in the development of lithium batteries. Currently, the existence of lithium is one of the main focuses for technological development in several developed countries.

In recent years, worldwide lithiummarket has a deficit value due to high lithium demand, especially for battery applications. When compared with other applications, lithium-ion battery has the largest consumption about 39% of total lithium. Globally, the major types of lithium resources could be obtained from minerals, brine water, and clay (an example of hectorite groups) with lithium reserves most often found in Argentina, Bolivia, and Chile. In minerals, commercial lithium is usually found in spodumene ($LiO_2.Al_2O_3.4SiO_2$), lepidolite ($KLi_{1.5}Al_{1.5}[Si_3O_{10}][F,OH]_2$), and petalite ($LiO_2.Al_2O_3.8SiO_2$) and others. Lithium is evenly distributed on hard rock (25%), Hectorite (7%),

Geothermal Brines (3%), Oilfield Brines (3%), Jaderite (3%). Indeed, the number of clays containing lithium is lower than hard rock, which is only about 7% of total production. However, it has a possibility for technological development of lithium recovery from clays source, especially Hectorite form [1].

The process of lithium recovering from clay resource has been studied since the discovery of hectorite and Li-Montmorillonite which contained 0.56% Li by Lithium Americas Corporation (Project of King Valley Lithium). Hectorite clays belong to the clay of smectite group that is as same as Montmorillonite because they have a typical 2:1 type of hydrous aluminosilicate. Their framework of layers are composed by one octahedral sheet (mainly AlO_6 and MgO_6) sandwiched between two tetrahedral sheets (mainly SiO_4 and AlO_4 units).On the geological formation of Montmorillonite, the isomorphous substitution (Al^{3+} by Mg^{2+} or Fe^{2+} on anoctahedral sheet) causes a negatively charged layer. It needs to be neutralized by exchangeable cations in the layers.The presence of inorganic cation substitution (eg Li^+, K^+, Na^+, Ca^{2+}, and Mg^{2+}) into interlayer structure causes different thermal layer characteristic [2]. Besides that, Li-Montmorillonite after heating at 250-300 °C caused the migration of the lithium cations from the interlayer space into the hexagonal and octahedral site with occupancies of 60 and 40%. According to this concept, the temperature is very influential in the migration of lithium to the interlayer position from Li-Montmorillonite when leaching is carried out by using water

In Indonesia, smectite groups found in Mud Volcano such as Lapindo Sidoarjo Clay (with contain 5 ppm Li) and Bledug Kuwu Clay (with contain 1,6 ppm). Bledug Kuwu contained Li-Montmorillonite phase. Rohmah (2018) found that solid/liquid ratio between Bledug Kuwu clays and distilled water affected lithium recovery, but, leaching time has no significant effect in the water leaching process [3].

On the previous research, lithium recovery increased with increasing temperature from 180°C until 250°C on the Egyptian Montmorillonite processing with H_2SO_4 leaching. So that, lithium recovery reached 90% at 250°C after 90 minutes [4]. So, in this research will aim to effect temperature on the lithium recovery for water leaching process.

2. Materials and Method

Raw material was the Montmorillonite-Containing Clay from Bledug Kuwu's Mud, Grobogan, Central Java. The Li-Montmorillonite phase of the clay determined by X-Ray Diffraction (XRD) and identified by Software of High Score Plus (HSP). The XRD result is shown in Figure 1. Bledug Kuwu's Mud is dominated by quartz compound and Li-Montmorillonite (40.1% semiquantitative with software Higscore). Based on Inductively coupled plasma - optical emission spectrometry(ICP-OES) test, Bledug Kuwu's Mud contains 0.0041% Li at sludge condition and 0.0029% at slurry condition [3].

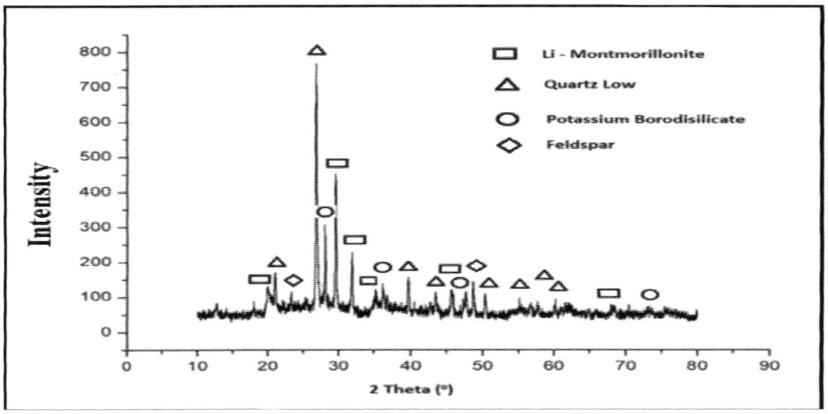

Figure 1. XRD pattern of Bledug Kuwu's mud [3]

Bledug Kuwu's Mud has done water leaching process with a variation of temperatures of 25°C, 30°C, 35°C, and 45 °C at the solid/liquid ratio between clay and distilled water was ½ for 3 hours. A total of 250 grams of wet sludge was dissolved into 500 ml of distilled water, then stirred by using Mixer Settler MX machine brand EYELA. After the stirring process, the solution was filtered to separate the filtrate and mud solid for 3 days. The filtrate was tested by ICP-OES analysis to determine the chemical composition of filtrate so that could be calculated the percentage of lithium recovery. Meanwhile, the leaching residue was dried at 100°C for 2 hours. The dried mud residue was tested by XRD and SEM-EDS analysis to confirm the change of compound and morphology in the Bledug Kuwu's Mud.

3. Results and Discussion

3.1 Effect of temperatures on the chemical composition of filtrates

The water leaching experiment was investigated to determine the optimum conditionof lithium recovery from Li-Montmorillonite at different temperatures. Temperature affected significantly on the dissolution of lithium and associated impurity from Li-Montmorillonite, as shown in Table 1.

Table 1. Chemical omposition of filtrate on the water leaching process (S/L ratio = 1:2 , t = 3 hours)

Element (ppm)	Temperature (°C)			
	25	30	35	45
Aluminum (Al)	0.004	0.042	0.033	0.037
Ferrous (Fe)	0.001	0.006	0.003	0.002
Magnesium (Mg)	22.45	29.76	8.69	12.49
Calcium (Ca)	62.72	72.01	30.78	38.92
Sodium (Na)	1360.26	1419.82	1726.07	1311.35
Potasium (K)	135.62	179.91	183.63	155.16
Lithium (Li)	15.78	15.11	4.36	9.29

Based on Table 1, the lithium recovery tends to decrease with increasing the leaching temperature. The highest acquisition of lithium is obtained at a leaching temperature of 25°C which is 15.75 ppm Li. As viewed in Table 1, the acquisition of Li at leaching temperatures of 25°C and 30°C is not significant or tends to be constant at 15 ppm Li. The recovery of lithium is seen to decrease at a temperature of 35°C that the lowest lithium content is approximately 4.36 ppm. At 45°C leaching temperature looks a recovery of lithium in the range of 9.29 ppm

The decrease of lithium recovery is due to the desorption competition of Na and K ions which tends to increase solubility with increasing the leaching temperatures. The acquisition of K and Na in filtrate reaches a maximum point at 35°C.This condition causes the exchangeable rate of Li with hydroxyl ions on the active side of clay would be disrupted. The Na and K ions might be in 2 side of the active surface (side and side edge) due to electrostatic repulsion so that the solubility of Na and K of Li-Montmorillonite in wateris increased (more than 1700 ppm Na and more than 180 ppm K) [5,6]. In addition, the simulation results of HSC Chemistry software proved that lithium solubility in water decreased as temperature increased. It was showed from data of log K values at 25°C, 30°C, 35°C, and 45 °C that were 70.07; 69.59; 68.44; 67.36; and 66.31, respectively. It can be explained that the lithium acquisition decreases ata warm temperature.

The trend of lithium acquisition almost isas same as Ca and Mg acquisition. However, it is different from K and Na acquisition. The warm temperature increases the adsorption level of the hydroxyl ion and monovalent element in the Li-Montmorillonite structure. It is caused by the bonds on the Li-Montmorillonite interlayer that will be very weak with the existence of competitive adsorption processes of Na and K ions[7].

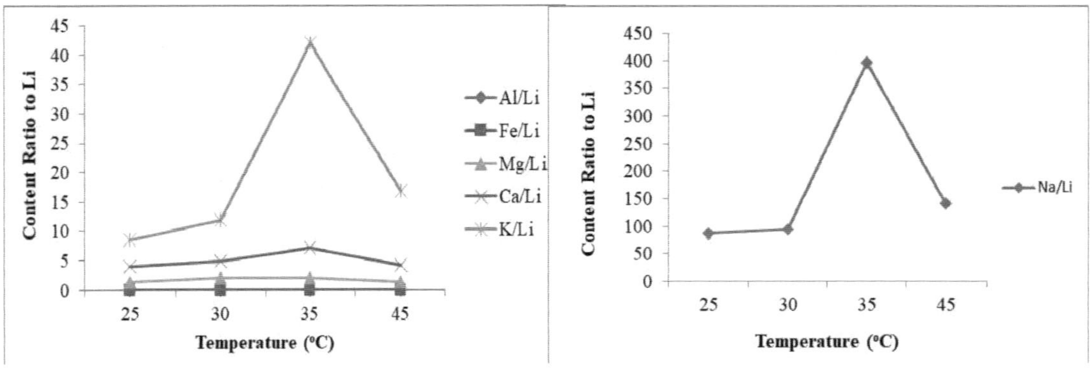

Figure 2. Content ratio of impurities to lithium at different temperatures

To investigate the effectiveness of the effect of temperatures on the optimum lithium recovery from the water leaching of Bledug Kuwu's Mud for 3 hours, it was necessary to calculate the selectivity ratio of impurities to lithium in the filtrates as shown in Figure 2. From Figure 2 can be seen that the optimum leaching conditions are obtained at a temperature of 25°C. This has the lowest Na/Li ratio of 86.2; K/Li of 8.59; Ca/Li of 3.97; and Mg/Li of 1.42. The following is the sequence of obtaining impurity levels as follows: Na/Li > K/Li> Ca/Li> Mg/Li> Al/Li = Fe/Li at water leaching of Bledug Kuwu's Mud for 3 hour.The lowest selectivity ratio of impurities to lithium shows that the filtrate produced by the water leaching process of Bledug Kuwu's mud at 25°C is rich in lithium content.

3.2 Effect of temperatures on the leaching residues

The effect of temperature on the residues from leaching Bledug Kuwu's Mud for 3 hours was investigated with the characterization of residues by using XRD analysis as shown in Figure 3.

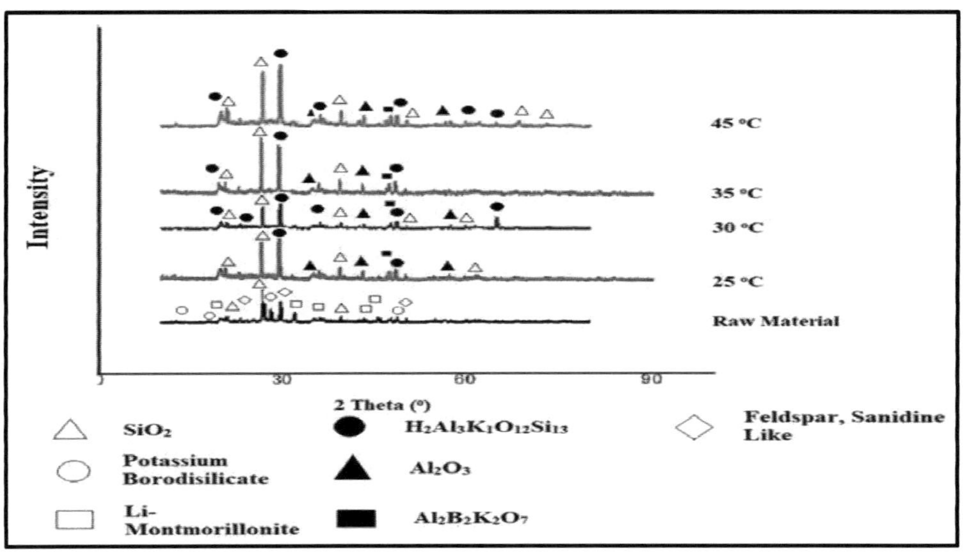

Figure 3. The XRD pattern of leaching residues of Bledug Kuwu's mud

According to Figure 3, there is phase changes between the initial raw material of Bledug Kuwu's mud and residues from water leaching processes. In initially Bledug Kuwu's mud appears the peaks with the intensity of phases of Li-Montmorillonite, SiO_2, feldspar and Potassium Borondisilicate that transforms phases into peaks with intensities of SiO_2, $H_2Al_3KO_{12}Si_3$, Al_2O_3, and $Al_2B_2K_2O_7$. The change of phase is indicated by the dissociation reactions between Li-Montmorillonite, feldspar, and Potassium Borondisilicate with water solvents at various temperatures of 25°C, 30°C, 35°C, and 45°C.

The leaching process of Li-Montmorillonite was very dependent on the interaction of the structure of clay with water or during the molecular hydration process. The initial cations in the Montmorillonite interlayer would react with water molecules so that the interlayer distance was wider. Ions with high hydration numbers tended to form outer sphere complexes that only attached to the surface of clay, such as Li and Na ions. While ions of K, Mg, Ca tended to form the inner sphere complex [8]. The Li-Montmorillonite leaching reaction in the Bledug Kuwu's mud using water medium was likely to be predicted as follows:

$$H_4Al_{3.21}Fe_{0.45}Li_{0.62}Mg_{0.54}O_{24}Si_{7.8(s)} + H_2O_{(l)} + 19H^+_{(aq)} \rightarrow 3.21Al(OH)^{4-}_{(s)} + 0.45Fe^{3+}_{(aq)} + 0.62Li^+_{(aq)} + 0.54\ Mg^{2+}_{(aq)} + 7{,}8HSiO^{3-}_{(s)} + 14OH^-_{(aq)} \quad (1)$$

Reaction (1) referred to the results of previous research by Takase et al (2004) [9] which explained the Montmorillonite reaction in a variety of neutral, acidic, and alkaline media. In other conditions, Hellman (2009) described that the dissolution of feldspar was influenced by pH, temperature, and solution composition [10]. Feldspar can be modified in distilled water media so that the possibility of a reaction that occurred as in Reaction (2). The leaching of feldspar also affected the content of compounds in the leaching residues. Therefore, the leaching residues tended to form Al-Si-K compounds.

$$Na_xK_yCa_zAl_{1+z}Si_{3-z}O_{8(s)} + 8H_2O_{(l)} \rightarrow xNa^+_{(aq)} + yK^+_{(aq)} + zCa^{2+}_{(aq)} + (1+z)[Al(OH)_4]^-_{(aq)} + (3-z)H_4SiO_{4(s)} \quad (2)$$

In an effort to investigate in detail the content of the compounds contained in the leachingresidues of the Bledug Kuwu's mud with water, the results of XRD analysis needed to be further calculated using the High Score Plus (HSP) software. The software could interpret quantitative of each phase of leaching residues as shown in Table 2

Table 2. Quantitative Analysis of phase of leaching residues in variation temperatures

Temperature	Phases of Residues			
	SiO_2 (ICDD no 98-002-9210)	$H_2Al_3KO_{12}Si_3$ (ICDD no 98-002-7170)	$Al_2B_2K_2O_7$ (ICDD no 98-040-9420)	Al_2O_3 (ICDD no 98-016-5594)
25°C	25%	39%	31%	5%
30°C	23%	33%	30%	14%
35°C	41%	29%	18%	12%
45°C	18%	47%	21%	17%

In Table 2 could be seen that leaching residues are dominated by SiO_2 and $H_2Al_3KO_{12}Si_3$ phases according to the XRD pattern in Figure 3. They have the highest peak intensity. Whereas the peaks showing the $Al_2B_2K_2O_7$ and Al_2O_3 phases also appear in the XRD pattern in Figure 3, but have a lower relative intensity. Therefore, to prove the existence of this phase, it is necessary to know the content of the chemical elements in the residues. The chemical composition of the residues with semi-quantitative can be further confirmed by EDS-SEM analysis.

3.3 Morphology of leaching residues

Figure 4 shows the morphological changes of the particles in the leaching residues. It looks that water leaching process at 25°C is capable of breaking the particle grains into the smaller size (4a). However, increasing temperature up to 45°C causesthe particles to re-agglomeration to become larger (4b-d). The change in particle size indicates an increasing of adsorption level in the monovalent cation. Related to Mustopa (2013), morphological changes in sludge has not affected by alkaline or acidic conditions but more affected by temperature changes [11]. Al-Ani et al (2008) showed that

IOP Conf. Series: Materials Science and Engineering **478** (2019) 012024 doi:10.1088/1757-899X/478/1/012024

temperature effected the ability of Li-Montmorillonite in the water adsorption and expantion of adsorption surface. The higher the temperature, the greater the capacity of water adsorption, but tended to decrease after the temperature reached up to 50°C and depended on the composition of the exchangeable cation in the interlayer of Li-Montmorillonite [7].

Figure 4. The morphology of leaching residues in variation temperatures of : (a) 25°C (b) 30°C (c) 35°C (d) 45°C (Magnification 500X)

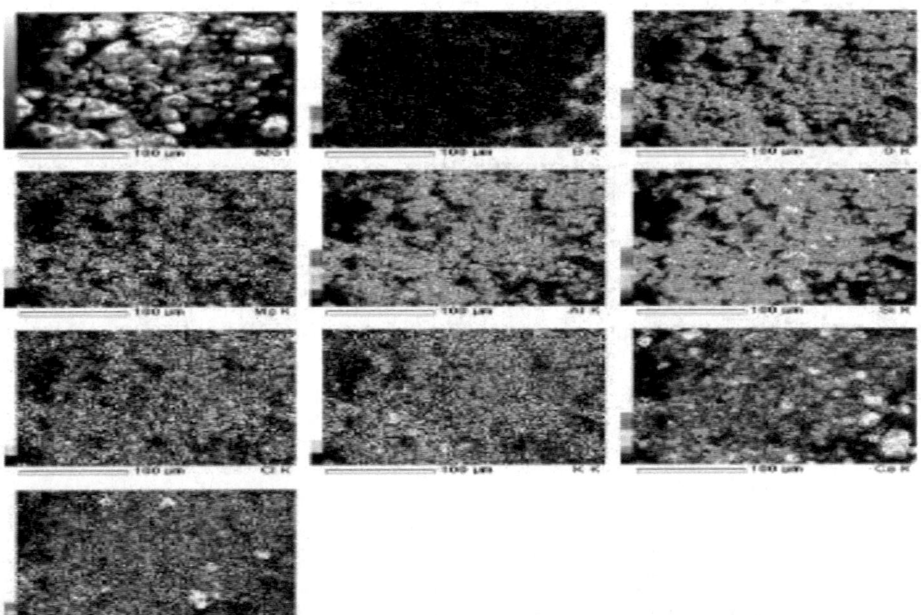

Figure 5. Elemental mapping of leaching residue at a temperature of 30°C(Magnification 500X)

Figure 6. Elemental mapping of leaching residue at a temperature of 45°C

In this discussion is only explained the elemental mapping of leaching residues at temperatures of 30°C (Table 5) and temperatures of 45°C (Table 6). It is clear that the particles of the leaching residue at room temperature at 30°C are relatively smaller compared to warm temperatures at 45°C. At room temperature of 30°C, the Bledug Kuwu's mud is likely to be dissolved in water so much that the particles break into smaller sizes. This indicates that Li-Montmorillonite, which is a constituent of Bledug Kuwu's mud, also dissolved in water. The larger solubility of Li-Montmorillonite in water at a temperature of 30°C causes lithium-rich filtrate compared to filtrate at a temperature of 45°C, as shown in Table 1. The same conditions may also occur in the leaching process of Bledug Kuwu's mudat a temperature of 25°C where the morphology in Fig. 4a shows particles of smaller residue compared to particles of residue at 30°C. If observed more clearly in Figure 5 and Figure 6, it appears that the dominant constituent elements present in the leaching residues are Al, Si, and O. These elements are indicated as SiO_2 and $H_2Al_3KO_{12}Si_3$ which are constituents of leach residues. These compounds are difficult to dissolve at low temperatures, as shown in Table 2.

Table 3. The Chemical composition of residue (EDX-SEM Test)

Element (%)	% wt			
	25°C	30°C	35°C	45°C
O	52.23	44.35	46.03	43.64
Na	1.35	1.36	1.07	1.34
Mg	1.31	1.42	1.29	1.37
Al	8.85	9.30	9.79	9.10
Si	20.28	21.72	22.06	21.51
K	1.30	1.39	1.45	1.69
Ca	10.01	12.19	12.07	13.15
Fe	4.67	7.27	6.25	7.29

According to Table 3, the alteration of chemical composition based on the EDX-SEM analysis occurs on Na, Mg, Al, and Si elements. The content of Na and Mg decreases with temperature rose from 25°C to 35°C. However, the levels of these two elements increase at 45°C. The decrease of Na content to 1.07% and Mg to 1.29% proves that the rising temperature can accelerate the Na and Mg adsorption level so that both of these elements would be dissolved largerin water as shown in Table 1. That was supported the acquisition of decreased Li content due to competitive adsorption between Li, K, and Na to hydroxyl ions[6].

The trend of the alteration composition in Al and Si are inversely related to the trend of alteration in Na and Mg. The composition of Al and Si on the residue tends to increase to a maximum at 35°C and then decreases at 45°C. The alteration of these two elements (Al and Si) explains that the temperature can accelerate the absorption of hydroxyl ions in the sludge resulting in significant structural changes in the octahedral (Al^{3+}) and tetrahedral (Si^{4+}) layers of Li-Montmorillonite which are aluminosilicate compounds [7]

3.4 Lithium Recovery

The lithium recovery is calculated by using equation (3). The recovery of Bledug Kuwu's mud leaching in distilled water medium as shown in Figure 7.

$$\%\, Recovery\ \mathrm{Li} = \frac{dis\mathrm{solved\ \ mass\ \ of\ Li\ in\ aquadest}}{Initial\ \ Mass\ of\ \mathrm{Li}} x\ 100\% \qquad (3)$$

Where dissolved mass of Li-ion (mg) which is the product of the concentration of Li (ppm) in the filtrate with the total volume of filtrate (ml). While the initial mass of Li (mg) is the product of the amount of Li (%) in the initial mass of the Bledug Kuwu's mud sludge (gram).

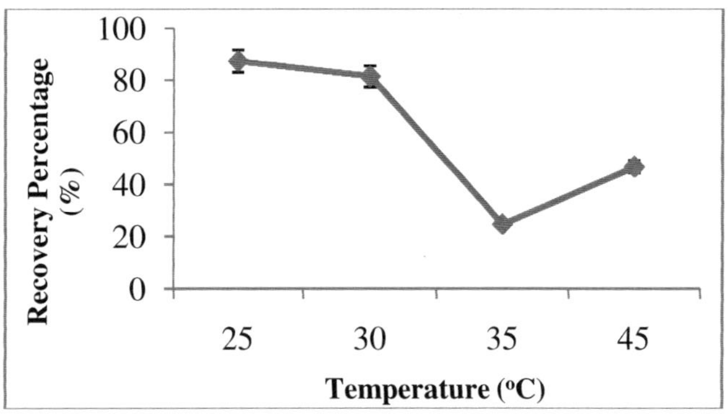

Figure 7. Effect of temperature on lithium recovery

Figure 6 shows that the temperature has a relation with the percentage of lithium obtained from the Bledug Kuwu's mud. The percentage of lithium recovery decreases with increasing temperature. Increasing temperature affects the adsorption level of H_2O in the clay structure. Bieski et al (2013) explained that higher temperatures could lead to higher surface area and modified the pore form of Li-Montmorillonite from crevice type to spheroidal, so that reaction ratewas higher for Fe removal in acid treatment. Leaching with temperatures below 30°C was not capable of dissolving Fe. Meanwhile, the temperature leaching above 30°C (i.e at 35°C) tend to substitute higher Ca^{2+}, Mg^{2+} and Na^{+} ions with oxonium ions [11]. Previous experiments conducted by Bieski et al (2013)[11] were different in conditions with this experiment where the increase in temperature caused a decrease in lithium recoveries from the water leaching of Bledug Kuwu's mud for 3 hours. The percentage of maximum lithium recovery obtained in this experiment was 87.2% at a temperature of 25°C. This is due to the temperature of 25°C, the solubility of Na and K ions is still relatively small in the water leaching of Bledug Kuwu's mud so that the adsorption rate of Li with hydroxyl ionsbecomes higher.

4. Conclusion

From this experiment were obtained some conclusions as follows: recovery of lithium from Li-Montmorillonite was strongly influenced by leaching temperature. Lithium levels from filtrate tended to decrease with increasing temperature. The increase in leaching temperature above room temperature caused the higher water adsorption in the Li-Montmorillonite. This caused the interlayer bond of Li-Montmorillonite that tend to weaken so that the exchangeable rate of Li with hydroxyl ions on the active side of clay would be disrupted with increasing solubility of Na and Kions from Li-Montmorillonite. The water leaching at a temperature of 25°C gave a high Li solubility of 15.78 ppm, while at a temperature of 45°C gave a Li solubility of 9.29 ppm. From the XRD analysis on the residue, it was seen that the Li-Montmorillonite phase did not appear in the residue at 25°C. This indicated that lithium from Li-Montmorillonite was almost perfectly extracted. The peak intensity of SiO_2 and $H_2Al_3KO_{12}Si_3$ dominated the residue with increasing temperature. The morphology of the residue appeared irregular particles and occurred agglomeration with rising temperature. This experiment succeeded in recovering lithium from Li-Montmorillonite (Bledug Kuwu'sMud) by water leaching at atmospheric pressure where the percentage of lithium recovery was 87.2% at 25°C, 81.27% at 30°C and 46.75% at 45°C.

Acknowledgments

This research was carried out at hydrometallurgy laboratorium in Research Center for Metallurgy and Material, Indonesian Institute of Sciences. The authors are grateful for the main financial supports from INSINAS budget 2018.

References

[1] Swain and Basudev 2017 *Separation and Purification Technology* **172** 388-403
[2] Qingze C, Runliang Z, Lingya M, Qing Z, Jianxi Z, and Hongping H 2017 *Applied Clay Science* **135** 129-135
[3] Rohmah M, Lalasari L H, Wahyuadi J, and Chrisayu N N 2018 *IEEE International Conference on Innovative Research and Development (ICIRD)*, Bangkok 1-5
[4] Amer A M 2008 *Journal of Metallurgy* **60(10)** 55-57
[5] Ikhsan and Jaslin 2005 *Memahami Proses Adsorpsi Ion Logam oleh Clay Mineral* Prosiding Semnas Penelitian, Pendidikan, dan Penerapan MIPA, FMIPA UNY, Yogyakarta
[6] Gournis D, Lappas A, Karakassieds M, Tobbens D, and Moukarika A 2008 *Physic and Chemical Minerals* **35** 49-58
[7] Al-Ani T and Sarapaa O 2008 *Clay and Clay Mineralogy: Physical - Chemical Properties and Industrial Uses.* Finlandia: Geological Survey of Finland
[8] Mitchel J K 1993 *Fundamental of Soil Behavior*, 2nd edition, John Wiley & Sons (Chapter 3)
[9] Takase H 2004 *Proceedings of International Workshop on Bentonite – Cement Interaction in Repository Environment*, Tokyo Japan, 14-16 April 2004
[10] Hellman R, Daval D, and Tisserand D 2009 *Comptes Rendus Geoscience* **342(7-8)** 676-684
[11] Bieseki L, Bertell F, Treichel H, Penha F G, and Pergher S B 2013 *Mat. Res.* **16(5)** 1122-1127

The effects of hot briquetting on the coke strength in the biocoke making process with coal blending method

E Yustanti[1], H Kusumawati[1], T Partuti[1], and A T Mursito[2]

[1]Department of Metallurgical Engineering, Sultan Ageng Tirtayasa University, Jl. Jenderal Sudirman KM.03 Cilegon 42435, Indonesia
[2]Reseacrh Center for Geotechnology, Indonesian Institute of Sciences (LIPI), Jl. Sangkuriang Komplek LIPI, Gd.70, Bandung 40135, Indonesia

Email: erlina.yustanti@untirta.ac.id

Abstract. The biomass waste of rice husk and coconut shell in Indonesia reaches more than 15.8 million tons/year. In addition, the utilization of biomass waste has not been performed optimally. Rice husk and coconut shell are the biomass that can be processed as briquette material. This research aims to find the maximum temperature of hot briquette to generate the maximum coke strength from blending. First, the biomass waste was performed by a carbonization process at 400°C with a coal composition of biomass 95:5 wt%. The production of coke briquette used molasses binder 15 wt%. The hot briquette temperature variation was at 100 and 150°C with a briquette pressure of 200 kg/cm^2. Coke briquette then was carbonized at 1100°C for 4 hours and followed by water quenching. The result indicates that hot briquette temperature depends on the type of biomass used. Coke briquette with a rice husk blend at 100°C and 150°C hot briquette temperature, respectively produced coke strength of 11.5 and 20.9 kg/cm^2. Meanwhile, coke briquettes with coconut shell charcoal blend at 100°C and 150°C hot briquette temperature respectively produced coke strength of 15.8 and 10.8 kg/cm^2.

1. Introduction

Coking coal limited availability limitation in Indonesia provides an opportunity for blending with biomass to produce biocoke. Biomass is a solid or liquid waste that can be utilized as a fuel source in blast furnace. This energy source contains the lowest sulfur relativity so it does not cause a huge air-pollution. Generally, biomass densification has several advantages, for example, it can increase the calories score per volume unit, is easily stored and transported, and has similar size and quality [1]. The availability of diverse biomass becomes a good alternative, especially the coconut shell and rice husk biomass. This condition occurs because the position of Indonesia as an agricultural country with a wide beach area. Obviously, both of coconut shell and rice husk availability are very abundant.

Based on the data from the Indonesian Agency for Agricultural Research and Development (2017), rice husk from rice milling in Indonesia in 2016 reached 15.8 million tons. Coconut shell is found almost in all areas of Indonesia [2]. The coconut shell is a part of coconut in the form of endocarp. It is hard and located inside the coconut after the husk containing a lot of lignin decreases about 30%. As biomass waste, coconut shell as a fuel is widely used either directly or indirectly as the carbonization process first [3]. When compared to other biomass charcoal, coconut shell charcoal has the highest carbon content.

Rice husk and coconut shell blending with coking coal is expected to overcome the limited coking coal in Indonesia and utilizing the abundant biomass as an alternative raw material in coke industry. Generally, using biomass directly as a fuel is less efficient. Therefore, a carbonization process is required. Biomass carbonization plays a very important role because it is related to the calories score produced. Agricultural waste can be processed with carbonization followed by briquetting to produce a good fuel. One of the ways to process agricultural waste [4]. Variation of briquette pressure affects the briquette thermal characteristic. The additional pressure will increase the value of coke compressive strength. In addition, the effect of briquetting temperature is the higher temperature will decrease the residual moisture content in coke briquette. The higher temperature has a positive effect on briquette strength [5].

2. Materials and methods

2.1. Coal Pulverizing and Sieving

The production of coke briquette in this research used a blend of coking coal from Central Borneo with rice husk and coconut shell biomass from Kebondalem, Cilegon. The size reduction process was conducted on coarse coal using pulverizer mill followed by sieving to obtain the size fraction of -40#. The more dominant use of point fraction -40 # will produce more compact coke briquettes because finer grains will fill in the cavity grains, so that it can increase density [6].

2.2 Biomass Carbonization

Biomass carbonization was conducted at temperature of 400°C and produced charcoal. Then, charcoal was followed by grinding and sieving (-60#). The process of biomass carbonization aimed to increase the calorific value and fixed carbon.The biomass carbonization process was conducted by using a mini rotary kiln and resulted about 20-30 %yield. The process of coconut shell carbonization in the temperature of 600°C with total bait of 1,946.6 gram produced 547 gram charcoal or about 28 %yield.

2.3 Coal Blending and Hot Briquetting Process

Blending between coal and biomass was conducted with the addition of adhesive in the form of molasses by 15 wt% of the total blend. Molasses as a binder composed with hydrocarbon bonds. It produced the higher fixed carbon by hot briquetting. In order to produce higher fixed carbon and increase the coke strength, it was compacted the briquetting pressure at 200 kg/cm^2. The suitable temperature of briquetting process for biomass was below 100°C and not more than 150°C. However, the temperature is also based on the type of biomass properties.

2.4 Re-carbonization Process of Briquette

The last process was briquette re-carbonization at 1100°C for 4 hours and followed by water quenching through giving a splash of water to a hot biocoke at room temperature. The variable for re-carbonization process was based on previous research [6]. Water quenching after 1100°C re-carbonization process led to the bond formation among the particles in coke briquette. The burn shrinkage of quenched coke was about 37-45%.

2.5 Proximate and Ultimate Analysis

The proximate analysis of coal and biomass charcoal was conducted to determine the content of fixed carbon to compare the difference between raw material and biomass charcoal. Proximate analysis was conducted in the Research Center for Geotechnology, Indonesian Institute of Sciences (LIPI), Bandung and ultimate analysis was conducted in Research Center and Development for Mining and Coal (TEKMIRA), Bandung.

3. Results and discussion

Biomass utilization provides many benefits such as renewable resources, relatively contain the lowest sulfur, provide an energy-efficiency and decrease the cost saving of waste treatment handling. Table 1 shows the results of raw materials proximate analysis. Preliminary analysis was performed after drying process of biomass at 110°C, which intended to facilitate the size reduction process using mini jaw crusher. Proximate analysis was performed to determine raw material quality and the content inside. The components based on proximate analysis consist of moisture, ash, volatile matter, and fixed carbon. Proximate analysis in this research was performed by applying LECO TGA 701 instrument based on ASTM D3172 standard.

Table 1. Proximate Analysis of Raw Material

Analysis (Air dried basis)	Coking Coal (%)	Rice husk (%)	Coconut shell (%)
Moisture	1.18	9.37	10.10
Volatile Matter	29.26	54.38	57.02
Ash	2.21	22.20	14.10
Fixed Carbon	67.36	14.05	18.60
Total Sulfur	0.69	0.02	0.03

In Table 1, we can see fixed carbon content of raw material. Rice husk contained lower bound carbon by 14.05% compared to fixed carbon of coconut shell by 18.60%. The biomass carbonization process can increase carbon content as shown in Table 2.

Table 2. Proximate Analysis of 400°C Charcoal

Analysis (Air dried basis)	Rice husk charcoal (%)	Coconut shell charcoal (%)
Moisture	7.44	5.03
Volatile Matter	7.73	33.09
Ash	45.85	1.33
Fixed Carbon	38.99	60.55
Total Sulfur	0.01	0.02

The biomass carbonization process aims to increase the carbon content. The carbon content is the amount of pure carbon contained in charcoal. The increasing temperature in carbonization process greatly affect the charcoal quality, including the carbon content [7]. The results of proximate analysis in Table 2 shows that the fixed carbon of rice husk charcoal increased by 278% from 14.05% to 38.99%, while the fixed carbon of coconut shell increased by 326% from 18.6% to 60.55%.

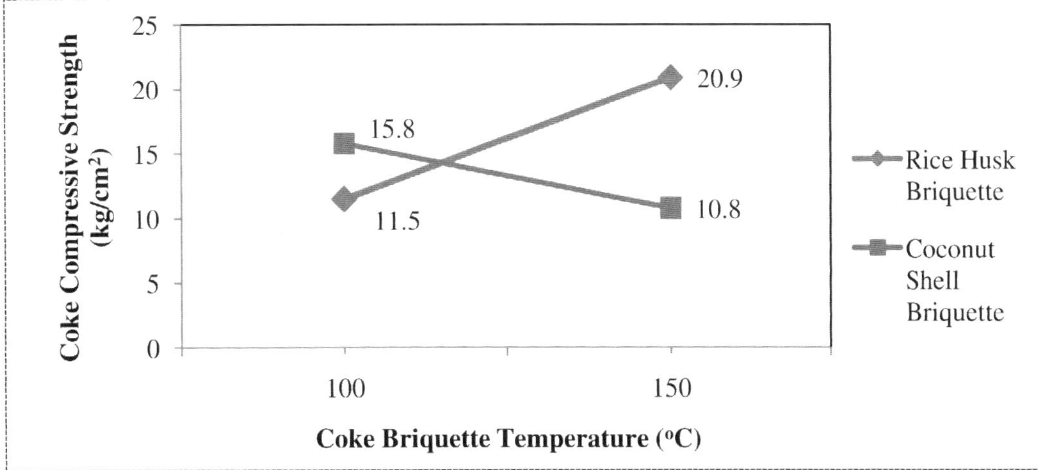

Figure 1. The Relationship between Hot Briquetting and Coke Strength

Next, we will discuss the effect of hot briquetting temperature on coke strength. Hot briquetting process in this research used a temperature variation of 100 and 150°C. Blending method was performed to blend between coking coal and biomass charcoal. The hydrothermal coal heating process can be performed through hot briquetting.

The hot briquetting process aims to increase coke strength. During this process, the release of volatile matter occurred, although it was just a little due to the heating. During the briquetting process, the molasses binder in the briquette burnt, so that the hydrocarbon chains inside were released. Consequently, the carbon content was expected to increase significantly. This condition allows for an increase in coke strength. Carbonized biomass charcoal at 400°C as blending material may provide effect on coke strength [8]. Figure 1 shows coke briquette with blended rice husk charcoal generates higher strength in hot briquetting process with temperature of 150°C equal to 20.9 kg/cm^2. In contrary, coke briquette with coconut shell charcoal generates higher strength in hot briquetting process with temperature of 100°C that is equal to 15.8 kg/cm^2. The metallurgical coke compressive strength for blast furnace was at the range of 28.12-103.34 kg/cm^2 [9]. For the next experiment, we suggest to increase compacting pressure when process briquetting until 4000 kg/cm^2 to reach more metallurgical coke compressive strength. The high pressure and temperature simultaneously act upon the mass, the cellular structures within the material release lignin, which binds individual particles into a compact unit-briquette. Strength of briquettes increases with the increase of the pressure to the strength limit (approx.150-250 MPa) of the compacting material. Increased strength causes low absorption of atmospheric humidity and thus increases the briquette durability [10].

Coke briquette with coconut shell charcoal blend has greater strength than briquette with rice husk charcoal blends, because the coconut shell charcoal generates tar during carbonization. With the carbonization process, wood tar resulted at temperature more than 400° C. It also has functions as a binder. Therefore, tar on coconut shell charcoal clearly provides the effect of increasing the strength value on coke briquette. On the hot briquetting effectiveness at 100°C on the blended biocoke with coconut shell charcoal, it was assumed that it occurred due to the tar from carbonization process. If hot briquetting temperature is increased to 150°C, then the vaporized tar may be much more than the hot briquetting at 100°C.

Table 3. The Nature of Proximate Coke

Material	M (%)	VM (%)	Ash (%)	FC (%)
BSP (HB 100°C)	0.7	2.15	12.14	85.01
BSP (HB 150°C)	0.38	1.77	11.93	85.92
BTK (HB 100°C)	0.89	3.02	20.75	75.34
BTK (HB 150°C)	1.27	1.69	19.06	77.98

BSP: Rice Husk Briquettes; BTK: Coconut Shell Briquettes; HB = Hot Briquetting

Coke is obtained from re-carbonization process of blending briquettes at high temperatures of 1100°C without oxygen, and it depends on the amount of carbon for 4 hours [11]. Stage of coal changes into coke starting at 375–475°C, briquettes experiencing decompose forms plastic layer around the wall. When the temperature reaches 475–600°C, tar fluid and hydrocarbon compound (oil) appear and it is followed by compaction of plastic mass into semi-coke. However, the re-carbonization process at temperature 800-900°C only produced semi-coke. And the suitable temperature for re-carbonization process was at 1100°C to produce the stabilized coke. Coking stage is strongly affected by the contraction of coke period, development of coke structure, and evolution of hydrogen at the end of coking.

As shown in Table 1, coke briquettes with a rice husk charcoal blend that had experienced hot briquetting process at 150°C generated fixed carbon of 85.92%, 0.91% higher compared to hot briquetting at 100°C that generated fixed carbon by 85.01%. Meanwhile, coke briquettes with the coconut shell charcoal blend that had experienced hot briquetting process at 150°C generated fixed carbon of 77.98%, 2.6% higher compared to hot briquetting at 100°C generating fixed carbon of 75.34%.

4. Conclusions

Based on the research results, it can be concluded that the hot briquetting depends on the type of biomass. Rice husk charcoal produced the coke strength 20.9 kg/cm^2 by hot briquetting at temperature 150°C, while coconut shell charcoal is more suitable with hot briquetting at 100°C with the coke strength of 15.8 kg/cm^2. The highest strength of coke briquette from blending with rice husk has not met the minimum metallurgical coke standard of 28.12 kg/cm^2. The quenching process may fix the proximate characteristic of briquette coke by increasing the density level.

Acknowledgment

The authors gratefully acknowledge the support of Reseacrh Center for Geotechnology Indonesian Institute of Sciences (LIPI) and Research Center and Development for Mining and Coal (TEKMIRA), Bandung for the research facilities. Research work is partially funded by the Ministry of research, technology and higher education under the research grant PDUPT with contract number: 095/SP2H/LT/DRPM/2018.

References

[1] Aqa S and Bhattacharya S 1992 *J. Energy* **17**(6) 575–8.
[2]Anonymous 2017 Sekam Padi Sebagai Sumber Energi Alternatif dalam Rumah Tangga Petani Agricultural Research and Development Agency, Ministry of Agriculture 5–7.

[3] Qistina I, Sukandar D, and Trilaksono 2016 *J. Kim Val Penelit dan Pengemb Ilmu Kim.* **2**(2) 136–42.
[4] Surono UB 2010 *J. Rekayasa Proses* **4**(1) 13–8.
[5] Subroto, Himawanto DA, and Sartono 2007 *J. Tek Gelagar* **18**(1) 73–9.
[6] Yustanti E, Pramudhita M, and Aghniya M 2014 *J. Tek Fak Tek Univ Sultan Ageng Tirtayasa* **10**(2) 103–13.
[7] Siahaan S, Hutapea M, and Hasibuan R 2013 *J. Tek Kim USU* **2**(1) 26–30.
[8] Ueki Y, Nunome Y, Yoshiie R, Naruse I, Nishibata Y, and Aizawa S 2014 *ISIJ Int.* **54**(11) 2454–60.
[9] Rahardjo BS 2010 *J. Ilmu Tek Energi* **1**(10) 58–69.
[10]Kaur A, Roy M, and Kundu K 2017 *Int J. Recent Sci Res.* **8**(10) 20561–8.
[11]Yustanti E 2012 *J. of Waste Manag Technol.* **15**(Suppl. Ed.) 15–30.

Iron bearing oxide minerals separation from rare earth elements (REE) rich coal fly ash

N D Nugroho[1], W Rosita[1], I Perdana[1], I M Bendiyasa[1], F R Mufakhir[2], and W Astuti[2]

[1]Advanced Material and Sustainable Mineral Processing Research Group, Chemical Engineering Department, Universitas Gadjah Mada, JlGrafika no 2, Yogyakarta, Indonesia
[2] Research Unit for Mineral Technology, Indonesian Institute of Science (LIPI), Jl. Ir.Sutami km 15, Tanjung Bintang, Lampung Selatan, Indonesia

Email: widyar@ugm.ac.id

Abstract. Recovery of Rare Earth Elements (REE) from waste is an important alternative to solve global REE crisis. Therefore, coal fly ash containing REE becomes a potential candidate as an alternative source of REE. Indonesia has a large potential of coal resources, however unlike other type of coals which is rich in SiO_2 and Al_2O_3, Indonesian coal contains significant amount of Fe_2O_3. As a consequence, fly ash resulted from burning Indonesia coal is also rich in iron oxides. The presence of high concentration of iron oxides will however reduce effectiveness of REE recovery from coal ash through hydrometallurgy process. Therefore, iron oxides in the coal fly ash needs, firstly, to separate preceding further REE separation. This study determined the optimal process conditions in the iron recovery process from coal fly ash. The variables studied were fly ash grain size, magnetic field strength and separation techniques. The iron recovery was determined by mean of the ratio of iron concentration after and before magnetic separation. Experimental results showed that increasing magnetic field strength and decreasing coal fly ash grain size in a wet separation will increase the recovery of magnetic iron component. Magnetic field strength of 4300 Gauss and grain size of -400 mesh were the optimal values in this study with iron recovery of 25.89%.

1. Introduction

Rare Earth Elements (REE) are elements of lanthanide group widely used in industries such as petroleum industry, electronics, information, and communication devices, even in green technology. REE play a key role in improving performance products, especially in clean energy technology and consumer goods [1][2][3]. Along with the extensive use of green technology to reduce global warming and improve energy efficiency, utilization of REE in various products is predicted to increase [4], between 2006 and 2035 with annual growth rate between 3.7% -8.6% [1]. Despite of its high annual demand rate the availability of REE is not enough, this condition is known as REE crisis [4][5][6]. One scenario to overcome this scarcity is to extract REE from waste containing REE including from coal fly ash.

Coal fly ash is known to contain REE and it becomes a candidate for REE alternative source. Several studies show that the amount of REE in coal fly ash (CFA) is relatively high so that several countries such as China, USA and South Africa consider coal fly ash as an alternative source of REE.

In 2015, Indonesian steam power plants had produced $4.38x10^6$ tons coal fly ash and bottom ash andit is predicted that the production will reach by $8.31x10^6$ tons in 2019 [7]. This information shows that Indonesia has an abundance resource of coal fly ash and coal bottom ash. During coal combustion process, concentration of trace elements in the coal fly ash will increase up to 30 times [8]. Therefore, utilization of coal fly ash as a source of REE might become more attractive in the near future. In addition, the recovery of REE from fly ash will become an alternative solution to reduce environmental pollution.

In recent years, there have been many studies in several countries related to REE recovery from coal ash. Acid leaching is usually used to extract REE element in coal fly ash. The results show that the dissolution process was highly depending on fly ash particle size, solution concentration and dissolution temperature [9][10].

In Indonesia, Anggara *et al.* (2018) conducted analysis of REE contain in coal from Banko coalfield, South Sumatera Basin using Inductive Couple Plasma Mass Spectroscopy (ICP-MS). It was found that the REE concentration in Banko coal is between 2.4 and 118.4 mg/(kg coal). When the coal was burned, the REE concentration in fly ash was predicted higher than 1000 ppm. This value was higher than world global average value[11]. In addition, the result of petrographic analysis of fly ash from South Sumatera coal combustion showed that coal fly ash before sieving has 12.7% unburned carbon, 41.5% glass, 15.3% quartz, 11.8% Iron oxide, 6.0% spinel, 12.7 % mullite [12].

The presence of iron oxides in coal fly ash can reduce the efficiency of REE recovery through leaching process (hydrometallurgy). Iron oxides can compete with REE to react with acid in leaching process. Therefore, to increase the REE recovery, iron oxides in the coal fly ash must be firstly eliminated. Magnetic separation and froth flotation are some techniques for reducing iron oxides from iron rich material. However, there are limited reports in case of reducing iron oxides content from REE rich coal fly ash. Lin *et al.* (2017) performed separation of iron oxides from USA coal fly using dry magnetic separation. They found that iron contents in magnetic component tend to decrease when magnetic field strength increases. In addition, REE are associated with aluminosilicates groups and are enriched in nonmagnetic component [13]. This REE analysis agrees with previous finding reported by Zhang *et al.* (2015) for United Kingdom coal fly ash, but there was no information whether the iron oxides are magnetic or nonmagnetic component [14]. In case of Indonesian coal fly ash, there is no report related to magnetic separation for reducing iron oxide content in REE rich coal fly ash. Even though , it has been known that coal fly ash from Indonesia contains relatively high iron oxides [12].The aim of this study is to investigate magnetic separation variables in order to remove iron oxides from fly ash. This will eventually increase REE content in the ash and therefore help improve the efficiency of REE recovery process.

2. Experimental method

Fly ash was produced from a commercial steam power plant which uses coal from South Sumatra as a source of energy. The first step was to sieve the samples in order to get the sample size +200 mesh and -400 mesh. The iron concentration in the fly ash samples were analyzed using ICP-MS performed by ALS Canada Ltd. (Vancouver, Canada) using fused bead method prior to acid digestion, followed by mineralogical analysis using X-Ray Diffarction (XRD, PANalytical type X'pert3 Powder). Subsequently the iron containing material was separated using magnetic separators developed by Research and Development Centre for Minerals and Coal Technology, Bandung. There were two kinds of magnetic separator, i.e dry magnetic separator and wet magnetic separator. Both magnetic separation techniques were used in this work. In wet separation, water was added to fly ash sample with a certain L/S then stirred until the mixture was well mixed. The mixture was then poured into magnetic separator with a specified magnetic strength. Magnetic components would be attracted to magnetic ball inside separator. The non-magnetic component flew to the bottom of separator and was then collected. Magnetic field strengths used in the wet separation study were 4300 Gauss and 10800 Gauss. In dry separation, fly ash particle was spread on a belt conveyor and passed into a magnetic separator. The magnetic component would be attracted to magnetic disc inside separator and non-magnetic component stayed on conveyor belt which is moving out of the separator. Fly ash sample in dry magnetic separator was set at code 1 for vibration to adjust the feed thickness. Magnetic field strength used in this study were 1100 Gauss, 1900 Gauss, 2650 Gauss, 4300 Gauss and 10800 Gauss. The amount of iron concentration in magnetic and non-magnetic components was analyzed using AAS (Perkin Elmer 3100).

The amount of iron collected as magnetic compound is calculated following this equation:

$$\alpha = \frac{m\ C_s}{M\ C_0}\ 100\%$$

where, $\boldsymbol{\alpha}$ = percentage of iron recovery (%),
$\boldsymbol{m}$ = mass of magnetic compounds (grams),
$\boldsymbol{C_s}$ = iron concentration in magnetic compounds (wt%),
$\boldsymbol{M}$ = initial mass of fly ash sample before separation (grams),
$\boldsymbol{C_0}$ = initial concentration of iron in fly ash before separation (wt%)

3. Results and discussion

3.1. Analysis of major compounds and mineralogy in coal fly ash

Table 1 shows major oxides in the coal fly ash before sieving using ICP-MS. The iron oxides content in this work is quite similar to result from petrology analysis previously reported by Besari (2018). As seen in the table, the concentration of iron oxides in the fly ash is 13.75%.This value is relatively higher compared to that of coal combustion from other countries, which is below 8% [15][16]. Therefore, reducing the iron oxides content from the fly ash is essential to increase efficiency of REE recovery through acid leaching process.

Table 1. The composition of major compounds in Coal Fly Ash before sieving

SiO_2 (%)	Al_2O_3 (%)	Fe_2O_3 (%)	CaO (%)	MgO (%)	Na_2O (%)	K_2O (%)
42.6	20.4	13.75	9.62	6.61	0.67	0.89

Previous study conducted by Besari (2018) found that unburned carbon is still present in coal fly ash [12]. The unburned carbon must be reduced because it could potentially interfere acid leaching process for REE recovery. The presence is indicated with the appearance of coarser grains with black colour. In this work, the unburned carbon was removed mostly as fraction of +200 mesh grain size.

Table 2 shows the iron concentration related to grain size. It is shown that the smaller size fraction contains higher concentration of iron.

Table 2. Iron concentration in fly ash fractions after sieving

Fraction	Iron Concentration (ppm)
+200 mesh	75512.3
-400 mesh	83229

In Table 3, it can be seen that in the fine fly ash, the fraction of hematite and magnetite are respectively 4.4% and 9.5%. In term of iron oxides, the content of both was 13.9%. Table 3 also shows the presence of amorphous glass fraction which amount reaches to46.6%. In comparison to petrographic data from previous study [12], the concentration in this work is relatively higher. Experimental result in this work shows that iron oxides and glass fraction can be concentrated in the finest fraction (-400 mesh).

Table 3. Mineralogical analysis on -400 mesh fraction using XRD

Quartz (%)	Hematite (%)	Magnetite (%)	Mullite (%)	Amorphous (%)
19.9	4.4	9.5	19.5	46.6

The increase of hematite, magnetite and glass fractions was also reported for coal fly ash from Poland and Kentucky [6][17]. Lower hematite and magnetite concentration in the coarser grain size is due to the presence of high concentration of organic compounds as well a slow pyrite and clay mineral concentration in the initial coal [6]. During combustion process, organic content partly becomes unburned carbon whereas pyrite and clay minerals become hematite, magnetite and glass fraction[18]. The formation of hematite indicated that boiler temperature possibly reached as high as 1400°C [19].

3.2. Iron oxide recovery from coal fly ash

The iron content in the fly ash within fraction of -400 mesh was then separated based on its magnetic properties using magnetic separator. This separation process is needed in order to reduce the iron content in the nonmagnetic components. The REE was predicted to concentrate in the glass fraction which is classified as nonmagnetic components of coal fly ash. Therefore, the iron content in nonmagnetic components is reduced in order to increase the efficiency of further REE recovery through acid leaching process. In this work, the influence of magnetic field strength, grain size and separation method were investigated.

3.2.1. *Effect of magnetic field strength.*

Hematite and magnetite have different magnetic properties. Hematite is paramagnetic while magnetite is ferromagnetic. Therefore, a greater magnetic strength is needed to separate hematite compared with the magnetite.

Table 4. Effect of Magnetic Field Strength on Iron concentration

Method	Grain Size, (mesh)	Magnetic Field Strength (Gauss)	Mass of Magnetic Components (gram)	Mass of Magnetic Components Percentage (wt%)	Iron Concentration in Magnetic Components (wt%)
Dry separation	+200	1100	57.68	59.35	7.47
		1900	70.98	72.77	7.82
		2650	75.65	77.35	9.63
		4300	80.01	82.37	9.71
Wet separation	-400	4300	9.49	18.99	13.78
		10800	17.98	35.83	12.40

Table 4 shows that increasing magnetic strength can increase iron concentration in the magnetic component and it also increases the amount of magnetic mass that is collected. Higher magnetic strength can attract more iron oxides as they are collected as magnetic component. However, fly ash also contains other minerals that might stick each other. This will cause that separation within magnetic field will also attracts others minerals. This is indicated by petrographic analysis data from Besari[12], as also shown in Figure 1, that some iron oxides were trapped in glass or quartz fraction. Therefore, an increase of magnetic strength is needed to attract iron oxide minerals as well as other minerals to magnetic fraction. However, experimental work in this study showed that fine fly ash

fraction is more hygroscopic and leads to entrainment among particles. The entrainment causes that non-iron components are likely to be also attracted to the magnetic components.

Figure 1. Petrographic analysis for iron oxide after Besari, 2018 [12]

3.2.2. *Effect of coal fly ash grain size*

Table 5 shows the effect of fly ash grain size on iron concentration in different separation techniques.

Table 5. Effect of Fly Ash Grain Size on Iron Concentration

Method	**Magnetic field strength (Gauss)**	**Particle Size (mesh)**	**IronConcentration in magnetic components (wt%)**	Mass of Magnetic Component Percentage **(wt%)**
Dry separation	1100	+200	7.47	59.35
		+325	8.94	55.04
		-400	Cannot be separated using this method	
	4300	+200	9.71	82.37
Wet separation	4300	+200	9.07	29.48
		-400	13.78	18.99

Table 5 shows that for both separation techniques decreasing particle size results the increase of iron concentration in the magnetic components. However, iron in the grain size fraction of -400 mesh cannot be separated using dry separation technique. This was due to entrainment phenomena that is more likely to occur for fly ash with particle size of -400 mesh so that non-iron components is also be attracted to the magnetic component.

As shown in Table 5, coarser grain size fraction contains lower iron concentration compared with the finer one. This is because the coarse coal size contains more unburned carbon resulted from coal combustion process. This result is in accordance with previous study which stated that separation using dry magnetic separator is limited for any material not finer than 75µm. For finer material, the effectiveness of separation would significantly reduce due to the effect of air current, particle-particle entrainment, and particle-device adhesion[20].

Based on results from both separation techniques, it is shown that the amount of separated magnetic components from wet separation was lower than that from dry separation. In the wet technique, particles are separated off in water so that the occurrence of entrainment can be avoided. Therefore, the iron separation process can be done more effectively and the concentration of iron is higher than that of dry separation technique.

Table 5 also shows the effect of separation techniques on the iron recovery with the same particle size fraction and magnetic field strength. Although, the iron concentration in magnetic components has slight differences, there is a significant difference in the total amount of magnetic components that can be separated. With dry separation technique, the collected magnetic components are 82.37% of coal fly ash. However, this result is not in a good agreement with the result of the mineralogy analysis, as reported in previous study [12]. This indicates that some non-iron components are likely being collected into magnetic components due to particle-particle entrainment. On the other hand, REE was predicted to present on the glass phase surfaces which are a non-magnetic component. The presence of non-iron components in the magnetic components will eventually decrease REE recovery. As a consequence, if dry separation technique is used, there will be significant amount of REE in the magnetic components and still mixed with iron components. This will cause further acid leaching process to recovery REE becoming more difficult and inefficient.

In this study, the most effective method to separate iron oxide from coal fly ash is using wet separation techniques with particle size of -400 mesh and 4300 Gauss magnetic field strength. In this condition the iron concentration is 13.78% with mass of magnetic components is 18.99%. So, the iron recovery will be 25.89%.

4. Conclusions

To separate iron components from fly ash, dry separation techniques should be used for fly ash with particle sizes larger than fraction of+200 mesh, while wet separation is for finer particles. The higher magnetic strength can increase iron concentration, glass, and mullite fraction in the magnetic component. The optimum value of magnetic field strength in this study is 4300 Gauss. Fly ash with fine particle size fraction has higher iron concentration in the magnetic component. The highest iron concentration can be achieved from fly ash having particle size fraction of - 400 mesh with an iron recovery of 25.89%.

5. Suggestion

It is recommended to investigate the effect of magnetic field strength between 4300 Gauss and 10800 Gauss in wet separation techniques for iron recovery in magnetic components.

Acknowledgement

The authors would like to thank Isyatun and Fandhy Rassyadi Putra for their help and support during the research and gratefully acknowledge for the support of the Chemical Engineering Department, Universitas Gadjah Mada.

References

[1] Alonso E, Sherman A M, Wallington T J, Everson M P, Field F R, Roth R and Kirchain R E 2012 *Environ. Sci. Technol.* **46** 3406–14
[2] Golev A, Scott M, Erskine P D, Ali S H and Ballantyne G R 2014 *Resour. Policy* **41** 52–9
[3] Simoni M, Kuhn E P, Morf L S, Kuendig R and Adam F 2015 *Waste Manag.* **45** 10–21
[4] Ahn J W, Thriveni T and Jegal Y 2015 *2015 World Coal Ash Conf.* 1–14
[5] Binnemans K, Jones P T, Blanpain B, Van Gerven T, Yang Y, Walton A and Buchert M 2013 *J. Clean. Prod.* **51** 1–22
[6] Blissett R, Smalley N and Rowson N A 2014 *Fuel* **119** 236–9
[7] Anonim 2016 International Coal Based Power Conferences 2016: Indonesia Electricity Development Plan and Indonesia Coal-Ash Management Implementation *Indones. Electr. Dev. Plan Indones. Coal-Ash Manag. Implement.*
[8] Meawad A S, Bojinova D Y and Pelovski Y G 2010 *Waste Manag.* **30** 2548–59
[9] Rajak D K, Guria C, Ghosh R, Agarwal S and Pathak A K 2016 *Int. J. Miner. Process.* **155** 106–17
[10] Roth E, Lin R, Howard B H, Bank T L, Granite E J and Soong Y 2017 *World of Coal Ash (WOCA) Conference* (Lexington)
[11] Anggara F, Amijaya D H, Harijoko A, Tambaria T N, Sahri A A and Asa Z A N 2018 *Int. J.*

Coal Geol. **196** 159–72
[12] Besari D A Y 2018 *Komponen Penyusun dan Model Keterdapatan Rare Earth Elements and Yttrium (REY) pada Fly Ash dan Bottom Ash* (Universitas Gadjah Mada) 71–79
[13] Lin R, Howard B H, Roth E A, Bank T L, Granite E J and Soong Y 2017 *Fuel* **200** 506–20
[14] Zhang W, Groppo J and Honaker R 2015 *World of Coal Ash (WOCA) Conference*(Nashville)
[15] Franus W, Wiatros-Motyka M M and Wdowin M 2015 *Environ. Sci. Pollut. Res.* **22** 9464–74
[16] Cao S, Zhou C, Pan J, Liu C, Tang M, Ji W, Hu T and Zhang N 2018 *Energy & Fuels* acs.energyfuels.8b01316
[17] Liu J, Dai S, He X, Hower J C and Sakulpitakphon T 2017 *Energy and Fuels* **31** 438–47
[18] Hower J C 2012 *Int. J. Coal Geol.* **92** 90–7
[19] Dai S, Zhao L, Hower J C, Johnston M N, Song W, Wang P and Zhang S 2014 *Energy and Fuels* **28** 1502–14
[20] Wills B A and Napier-munn T 2006 *Mineral Processing Technology: An Introduction to the Practical Aspects of Ore Treatment and Mineral Recovery* (Elsevier Science & Technology Books) **Chapter 13** 360

Effects of carbonization duration on the characteristics of bio-coal briquettes (coal and cane waste)

H E Handayani, Y B Ningsih, and M S Meriansyah

Department of Mining Engineering, Universitas Sriwijaya

Email: harminuke@ft.unsri.ac.id

Abstract. One of the technologies used to utilize low quality coal is by making bio-coal briquettes. Bio-coal briquettes are briquettes made of a mixture of coal and biomass. In this study, bagasse biomass is used as a mixture material of the briquettes. The objective of this study is to disclose the effects of the duration of carbonization on the characteristics of the briquettes. The process of carbonization is conducted to reduce water content and the level of volatile substances and increase the calorific value of the raw materials, so that the resulting briquettes will be of higher quality. This study uses the variations of carbonization duration of 30, 60, 90, 120, and 150 minutes. The result of the study reveals that the best duration of carbonization of raw materials in making bio-coal briquettes is 150 minutes. The bio-coal briquettes made with the duration of carbonization of 150 minutes has moisture content of 10.25%, ash content of 12.91%, volatile substance content of 31.88%, fixed carbon content of 44.96%, calorific value of 5,897 cal / gram, initial ignition time of 6.85 minutes, duration of combustionof 20.26 minutes, CO gas emission of 395 mg / Nm^3, and NO gas emission of 5.6 mg / Nm^3.

1. Introduction

The data obtained from the Directorate of Energy and Mineral Resources of the Year 2016 revealed that Indonesia had a low quality coal reserve of 50% of the total coal reserves. However, if the low-quality coal is directly used as a fuel, it will produce considerable emissions. Therefore, it is necessary to have a technology to utilize low quality coal as an environmentally friendly alternative energy. One of the technologies for utilizing low quality coal is the manufacture of bio-coal briquettes. Bio-coal briquette is a type of briquette product that uses raw materials of coal particles and biomass, both with or without binder or other added ingredients [1].

One of the biomass that can be used as a mixture in making bio-coal briquettes is bagasse. Bagasse is a by product of the process of extracting (milking) sugar cane in the manufacture of sugar and Sugar Cane Ice. Bagasse is usually used as fuel, material for paper pulp, organic fertilizer, and animal feed [2]. However, in reality,the traders of Sugar Cane Icesimply discard the bagasse,so that it can be classified as a waste product. This will certainly add to the environmental problem, whereas bagasse can be recycled.

Several studies have been conducted to utilize bagasse as an energy source in the form of briquettes. However, in general the briquettes only consist of bagasse or added with other biomass (Bio-briquettes). Many factors influence the quality of bio-coal briquettes, one of which is the duration of carbonization factor of raw materials that can affect the physical and chemical properties of briquettes produced.

High level of water content and volatile substances both in low quality coal and bagasse will result in a low calorific value of the bio-coal briquettes produced. To reduce the moisture content and the content of volatile substances in bio-coal briquettes, carbonization process can be carried out on coal and bagasse before being made into briquettes. In making bio-coal briquettes, the longer the duration of carbonization, the lower the content of water and volatile substances and the higher the calorific value in the briquettes will become [3].

One of the criteria of good briquettes is that the exhaust emissions produced from the combustion process do not contain toxins. The emissions produced from burning coal such as NO and CO will increase if more coal is used [4]. Biomass has low emissions so that if added to coal it will reduce emissions produced in the combustion process. Briquettes made by mixing coal and biomass produce better fuel properties and lower emissions. In addition, the addition of biomass can accelerate the ignition time in the briquetting process [5].

The aforementioned problem is the background for conducting this study on the effects of the duration of carbonization in the manufacture of bio-coal briquettes of coal and bagasse on the characteristics of the briquettes. The objective of this study is to obtain an optimal duration of carbonization that can produce the best quality bio-coal briquettes based on the quality standards of coal briquettes.

2. Research method

This study is a laboratory-scale study. This study uses low quality coal and bagasse as the raw material for making briquettes. The adhesive material used is tapioca flour. The variations of duration of carbonization used are 30, 60, 90, 120, 150 minutes.

The process of making bio-coal briquettes begins with the preparation of the raw materials. Large sized coal is crushed first using a jaw crusher and ball mill to reduce the size of the coal. After being ground, then the sieving process is carried out on the coal. The sieve used measures 30 mesh and 50 mesh. Ground coal sieving aims to obtain coal size of -50 mesh. The flowchart of the process of bio-coal briquette manufacturing with overall variations of duration of carbonization can be seen in Figure 1.

The coal that has undergone the process of size reduction and subsequent sieving is carbonized using the Nabertherm Carbolite Furnace, 30-3000^{0}C with a temperature of 500^{0}C. The duration of carbonization varies, namely 30, 60, 90, 120, 150 minutes. After carbonization, the coal cannot be directly removed from the furnace until the temperature reaches a room temperature. This process takes 5-6 hours or is best left for 1 night. The same process is also carried out in bagasse preparation process. However, the bagasse is first carbonized with the same temperature and variations of duration of carbonization in the carbonization process of the coal before sieving. The size of the bagasse used in this study is the same as the size of the coal which is -50 mesh.

The raw materials that have been prepared are then mixed by adding adhesive material in the form of tapioca flour. This mixing process uses a composition percentage of 50: 50 between coal and bagasse. The adhesive material used is 3.75 grams of the weight of one briquette of 15 gram. The ingredients and adhesives that have been mixed are then made into bio-coal briquette mixture. The dough is then molded using a manual molding tool with a compressive strength of 470 kg / cm^2.

The molded briquettes are then dried to reduce the water content that is still contained in bio-coal briquettes. The drying process is carried out using an oven at 50^{0}C for 1 hour. After being dried the briquettes cannot be removed directly from the oven until the temperature inside the ovens is the same as the room temperature. This process takes 5-6 hours and is best left for 1 night.

The dried briquettes are then analyzed to obtain the optimal duration of carbonization out of the existing variations of carbonization duration. The analysis conducted includes proximate tests (moisture content, ash content, volatile substance content, fixed carbon content, calorific value), initial ignition time, duration of combustion and CO and NO emissions produced from burning bio-coal briquettes.

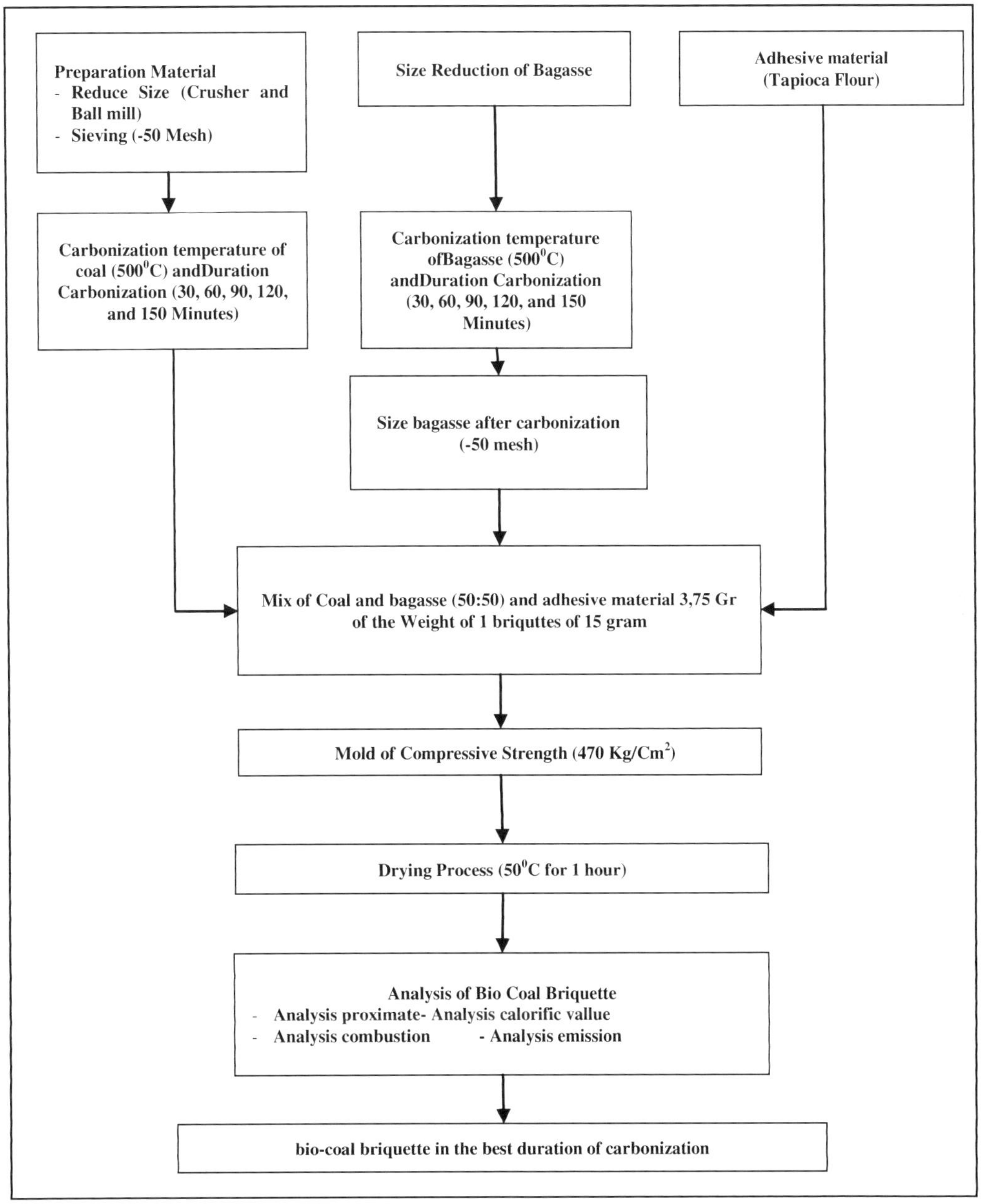

Figure 1. Flow chart of the study

3. Results and discussion

The raw materials for making briquettes in the form of low quality coal and bagasse were first tested to disclose the initial chemical characteristics of the raw materials. The initial chemical characteristics of the raw materials tested were the water content (Inherent Moisture) with the method of ASTM D. 3173-11, the ash content with the method of ASTM D. 3174-11, the volatile matter content with the method of ISO-562: 2010 (E), the content of the fixed carbon with the method of ASTM D. 3172-02 (11), and the calorific value with the method of ASTM D. 5865-11a. The test results of the initial chemical characteristics of the raw materials can be seen in Table 1. Then, the raw materials are carbonized with the variations of duration of carbonization of 30, 60, 90, 120, 150 minutes.

The results of proximate analysis and calorific value of bio-coal briquettes using variations of duration of carbonization can be seen in Table 2. The results of the analysis of ignition time and the duration of bio-coal briquette combustion using the variations of duration of carbonization can be seen in Table 3. In addition, the results of the analysis of CO and NO emissions of bio-coal briquettes using the variations of duration of carbonization can be seen in Table 4.

Table 1. Proximate Test and Calorific Value of Raw Materials

No	Raw Materials	Inherent moisture Content (% adb)	Ash Content (% adb)	Volatile Matter (% adb)	Fixed Carbon Content (% adb)	Calorific Value (cal/gram)
1	Coal	28.84	4.52	33.76	32.88	4348
2	Bagasse	21.00	8.40	65.55	5.05	3255

Table 2. Proximate tests and calorific value of briquettes using variations of duration of carbonization

No	Duration of Carbonization (minute)	Inherent Moisture Content (% adb)	Ash Content (% adb)	Volatile Matter Content (% adb)	Fixed Carbon Content (% adb)	Calorific Value (cal/gram)
1	30	14.68	7.46	36.36	41.50	5508
2	60	12.75	9.91	33.97	43.37	5673
3	90	12.36	10.69	33.28	43.67	5698
4	120	10.72	12.15	32.76	44.37	5734
5	150	10.25	12.91	31.88	44.96	5897

Table 3. Initial ignition time and duration of briquette combustion

No	Duration of Carbonization (minute)	Initial Ignition Time (minute)	Duration of Combustion (minute)
1	30	10.2292	14.4500
2	60	9.7875	16.0167
3	90	9.3625	17.1583
4	120	8.6292	18.2833
5	150	6.8458	20.2584

Table 4. Results of analysis of emission test using variations of duration of carbonization

No	Duration of Carbonization (minute)	NO(mg/Nm^3)	CO(mg/Nm^3)
1	30	9.4	3155
2	60	7	1478.75
3	90	6.15	1280
4	120	6	1043.75
5	150	5.6	395

The relationship between the duration of carbonization and the water content, the ash content, the volatile substance content, the fixed carbon content, the calorific value, the initial ignition time, the duration of combustion and the combustion emissions can be seen in Figure 2-10.

The moisture content indicates the amount of water contained in the structure of a solid fuel [6]. The higher the moisture content, the smaller the calorific value of the briquettes will be [7]. The effects of the duration of carbonization on the moisture content of the bio- coal briquettes generated in this study are shown in Figure 2.

Ash is the amount of residue after the organic matter is burned [6]. The effects of the duration of carbonization on the ash content of the bio-coal briquettes produced in this study are shown in Figure 3. Volatile matter is one of the chemical characteristics of a briquette. High content of volatile matter in bio-coal briquettes will make the briquettes easy to ignite and burn, so that the initial ignition rate in the briquetting process is faster [6]. The effects of the duration of carbonization on the volatile substance content of the bio-coal briquettes produced in this study are shown in Figure 4.

Fixed carbon content or can also be called bonded carbon content is a component found in solid fuels in the form of char [8]. The bonded carbon content shows the number of tethered carbon elements contained in the briquettes. The effects of the duration of carbonization on the level of bonded carbon of the bio-coal briquettes produced in this study are shown in Figure 5.

Calorie value shows the energy content contained in a fuel. High calorific value will produce good quality briquettes [9]. The effects of the duration of carbonization on the calorific value of the bio-coal briquettes produced in this study are shown in Figure 6.

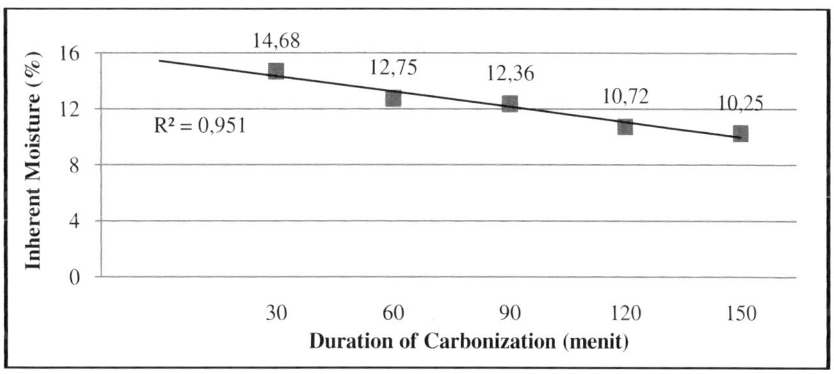

Figure 2. The graph of the effects of the duration of carbonization of the raw materials on the inherent moisture of the briquettes

Figure 2 shows that the duration of carbonization affects the moisture content of the briquettes. The longer the duration of carbonization of raw materials for the briquettes, the lower the moisture content in the briquettes will become. The results of the analysis showed that the duration of carbonization of 150 minutes had the lowest moisture content of 10.25%. This is consistent with the theory that states that the longer the duration of carbonization the less the moisture content of the briquette will become [10]. The long duration of carbonization makes the pores of the charcoal more open resulting in the release of moisture content contained in the materials [10]. The longer the duration of carbonization, the more pores that will open and more moisture content is released. The results of the study showed that the highest moisture content produced was 14.68%. Based on the standards, the maximum moisture content in the bio-coal briquettes is 15% [1]. The results of the analysis show that the moisture content of the all samples produced still meet the standards.

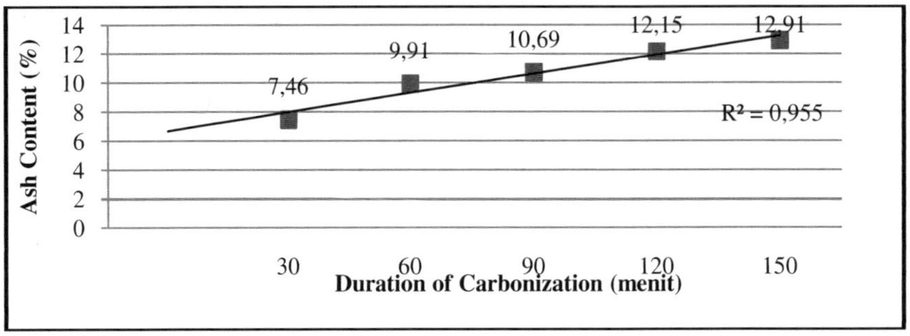

Figure 3. The graph of the effects of the duration of carbonization of the raw materials on the ash content of the briquettes

Figure 3 showed that the duration of carbonization affects the ash content of the briquettes. The longer the duration of carbonization of the briquette raw materials the the more ash content of the briquettes will become. The results of the analysis showed that the duration of carbonization of 150 minutes produced the highest ash content of 12.91%. The duration of carbonization of 30 minutes produced the lowest ash content of 7.46%. The duration of carbonization affects the ash content contained in the briquettes. The longer the duration of carbonization the more the ash content of briquettes will be. [11]. It is because the carbon will burn out and leave the ash which is the result of the combustion. The ash content can affect the calorific value of the briquettes. High level of ash content can reduce the calorific value of the briquettes so that the quality of the briquettes will decrease. The ash content contained in the briquettes must be as low as possible so that the quality of the briquettes is maintained [12]. The results of the study showed that the highest ash content produced was 12.91%. Based on the standard, the maximum ash content in bio-coal briquettes is 10% [1]. It shows that the ash content of the samples with the variations of duration of carbonization of 90, 120, and 150 minutes have not met the standards yet.

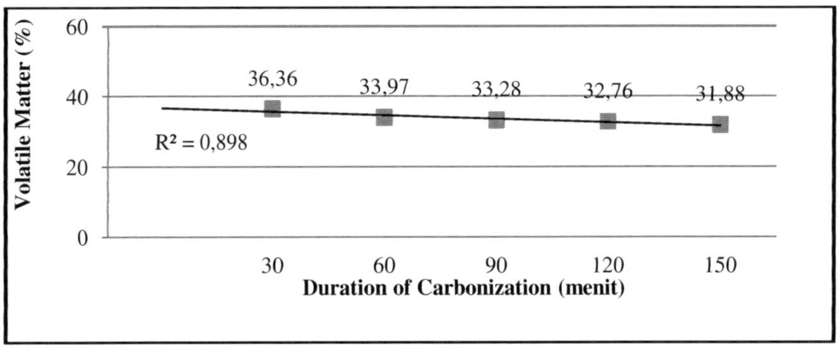

Figure 4. The graph of the effects of the duration of carbonization of the raw materials on the content of volatile matter of the briquettes

Figure 4 showed that the duration of carbonization affects the content of the volatile matter of the briquettes. The longer the duration of carbonization of the raw materials for the briquettes is the lower the content of volatile substances in the briquettes. The content of volatile substance tends to decrease as the duration of carbonization increases. So that the longer the duration of carbonization the lower the content of the volatile matter of the briquettes will become. The results of the analysis showed that the duration of carbonization of 150 minutes produced the lowest content of volatile substance of 31.88%, while the duration of carbonization of 30 minutes produced the highest content of volatile substance of 36.36%. Based on the standards, the maximum content of volatile substance in the bio-coal briquettes depends on the raw materials used [1]. The maximum volatile substance level is 65.55%. This shows that the content of volatile substance of the all samples produced still meet the standards.

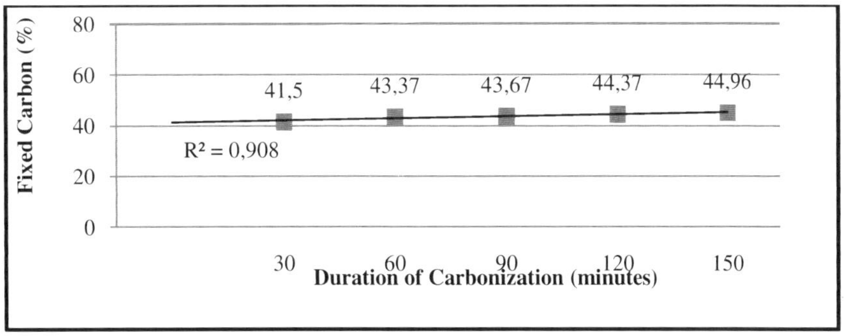

Figure 5. The graph of the effects of the duration of carbonization of the raw materials on the content of the fixed carbon of the briquettes

Figure 5 showed that the duration of carbonization affects the content of the fixed carbon in the briquettes. The longer the duration of carbonization of the briquette's raw materials, the higher the content of the fixed carbon in the briquettes produced. The results of the analysis showed that the duration of carbonization of 150 minutes produced the highest content of fixed carbon of 44.96%, while the duration of carbonization of 30 minutes produced the lowest content of the fixed carbon of 41.5%. The longer the duration of carbonization the more fixed carbon in the briquettes produced will become. The longer the duration of carbonization, the more perfect the carbon formation will be and will produce high fixed carbon content in the briquettes produced. In addition, the content of the fixed carbon is also influenced by the content of volatile substances in the briquettes. The lower the content of the volatile substances is the higher the content of the fixed carbon in the briquettes and vice versa [6].

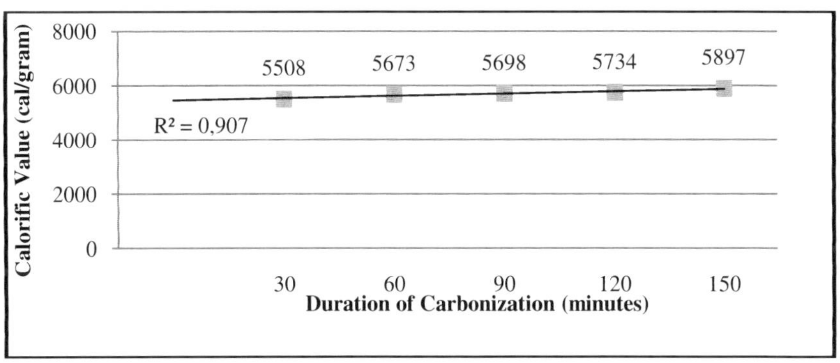

Figure 6. The graph of the effects of the duration of carbonization of the raw materials on the calorific value of the briquettes

Figure 6 showed that the duration of carbonization affects the calorific value of the briquettes. The longer the duration of carbonization of the raw materials for the briquettes, the more calorific value of the briquettes will be obtained. The calorific value is affected by the water content of the briquettes.

The higher the water content, the smaller the calorific value of the briquettes will be [7]. In addition, the calorific value of briquettes is also influenced by the content of volatile substances. The high content of volatile matter will reduce the calorific value of the briquettes and will reduce the quality of the briquettes [13]. The results of the analysis showed that the duration of carbonization of 150 minutes produced the highest calorific value of 5897 cal / gram. This is because the duration of carbonization of 150 minutes produced the lowest water content and also the lowest volatile substance content. The duration of carbonization of 30 minutes produced the lowest calorific value of 5508 cal / gram. This is because the duration of carbonization of 30 minutes produced the highest water content and also the highest volatile substance content. The results of the study showed that the lowest calorific value produced was 5508 cal / gram. Based on the standard, the minimum calorific value in the bio-coal briquettes is 4400 cal / gram [1]. This shows that the calorific value of all briquettes produced still meets the standards.

The content of water and volatile substances contained in the briquettes will affect the length of time of the initial ignition of the briquettes. The lower the water content, the faster the briquettes will light up during the combustion process. In addition, the higher the content of the volatile substance is the faster the process of ignition of the briquettes [14]. The duration of combustion is affected by the calorific value contained in the briquettes. High calorific values will make burning longer and can save the briquettes [14]. The effects of the duration of carbonization on the time of the initial ignition and the duration of the combustion of bio-coal briquettes in this study are shown in Figures 7 and 8.

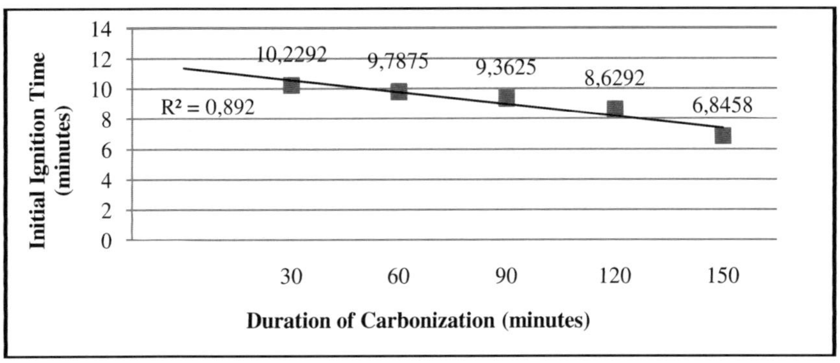

Figure 7. The graph of the effects of the duration of carbonization of raw materials on the initial ignition time of the briquettes

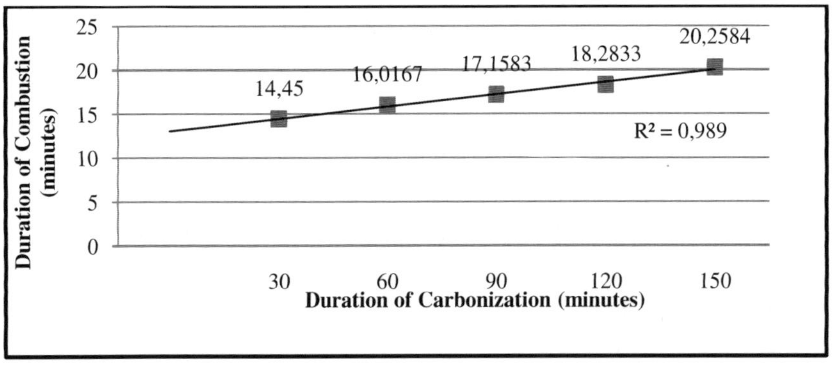

Figure 8. The graph of the effects of the duration of carbonization of raw materials on the duration of the combustion of the briquettes

Figure 7 shows that the duration of carbonization of the raw material influences the initial ignition time produced by the briquettes. The longer the duration of carbonization is the faster the initial ignition of the briquettes. This is because the longer the duration of carbonization of the briquettes, the lower the content of water in the in the briquettes, so that the initial ignition time of briquettes becomes faster [14]. In addition, the content of the volatile substance of the briquettes also affects the initial ignition time. The high content of volatile substances in the briquettes will make the initial ignition time of the briquettes become faster. The results of the study showed that the duration of carbonization of 30 minutes produced the longest initial ignition time, namely 10.2292 minutes. It is because the briquettes with that duration of carbonization has a high water content. The high water content will prolong the initial ignition of the briquettes because it takes time to drain the water content before the volatile substance can burn. The duration of carbonization of 150 has the fastest initial ignition time of 6.8458 minutes. It is because the briquettes with that composition has a low water content and high content of volatile substances. The low water content will accelerate the initial ignition of the briquettes because it does not require a long time to drain the existing water content before the volatile substance burns.

Figure 8 shows that the duration of carbonization of raw materials affects the length of combustion produced by the briquettes. The longer the duration of carbonization, the longer the combustion of the briquettes will be. The length of time of the combustion of the briquettes is also influenced by the content of the fixed carbon in the briquettes. The high content of the fixed carbon will make the combustion become longer and can save the use of the briquettes [14]. The results of the study showed that the duration of carbonization of 150 minutes has the highest combustion time of 20.2584 minutes. It is because the briquettes with that duration of carbonization has a high content of fixed carbon so the combustion time is long enough. The duration of carbonization of 30 minutes has the shortest combustion time of 14.45 minutes.It is because the briquettes with that duration of carbonization has a low content of fixed carbon resulting in shorter combustion time of the briquettes.

Emissions are residual gases of the combustion of a fuel. The test of briquette's gas emissions is needed to disclose the types of gases produced by the combustion, so as to prevent air pollution as a result of burning briquettes. Theoretically, burning fuel consisting of hydrocarbon elements will only produce H_2O and CO_2. However, in reality, there are other substances produced by the combustion process. These substances can be CO (Carbon Monoxide), NO (Nitrogen Monoxide) and other harmful substances [15]. The effects of the duration of carbonization on the emissions of bio- coal briquettes in this study are shown in Figure 9 for CO gas emissions and Figure 10 for NO gas emissions.

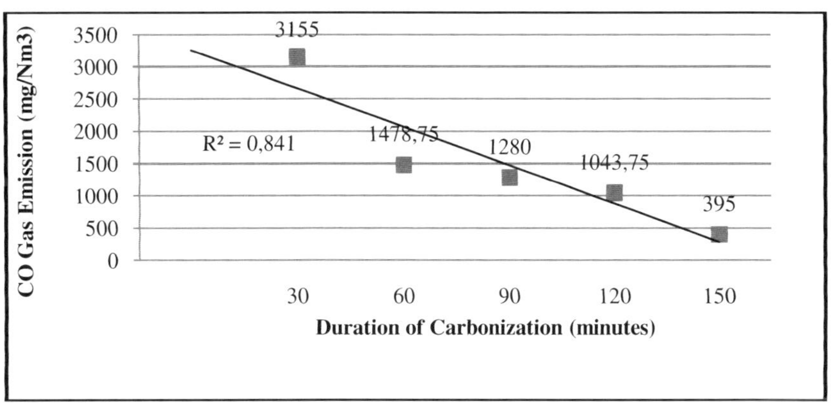

Figure 9. The graph of the effects of the duration of carbonization of raw materials on co gas emissions of the briquettes

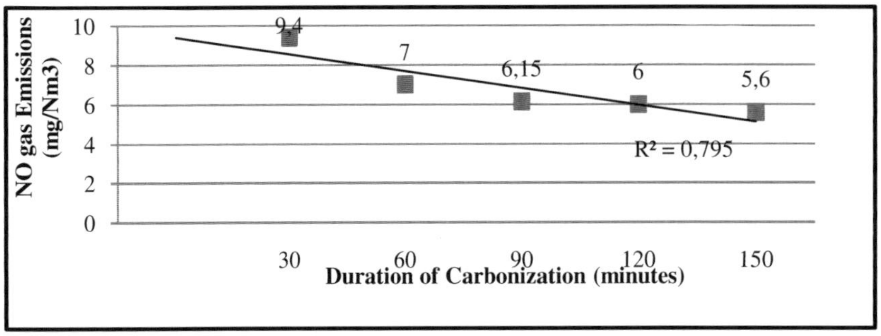

Figure 10. The graph of the effects of the duration of carbonization of raw materials on the no gas emissions of the briquettes

Figures 9 and 10 show that the duration of carbonization affects the emissions produced by the briquettes. The longer the duration of carbonization of the briquette's raw materials is the lower the emissions of CO and NO gas of the briquettes. The high content of volatile matter in the briquettes will produce more smoke when the briquettes are burned [15]. A lot of smoke will produce high CO and NO emissions so this will have a negative impact on the health and the surrounding environment. The duration of carbonization will affect the content of the volatile substances of the briquettes. The long duration of carbonization will reduce the content of volatile substances [10]. The results of the study revealed that the briquettes with the duration of carbonization of 150 minutes have the lowest level of CO and NO gas emissions. This is because the briquettes with that duration of carbonization have low level of volatile substances. The results of the study showed that the lowest CO gas emissions produced were 395 mg / Nm^3, while the highest NO gas emission was 9.4 mg / Nm^3. Based on the standard, the maximum CO gas emission is 726 mg / Nm^3, whereas NO gas emission is 140 mg / Nm^3 [1]. This indicates that only the briquettes with the duration of carbonization of 150 minutes have met CO gas emission standards. In addition, NO gas emission of all the produced briquettes has met the standards.

4. Conclusions

Based on the results of the testing and the data analysis conducted in this study, the following conclusions are drawn:

1. The duration of carbonization in the process of making bio-coal briquettes has an effect on the characteristics of the bio-coal briquettes produced. The results of the study disclosed that the longer carbonization duration will reduce the moisture content, volatile matter, initial ignition, CO and NO gas emissions. On the other hand, the longer carbonization duration will increase ash content, fixed carbon, calorific value, and burning time of the briquettes.
2. The best duration of carbonization of the raw materials for bio-coal briquette manufacturing is 150 minutes. The bio-coal briquettes produced with the duration of carbonization of 150 minutes has a moisture content of 10.25%, ash content of 12.91%, volatile substance content of 31.88%, fixed carbon content of 44.96%, calorific value of 5897 cal / gram, initial ignition time of 6.8458 minutes, duration of combustion of 20.2584 minutes, CO gas emission of 395 mg / Nm^3, and NO gas emission of 5.6 mg / Nm^3.

References

[1] Ministry of Energy and Mineral Resources 2006. Regulation of the Minister of Energy and Mineral Resources No. 047/2006
[2] Li-An'Amie N L 2014 *Jurnal tingkat sarjana senirupa dan desain.* **3 (1)** 1-7

[3] Junari E, Julham P P, and Herlina N *2015 Jurnal Teknik Kimia USU* **4(2)**
[4] Nukman 2010 *Jurnal Rekayasa Sriwijaya*, **19 (1)** 34–38
[5] Adekunle J, Ibrahim J and Kucha E 2015 *British Journal of Applied Science &Technology* **7(1)**, 114–123. doi: 10.9734/BJAST/2015/15154
[6] Sudiro and Suroto S 2014 *Jurnal Sainstech Politeknik Indonesia Surakarta.* **2 (2)** 1-18
[7] Rasyidi F A, Tuti I S, Arco Y D, Jasril N 2012 *Prosiding Seminar Nasional Perkembangan Riset dan Teknologi di Bidang Industri ke 16* (Inderalaya: Department of Chemical Engineering UNSRI)
[8] Nugraha J R 2013 *Karakteristik termal briket arang ampas tebu dengan variasi bahan perekat lumpur lapindo*. Skripsi (Jember : University of Jember)
[9] Patabang D 2009 *Majalah Ilmiah Mektek* **Year XI No. 3**
[10] Siahaan S, Hutapea M, and Hasibuan R 2013 *Jurnal Teknik Kimia USU* **2(1)**
[11] Yudanto, Angga, and Kartika K 2009 *Pembuatan Briket Bioarang dari Arang Serbuk Gergaji Kayu Jati* Final Paper (Skripsi) (Semarang : Department of Chemical Diponogoro University)
[12] Masturin A 2002 *Sifat Fisis dan Kimia Briket Arang Dari Campuran Arang Limbah Gergajian Kayu* Undergraduate Final Paper (Skripsi) (Bogor: Bogor Agricultural Institute)
[13] Wahyu K , Sarwono, Ronny D N 2013 *Jurnal Teknik Pomits* **2(1)**
[14] Jamilatun S 2008 *Jurnal Rekayasa Proses* **2 (2)** 34-41.
[15] Sulistyanto A 2006 *Media Mesin* **7(2)** 77-84

IOP Conf. Series: Materials Science and Engineering **478** (2019) 012028 doi:10.1088/1757-899X/478/1/012028

Effect of inlet speed on gravitational air separator for cenospheres accumulation from fly ash: modeling using computational fluid dynamics (CFD)

Sutijan, S W Sandiwan, K I Saputra, M H Mahdi, M S Albarkan, A Prasetya, and H T B M Petrus

Department of Chemical Engineering (Sustainable Mineral Processing Research Group), Faculty of Engineering, Universitas Gadjah Mada, Jl. Grafika No. 2 Yogyakarta 5528, Indonesia

E-mail: bayupetrus@ugm.ac.id

Abstract. Dry gravity separation is widely used in industries due to its low operational cost. Furthermore, it does not require water to perform its separation process. Commonly used gravity separator is cyclone separator. Unfortunately, this type of separator is unable to yield various purities for its products. A modified gravity separator as an alternative relies on terminal velocity difference for different diameter and density is evaluated in this paper. The separated particles are coal fly ash and cenosphere which are the results of coal combustion where the content of cenosphere is generally 0.1% of total mass. Separated cenospehere can be used for various applications in the field of materials. In this simulation we use a particle model with a ratio of fly-ash to cenosphere mass of 1:1. Particles behavior in this gravity separator is affected by terminal velocity, linear speed, as well as rheology. Modeling with CFD (Computational Fluid Dynamics) using the Discrete Phase Model can take into account the behavior of particles that are affected by these factors. This model assumes that the simulated particles are perfectly round in shape. It uses the Euler-Lagrange framework in which the perfectly rounded particles (discrete phases, Lagrange frames) interact with air (continuous phase, Euler framework) and performed with ANSYS Fluent software. This study evaluates the effect of speed variation on Newton efficiency (NE) on a given geometry. Simulations were performed with variations of inlet velocity: 0.1 m/s; 0.25 m/s; 0.5 m/s; 1 m/s; 2 m/s; and 3 m/s. It was found that the lower the speed, the higher the NE and the optimum NE of 0.41 is obtained with 0.25 m/s inlet velocity.

1. Introduction

Developing countries like Indonesia rely on coal-fired power plants in large quantities to fulfill their energy. It is because of the low price and abundance of mass energy source of coal that could not be found in any other energy resources. In a case of Indonesia, the need of energy is increasing along with the increase of its population. Thus, the Indonesian government through Presidential Decree No. 22 year 2017 determines the energy supply management to result 135.5 GW at which coal poseses 30% of the national energy consumption until the year of 2025 as it can be seen in Figure 1.

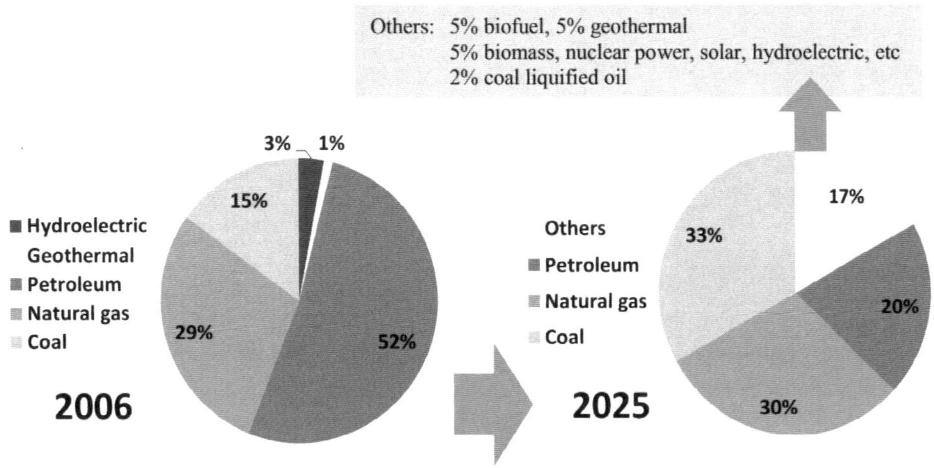

Figure 1. Model of the National Supply and Demand Energy in the year of 2025 [1]

In line with the increase of coal needs up to the year of 2025, there will be also fly ash resulted from the coal combustion with the estimated amount of 200 tonnage per year. It will surely pose problems to the environment, if the fly ash is not treated carefully. Exiting fly-ash from coal combustion process has no selling value and usually only disposed in landfill while in the developed countries coal-fly ash is used in cement industries [2]. However, fly-ash actually contains a type of micro particle called a cenosphere mainly consists of quartz, alumina, Calcium Oxide and Hematite [3] which has a hollow spherical structure and has many applications in the industry due to its superior nature. Cenosphere is made by a process similar to a glass-making process that makes cenosphere has a distinctive properties [4]. This material has low bulk density, high thermal resistance, and high mechanical strength [5].

The commonly used separation technique is based on the density difference at which the separation provides high efficiency to separate cenosphere from fly ash [6]. This separation process is effective but it currently consumes large amount of water and add to water pollution due to leaching of toxic material from fly-ash. Thus, the exploration of dry separation processes have been conducted by some researchers to minimize the propensity of water pollution [7,8]. One of the commonly used tools is cyclone, but it cannot separate different fly-ash diameters at once needing multistage of cyclones. Therefore, a new separation device is proposed as shown in section 3.1. This geometry is expected to separate fly-ash into several different product fractions.

Before running the CFD calculation, two models have been selected which are turbulence model and particle model. Standard k-epsilon model with standard wall treatment are used for turbulence condition because this method is easier to use in the first simulation [9]. Discrete Phase Model (DPM) is used for particle model approach.

2. Mathematical Modeling

2.1. Governing Equation

The main equations govern this modeling are based on the continuity law (equation 1) and momentum conservation (equation 2, 3, and 4). The latter equation is also known as Navier-Stoke equation.

$$\rho \frac{\partial \rho}{\partial t} + \nabla(\rho \bar{V}) = 0 \quad (1)$$

Momentum conservation in x direction

$$\frac{\partial(\rho u)}{\partial t} + \nabla(\rho u\bar{V}) = -\frac{\partial p}{\partial x} + \frac{\partial \tau_{xx}}{\partial x} + \frac{\partial \tau_{yx}}{\partial y} + \frac{\partial \tau_{zx}}{\partial z} + \rho f_x \quad (2)$$

Momentum conservation in y direction

$$\frac{\partial(\rho v)}{\partial t} + \nabla(\rho v\bar{V}) = -\frac{\partial p}{\partial y} + \frac{\partial \tau_{xy}}{\partial x} + \frac{\partial \tau_{yy}}{\partial y} + \frac{\partial \tau_{zy}}{\partial z} + \rho f_y \quad (3)$$

Momentum conservation in z direction

$$\frac{\partial(\rho w)}{\partial t} + \nabla(\rho w\bar{V}) = -\frac{\partial p}{\partial z} + \frac{\partial \tau_{xz}}{\partial x} + \frac{\partial \tau_{yz}}{\partial y} + \frac{\partial \tau_{zz}}{\partial z} + \rho f_z \quad (4)$$

2.2. Standard k-epsilon Model (Standard k - ε Model)

Two-equation turbulence models allow the determination of both turbulent length and time scale by solving two separate transport equations. The standard k - ε model in ANSYS Fluent falls within this class of models and has become the workhorse of practical engineering flow calculations in the time since it was proposed by Launder and Spalding. Robustness, economy, and reasonable accuracy for a wide range of turbulent flows explain its popularity in industrial flow and heat transfer simulations. It is a semi empirical model, and the derivation of the model equations relies on phenomenological considerations and empiricism. The standard k - ε model is a model based on model transport equations for the turbulence kinetic energy (k) shown by equation (5) and its dissipation rate (ε) shown by equation (6). [10].

$$\frac{\partial}{\partial t}(\rho k) + \frac{\partial}{\partial x_i}(\rho k u_i) = \frac{\partial}{\partial x_j}\left[\left(\mu + \frac{\mu_t}{\sigma_k}\right)\frac{\partial k}{\partial x_j}\right] + G_k + G_b - \rho\varepsilon - Y_M + S_k \quad (5)$$

and

$$\frac{\partial}{\partial t}(\rho \varepsilon) + \frac{\partial}{\partial x_i}(\rho \varepsilon u_i) = \frac{\partial}{\partial x_j}\left[\left(\mu + \frac{\mu_t}{\sigma_\varepsilon}\right)\frac{\partial \varepsilon}{\partial x_j}\right] + C_{1\varepsilon}\frac{\varepsilon}{k}(G_k + C_{3\varepsilon}G_b) - C_{2\varepsilon}\rho\frac{\varepsilon^2}{k} + S_\varepsilon \quad (6)$$

The turbulent (or eddy) viscosity, μ_t, is computed by combining k and ε as follows :

$$\mu_t = \rho C_\mu \frac{k^2}{\varepsilon} \quad (7)$$

2.3. Discrete Phase Model

This model assumes individual particle is perfectly spherical and uniform in its mass distribution throughout the sphere. Trajectory of particle/droplet are computed in a Lagrangian frame and coupled with the momentum of Eulerian frame by integrating equation (8). This model is preferred because it is able to track particles individually and hence giving off information of a particle's detail such as diameter and mass at the selected pressure-outlet.

$$\frac{\partial u_i^p}{\partial t} = F_D(u_i - u_i^p) + g_i\frac{(\rho_p - \rho)}{\rho_p} + \frac{F_i}{\rho_p} \quad (8)$$

3. Results and discussion

3.1. Geometry Specification and Nomenclature

In order to be able to separate cenosphere from fly ash with different fraction based on their diameter, the air separator has been developed with geometry as it can be seen in Figure 1 and 2.

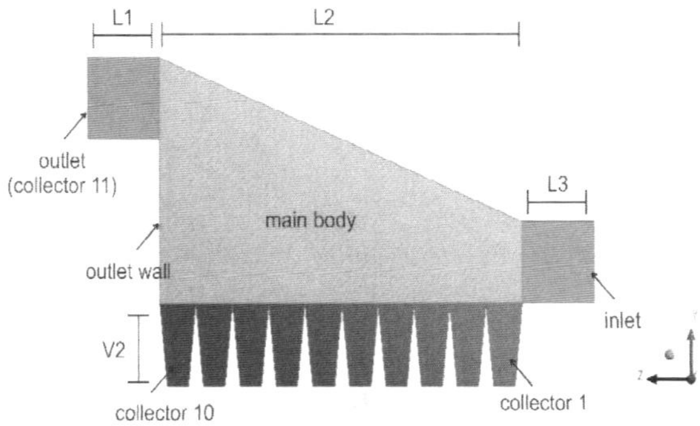

Figure 2. Side view of geometry

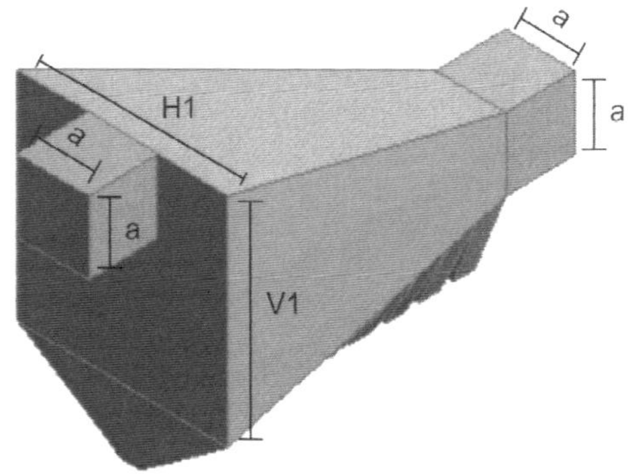

Figure 3. Isometric view of geometry

By having such an air separator as shown in Figure 2 and 3, the terminal velocity of the cenospheres will drop that the classification can be conducted along the fractionated collectors. The dimension of the separator is shown in table 1.

Table 1. Geometry of the Air Separator

Marker	Dimension
L1	0.1 m
L2	0.5 m
L3	0.1 m
V1	0.3 m
H1	0.3 m
V2	0.1 m
A	0.1 m

3.2. Model Particle

Real cenosphere concentration commonly found in coal-fueled power plant is less than 1% [6] which is very small. This study uses a model particle with a total mass of 2e-6 kg and a cenosphere : fly-ash mass ratio of 1:1 in order to fully map every diameter distribution and still uses a reasonable computing power. Particle distribution is based on particle distribution found in [7]. It is then fitted for Rosin-Rammler particle distribution which is shown by equation (9) and mapped for every diameter. Result is shown on figure 3.

$$Y_d = e^{-(\frac{d}{d_{mean}})^\wedge n} \quad (9)$$

where:

Y_d = mass fraction of particle with the diameter greater than d

d_{mean} = size constant
n = size distribution parameter

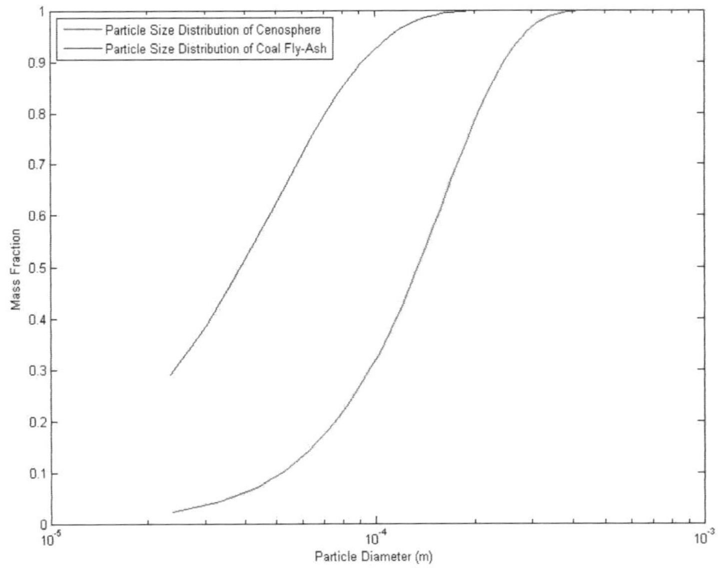

Figure 4. Particle size distribution for model particle of raw material

3.3. Particle Trajectory

The separation of cenosphere from fly ash will be highly affected by the rheology inside the air separator. Providing the trajectory of the air flow as a function of inlet velocity as it can be seen in Figure 4-9 will underline the separation mechanism that will be further represented by Newton efficiency.

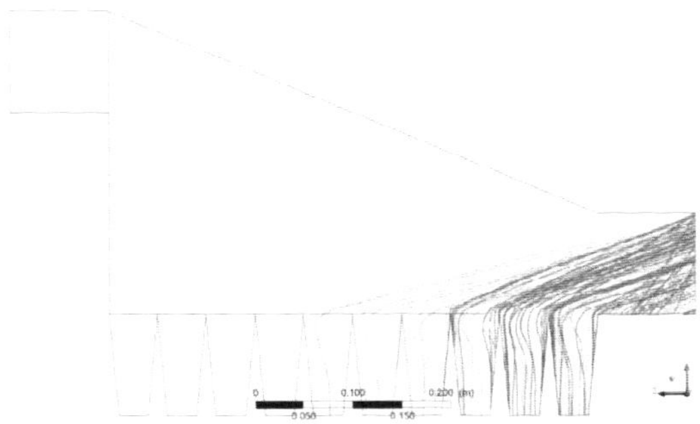

Figure 5. Particle Trajectory for 0.1 m/s Inlet Speed

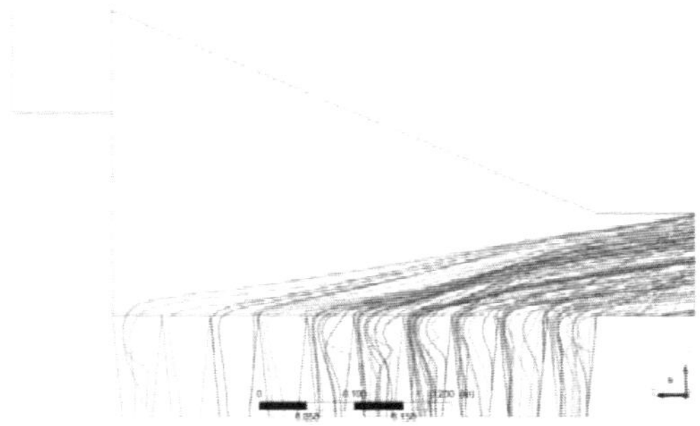

Figure 6. Particle Trajectory for 0.25 m/s Inlet Speed

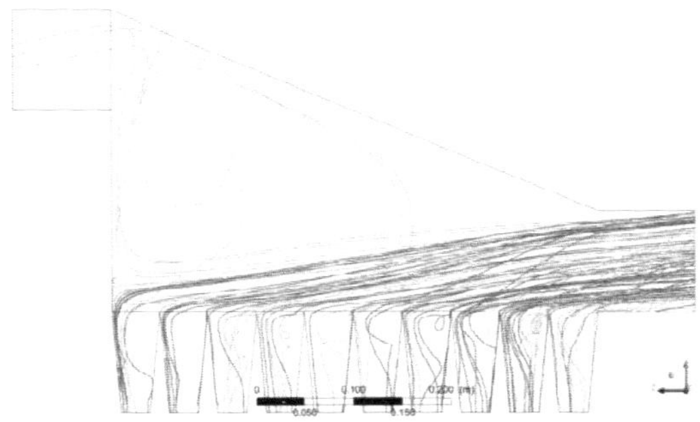

Figure 7. Particle Trajectory for 0.5 m/s Inlet Speed

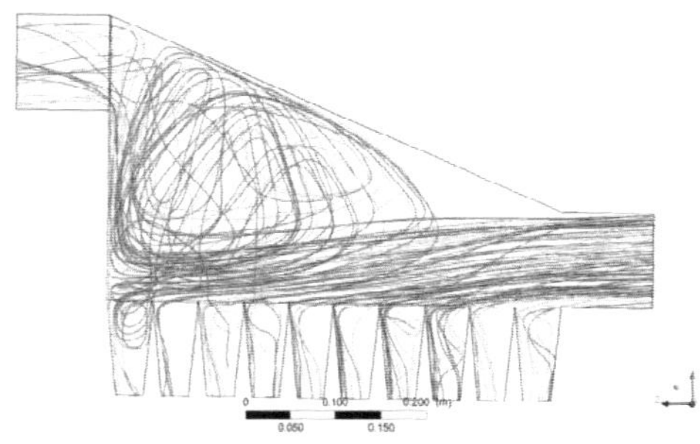

Figure 8. Particle Trajectory for 1 m/s Inlet Speed

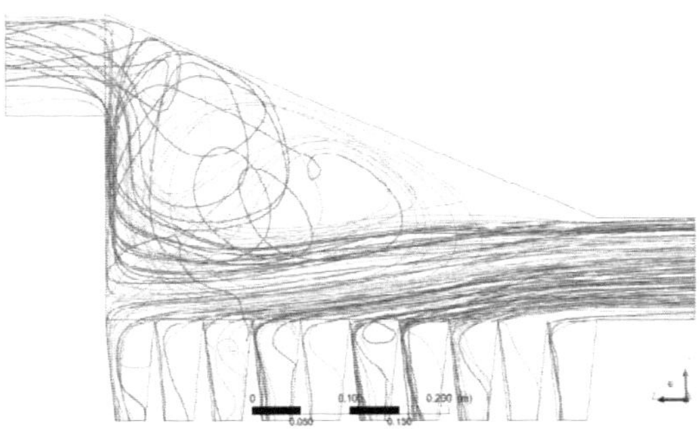

Figure 9. Particle Trajectory for 2 m/s Inlet Speed

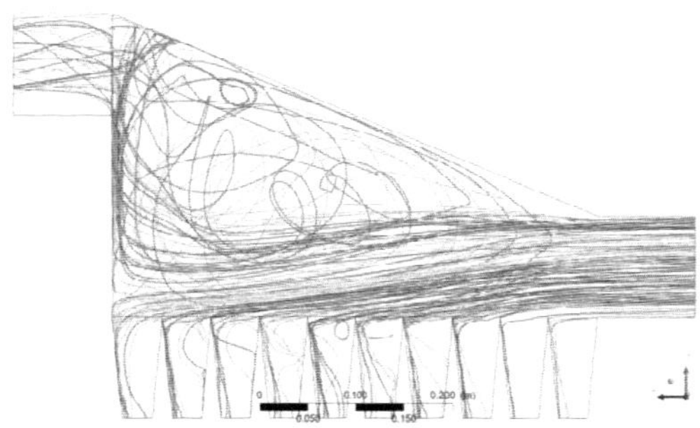

Figure 10. Particle Trajectory for 3 m/s Inlet Speed

Figure 5 to Figure 10 show that particle trajectory is heavily affected by inlet speed variation. Particles are more likely to spread further from the inlet position with the increase of inlet speed. There is a backflow starting from 0.5 m/s velocity variation which is caused by stream and particle collides with the outlet wall. This phenomenon causes the separation efficiency of this gravitational separator is not merely controlled by linear velocity and terminal velocity.

3.4. Newton efficiency (NE) of each collector

Newton efficiency of each collector is calculated by assuming that each collector is a cenosphere concentrated product and the other 10 collectors are fly-ash concentrated product for every collector's NE calculation. This method is described in equation (10) found in Hirajima et.al. (2010) [7].

$$NE\ (i) = \frac{mc\ (i)}{mct} - (1 - \frac{mft - mf\ (i)}{mft}) \quad (10)$$

Rearranging equation (10) above yields :

$$NE\ (i) = Rc(i) - Rfa(i) \quad (11)$$

where :

(i)= index of collector
NE = newton efficiency of specific collector
mc = mass of cenosphere in a specific collector
mct = total mass of cenosphere
mft = total mass of fly-ash
mf = mass of fly-ash in a specific collector
Rc = recovery of cenosphere in a specific collector
Rfa = recovery of fly-ash in a specific collector

Detailed calculation result will be shown for inlet speed = 0.1 m/s in Table 2, while Newton efficiency for each collector for every inlet speed variation will be visualized in Figure 11 to figure 16.

Table 2. Detailed Calculation for Eah Collector at Inlet Speed = 0.1 m/s

(i)	1	2	3	4	5	6	7	8	9	10	11
Rc	0,498772	0,22608	0,10672	0,057416	0,040439	0,017048	0,014098	0,011061	0,007267	0,009973	0,011126
Rfa	0,308726	0,143193	0,120941	0,100971	0,073772	0,059547	0,041558	0,03328	0,032026	0,041275	0,044711
NE	0,190047	0,082887	-0,01422	-0,04356	-0,03333	-0,0425	-0,02746	-0,02222	-0,02476	-0,0313	-0,03359

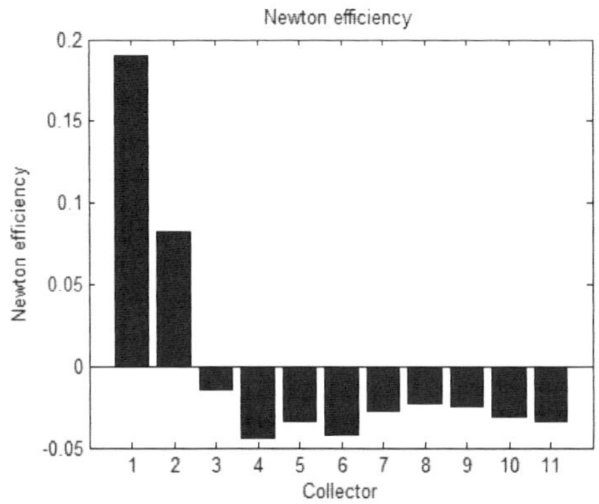

Figure 11. NE(i) at inlet speed=0.1 m/s

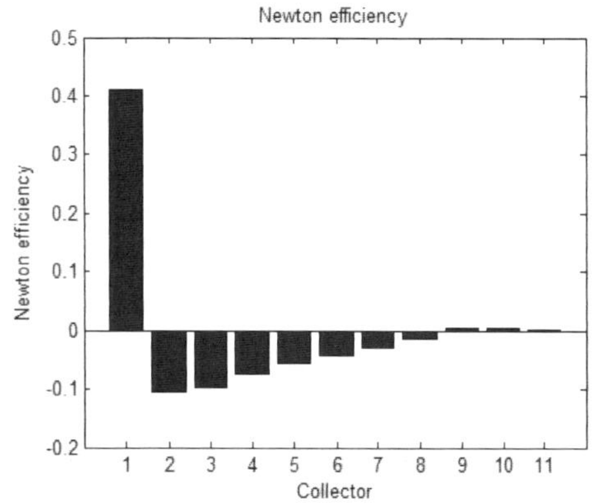

Figure 12. NE(i) at inlet speed=0.25 m/s

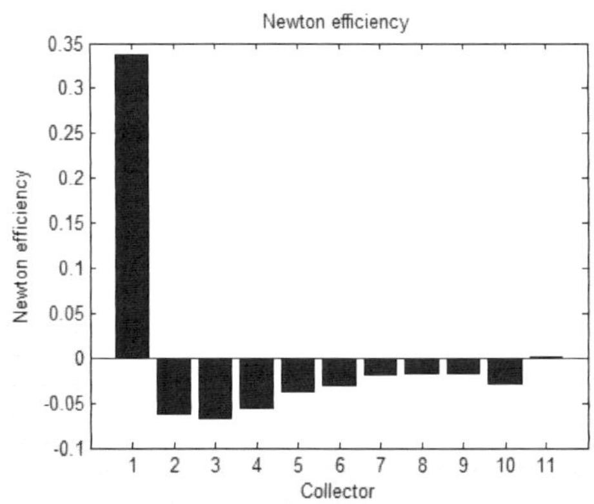

Figure 13. NE(i) at inlet speed=0.5 m/s

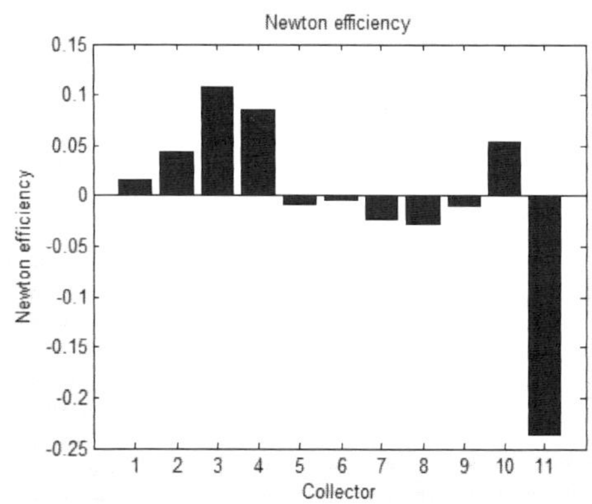

Figure 14. NE(i) at inlet speed=1 m/s

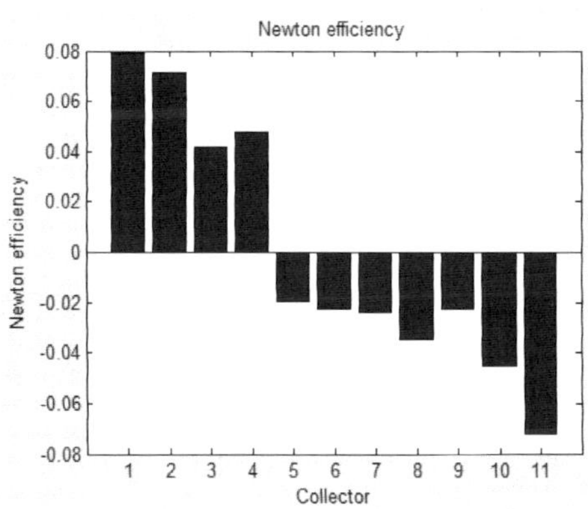

Figure 15. NE(i) at inlet speed=2 m/s

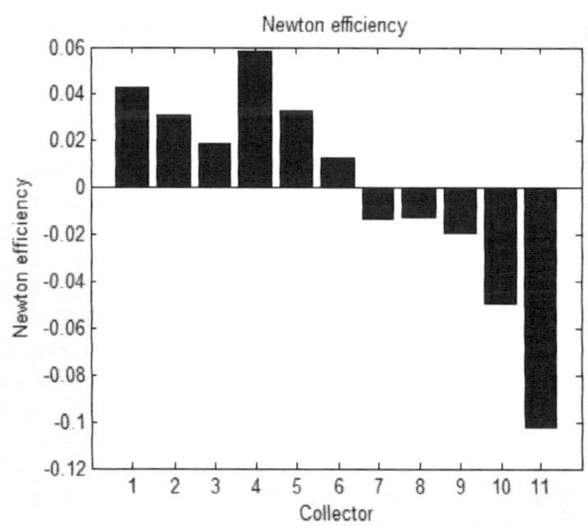

Figure 16. NE(i) at inlet speed=3 m/s

The NE distributions shown above effectively shows the cenosphere and fly-ash distribution across the geometry. The higher the inlet speed, the more scattered cenosphere is. This is mainly due to the size distribution of cenosphere and fly-ash. Fly-ash mainly consist of small particles with 20% of its mass fraction is under 2.5e-5 m in diameter as shown in section 3.2. Fly-ash particle distribution is more evenly distributed compared to cenosphere particle distribution. This unique particle distribution causes the cenosphere to be concentrated at the 1st and 2nd collector at low speed (0.1 m/s ; 0.25 m/s ; 0.5 m/s). Even at low inlet speed variations, small fly-ash particles is blown further than cenosphere particles and causes the 1st and 2nd collector to be less occupied by fly-ash particles. This phenomenon is further confirmed in Figure 17 and 18. It can be seen that particles with the biggest diameter are accumulated in collector-1 for both fly-ash and cenosphere.

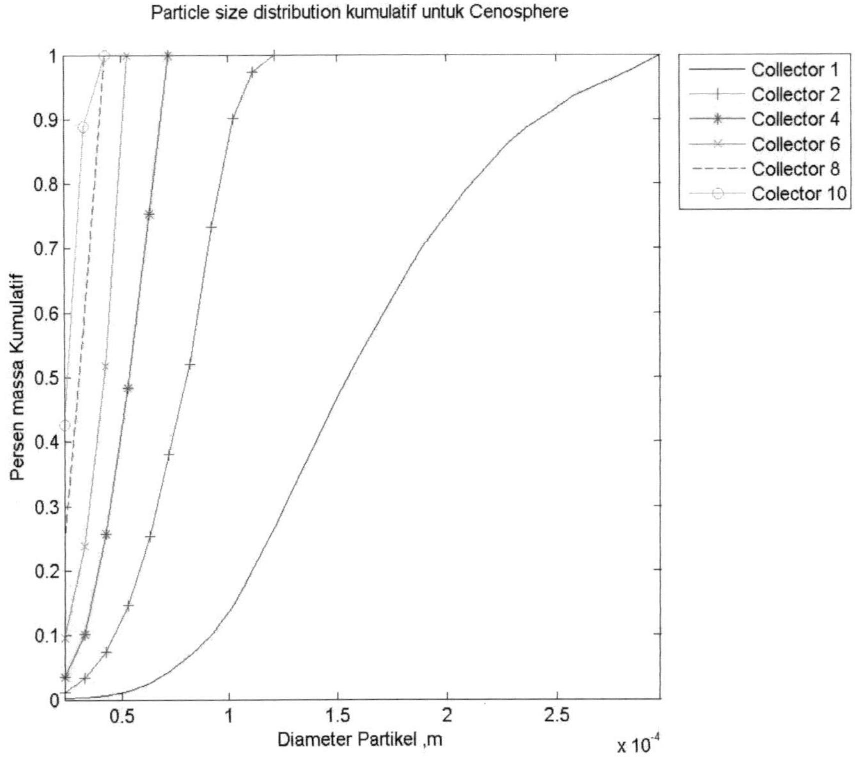

Figure 17. Cumulative particle size distribution for cenosphere in every collector at inlet speed =0.5 m/s

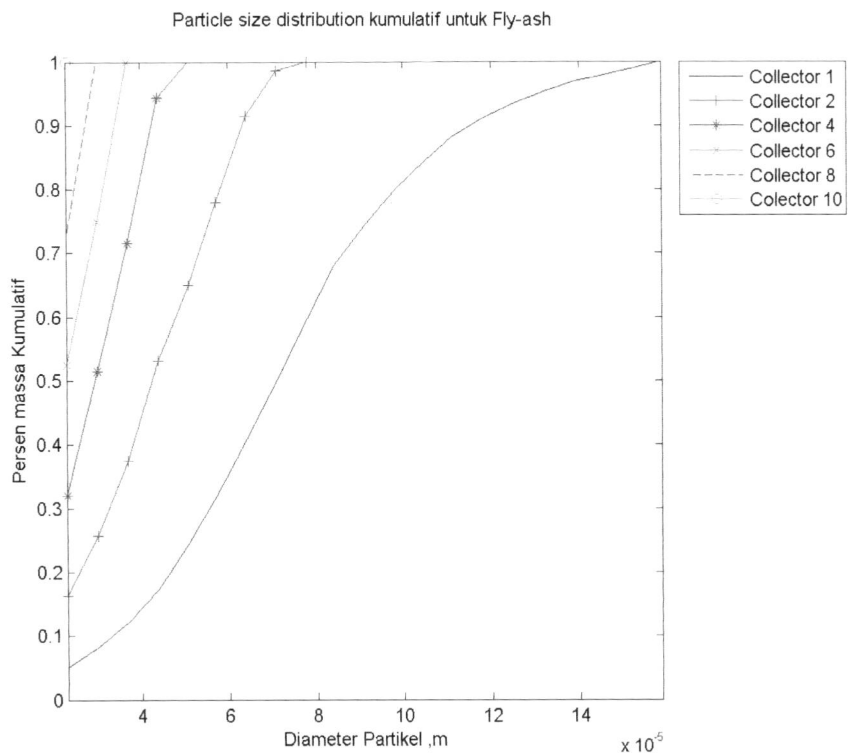

Figure 18. Cumulative particle size distribution for fly-ash in every collector at inlet speed =0.5 m/s

3.5. Overall efficiency

Collector(s) with positive value of NE (>0) described in section 3.4 are considered as cenosphere concentrated product and collector(s) with negative value of NE (<0) are considered as fly-ash concentrated product thus yielding a single parameter to measure the separation performance for every inlet speed variation. This method is described in equation (12)

$$NET = \frac{mcs}{mct} - (1 - \frac{mft - mfs}{mft}) \quad (12)$$

Where:
NET= Overall newton efficiency
mcs = mass of cenosphere in the cenosphere concentrated product
mct = total mass of cenosphere
mft = total mass of fly-ash
mfs = mass of fly-ash in the fly-ash concentrated product

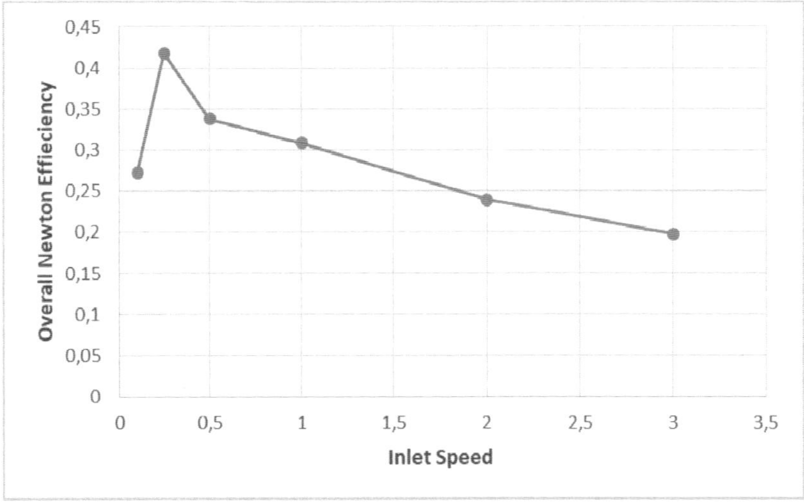

Figure 19. Inlet speed vs overall newton efficiency

Optimum speed is obtained at inlet speed= 0.25 m/s. This phenomenon is explained at section 3.4 where cenosphere is concentrated at the 1st collector thus yielding the largest value of newton efficiency by retaining most of injected cenosphere in a single collector resulting in a newton efficiency of 0.41.

4. Conclusion

Gravitational separation is an environmentally friendly and practical method for separating cenosphere and fly-ash. It uses no water as its medium of separation while offering the reasonable separation efficiency. Computational Fluid Dynamics (CFD) can provide detail information related to the potential of the modified air separator studied in this paper which offers the reasonable separation efficiency (Newton efficiency) up to 0.41.

Acknowledgment

Highly appreciation to the Minister of Higher Education of Indonesia for the financial support to conduct this project under PTUPT scheme with contract number 1860/UN1/DITLIT/DIT-LIT/LT/2018.

References

[1] Kementerian ESDM 2017 *Perpres RI No. 22 Tahun 2017 tentang Rencana Umum Energi Nasional* Jakarta
[2] Acar I and Atalay M U 2016 *Fuel* **180** 97-105
[3] Sen S 2014 *Physical properties of cenosphere* Diss
[4] Żyrkowski M 2016 *Fuel* **174** 49-53
[5] Ranjbar N and Kuenzel C 2017 *Fuel* **207** 1-12
[6] Hirajima T, Oosako Y, Nonaka M, Petrus H T B M, Sasaki K and Ando T 2008 *Journal of MMIJ* **124(12)** 878-884
[7] Hirajima T, Petrus H T B M, Oosako Y, Nonaka M, Sasaki K and Ando T 2010 *International Journal of Mineral Processing* **95(1-4)** 18-24

[8] Petrus H T B M, Hirajima T, Oosako Y, Nonaka M, Sasaki K, and Ando T 2011 *International Journal of Mineral Processing* **98(1-2)** 15-23
[9] Wasilewski M 2017 *Separation and Purification Technology* **179** 236-247
[10] Soe T M and Khaing S Y 2017 *Int J Sci Res Publ* **7(8)**

IOP Conf. Series: Materials Science and Engineering **478** (2019) 012029 doi:10.1088/1757-899X/478/1/012029

Pre-heating of multi-axial forging (MAF) on aluminum based composites

A Pramono, A Yolanda, and A A Alhamidi

Department of Metallurgy Engineering, Universitas Sultan Ageng Tirtayasa. Cilegon, 42435 Banten, Indonesia

Email: agus.pramono@untirta.ac.id

Abstract. Aluminum is one of the materials used for automotive, almost 90% of aluminum scrap from automotive cars can be recycled. The properties of aluminum are light and have a high ductility. The addition of SiC as reinforcement to aluminum based composite materials by multi-axial forging (MAF) is one of the newly developed SPD processes. Addition of SiC is to increase strength greater than the base material. The purpose of this research is to know the micro structure especially spread of distribution of precipitate evenly in matrix and change of mechanical properties after MAF process. Therefore, variation of weight fraction of SiC and pre-heating temperature were conducted. Kind of aluminum type AA1100 shaped plate as matrix added SiC as reinforcement, using variation 0.2%:0.4% and 0.6% wt. As for pre-heating temperature variation of 350°C, 450°C and 550°C then done MAF process with every change of pass rotate 90°. The results obtained from aluminum-based composite study by MAF, that with the addition of SiC on microstructure observation there is a change of grain boundary shape from equaxial to lamellar, there are grains of precipitate on the matrix that spread evenly and tensile test results and microhardness test increased along with the addition SiC fraction at a temperature of 350°C while for temperatures of 450°C and 550°C there is a decrease in hardness value, tensile strength value on Al/SiC.

1. Introduction

The Indonesian automobile manufacturers association (Gaikindo) showed that motor vehicle sales in Indonesia amounted til 86,234 units, a year-on-year increase of 17.7%. Gakindo predicts sales will reach 1.1 million units seen from infrastructure and increase commodity prices, will help increase demand for cars [1]. The use of aluminum in automotive cars has many advantages including a fast, safe, and cost-effective process, in addition, the benefits of using aluminum in the car automotive is environmentally friendly. Almost 90% of aluminum scrap from automotive cars can be recycled. Recycling 1 ton of aluminum can save energy equivalent to 21 barrels of oil. The energy department at Oak Ridge National Laboratory found that vehicles with aluminum materials could reduce 20% of total energy cycle consumption and reduce 1% of CO_2 emissions [2]. Properties of aluminum are: high ductility, light weight, corrosion resistance and low strength so easily formed. AA1100 has a very high ductility with adequate strength. To increase the power up to 290 MPa in automotive applications it is necessary to add silicon carbide (SiC) by severe plastic deformation (SPD) [3]. Some SPD processes: equal channel angular pressing (ECAP), accumulative Roll Bonding (ARB), high pressure torsion (HPT) and multi-axial forging (MAF) are the development of conventional forming. MAF processing on automotive component are more appropriately selected for the process. The advantages of MAF method when compared to other methods in SPD, this process can be used for large materials and can be combined for mass production in industries with relatively low cost [4-6]. In most studies, the application of MAF is used for metals and alloys.

In this study, MAF was applied to the processing of aluminum-based composite with SiC as reinforced, variation of 0.2, 0.4, and 0.6% SiC weight fraction and using 350, 450 and 550°C processing temperatures, the results obtained were analysed to obtain ideal mechanical properties as recommendations for automotive components.

2. Experimental Methods

First step cutted aluminum plate AA 1100 with size 135 mm x 25 mm x 3 mm and 3 mm of thickness. The sample brushed using a wire brush with the same direction. Before the grinding process was done, the material is first cleaned using acetone solution. The acetone to prevent the material from getting dirt so no oxidation and then holes on each edge of the sample, The next step, done grinding and then sprinkling SiC powder 0.2% wt, 0.4% wt, 0.6% wt surface plate to bind the end plate used wire or rivet process. The pre-heating process was done in the muffle furnace used 3 time variables of choice: 350°C, 450°C and 550°C holding time of 1 hour. After preheating process aluminium directly done by pressured of MAF process.

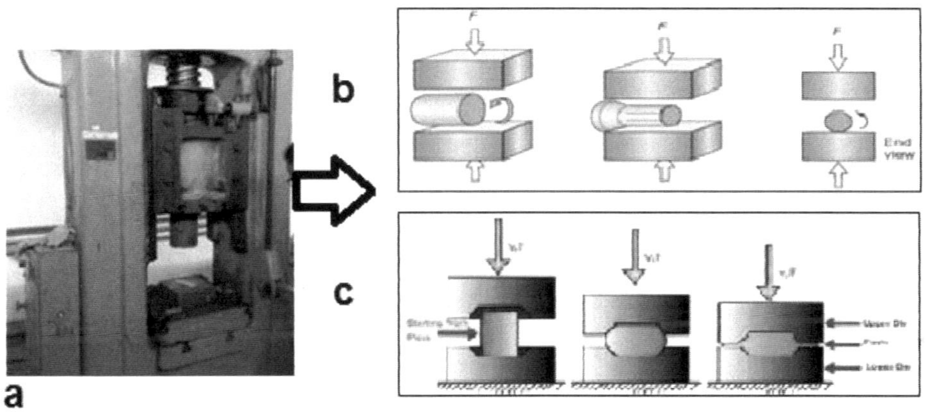

Figure 1. a) Multi-axial forging machine, b) Scheme of open die drop Multi axial forging process, c) Scheme of close die drop multi-axial forging process.

Figure 1 presented MAF machine were used to process aluminum-based composite materials, there are two types of division that are open and close dies of MAF. In this experiment use type open die MAF. The material is placed into the hydraulic press tool using a clamp as a tool, to pick up and remove the sample from the muffle furnace to the hydraulic press machine, as well as the handle while the sample is in forging. How to use this machine is by lifting the lever first to raise the load forging up to 100 kN. MAF process is done as much as 5 passes of pressed with direction rotation 180° final reduction reach 50%. The initial thickness of the material before the MAF process is 6 mm and the final thickness after MAF process reaches 3 mm.

3. Result and Discussion

3.1. Microstructure analysis

Observation of microstructure was conducted using optical microscope 200x magnification and image-J software for calculate of grain size, this is done to see and know the microstructure formed especially formation of bond formed between aluminum as a matrix with SiC particles. The aluminum microstructure as received shows the grain boundary in the shape of equiaxial and the grain size at the initial condition is still large and rough with the average grain size of 10.035 μm, shown in Figure 2a. The aluminium based composite microstructure with 0.6 wt% SiC at 350ºC for 5 pass formed precipitate grains that spread evenly within the matrix with lamellar grain boundaries and had an average grain size of 1.197 μm. Temperature 350ºC in Figure 2 b. it is seen that the sample has a bonding interface, where the interface bonding area contains SiC particles trapped in the interface bonding area. Figure 4.2 (c) shows a temperature condition of 450ºC with a 0.6% wt SiC fraction formed of lamellar granules with an average grain size of 1.81 μm where the number of precipitate grains is not as highest on 350ºC, the spread of precipitation grains at this temperature is spread evenly across in the matrix of this sample the grains begin to shrink and smooth. Figure 4.2 (c) shows a larger bonding interface when compared to the results of Fig. 2. b. Figure 4.2 (d) shows an optical microscope at 550ºC with a fraction of 0.6% wt SiC. The bulb is lamellar with an average buitr size of 1.987 μm, where at this temperature there is a fine prep filipipate and not all parts of the matrix are filled by precipitate grains. this indicates the presence of bonding interfaces and more SiC particles trapped in the interface area.

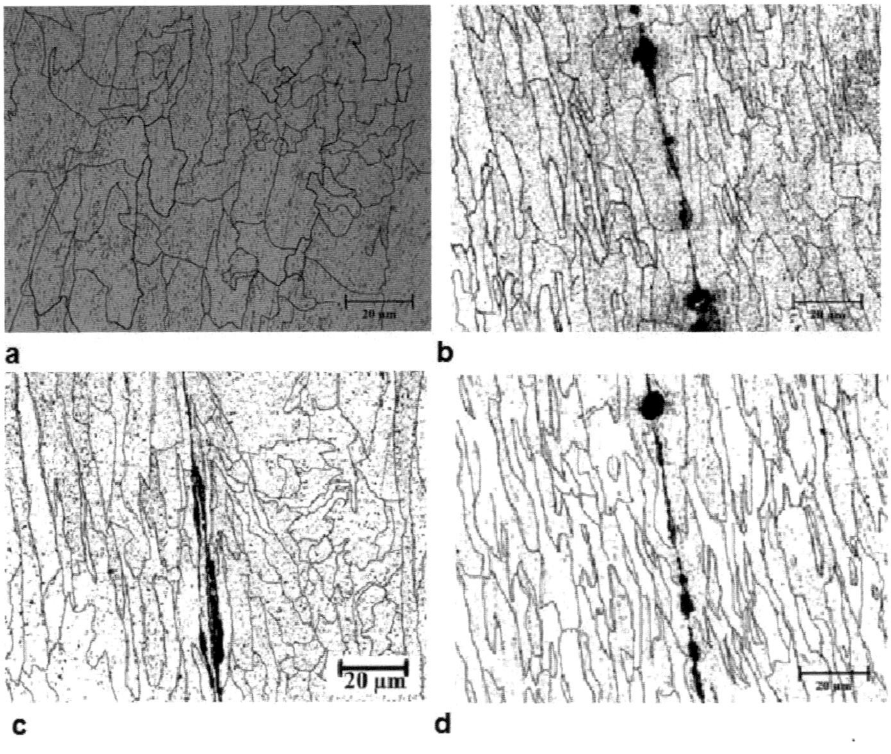

Figure 2. Result of microstructures a) Al as received, b) Al based composites by MAF pre-heating 350°C, c) Al based composites by MAF pre-heating 450°C, and d) Al based composites pre-heating 550°C.

The result of optical microscope in each temperature still has a bonding interface. Addition of SiC on aluminum fraction can increase the strength and hardness value, that can be seen from the grain size produced. Increase of mechanical properties occurs because the reduction of thickness in each sample causes the particles in the interface layer to move, this will result in changes in grain boundary shape from equaxial to lamellar after MAF process. At the same time, the aluminum matrix moves between particles and causes the particles to separate from each other. The mechanical properties of the composites are strongly influenced by the presence of bonding interfaces, the presence of SiC particles in the aluminum layer interface causes a reduction in bond strength and this stage of the particles is not distributed in the matrix. Increasing the number of cycles in the SPD gives the effect of spreading SiC particles from the interface to most of the aluminum layers, all discontinuities in the interface disappear as a result of porosity closure. With the cyclical emphasis the thickness of the layer interface and SiC particles are scattered across most of the aluminum matrix. It show the dispersion of SiC particles in aluminum matrix becomes more uniform [7-9]. In the process of MAF 5 pass shows the result of spreading the grain structure evenly. The grain forming mechanism is part of dynamic recovery, under which conditions small grain sizes remain constant and incorrectly oriented granules will increase. There are two factors supporting the dynamic recovery of Al/SiC composite which is the shear deformation and micro particles size. When composites are exposed to deformation, the presence of non-deformable particles induces a stretch around the region causing the particles to form a new boundary due to the high density of dislocations [10-11].

3.2. Microhardness

The hardness as received was 34.6 HV. The highest hardness 45.62 HV produced by pre-heating process 350°C with fraction 0.6% wt of SiC. While for lowest hardness value reach 37.12 HV at temperature 550°C, with fraction of SiC 0.6 % wt. The results of Al/SiC composite hardness in the MAF process can be seen in Table 1.

Table 1. Formatting sections, subsections, and subsubsections

***Pre Heating* (°C)**	**Percent SiC (% wt)**	**Microhardness (HV)**
As Received	**No**	34.6
350°C	0.2	43.04
	0.4	43.08
	0.6	45.62
450°C	0.2	41.40
	0.4	40.72
	0.6	37.86
550°C	0.2	40.66
	0.4	40.74
	0.6	37.12

Several studies had shown an association between microhardness and microstructure, generated through various SPD processes. To determine the uniformity of deformation, the value of microhardness was measured at cross-section with the distribution that can be expected occured homogeneous microstructure at various SiC fractions. Samples before treatment and after MAF process had a mean hardness of 30-40 HV [7]. At 350°C the mechanical properties are affected by the addition of SiC fraction. In accordance with [10] the addition of SiC plays a very important role in increasing the hardness of a material because the SiC composite has a role as an reinforcement. As the SiC fraction increases, the resulting hardness value decreases. Pre-heating exceeds the optimal limit where the optimum temperature during heating for AA1100 is 0.3Tm – 0.5Tm [12] so that recovery time is saturated where the balance between hardening and softening has been reached. The increase of grain boundaries at low temperatures becomes a barrier where the area is considerably increased rather than coarse grain. Therefore, nanocrystalline metal alloys are expected to have higher mechanical properties than polycrystalline materials of the same composition [13]. Some of factors can be affect for Al based composite when the samples was dry and brushing before it is attached, strain hardening due to friction between the load and the specimen, grain refinement by dynamic recovery. Microhardness increases during initial pass due to strain hardening (based on dislocation density and dislocation effect with each other [14]. Dynamic recovery at high voltage can cause absorption of dislocations to more grain boundaries. Density dislocations of aluminum will decreases with increasing pressure and some free grain from dislocations after large deformations. Hardness of Al/SiC was greater than aluminum samples without the addition of SiC particles, because of resistance in SiC particle movement and strength of the bond between matrix and SiC particles increases. The pass number on SPD produces a more uniform distribution and reduces porosity content [15-16].

3.3. Tensile strength

Temperature of 450°C and 550°C there was a decrease in the tensile strength value of the fraction 0.2 wt to 0.6 wt. The lowest strength value of the temperature is 550°C with SiC fraction 0.4% at 99.804 MPa. Previous researchers have proved that increasing the amount of SiC in the aluminum matrix can increase the tensile strength. It can be seen at 350°C which shows an increase in tensile strength. This is related because SiC acts as an amplifier in the aluminum matrix. The difference in results at 350°C with a temperature of 450°C and 550°C is due to the influence of the preheating temperature where temperatures of 450°C and 550°C experience maximum heating (overheating).

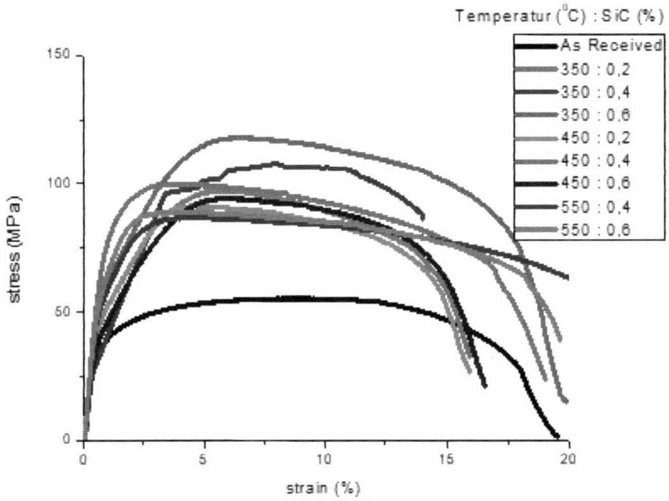

Figure 3. Diagram Stress-Strain Al based composite by MAF.

Temperature on 400°C for composite by SPD is a dynamic recrystallization temperature where the stored energy will produce new nucleation of granules, nucleation starts from the beginning of the formation of dendrites as grain growth [17]. The highest UTS value at 350°C with a SiC fraction of 0.6%. It was explained that increasing the SiC fraction can affect the material strength of the ARB process due to ultrafine grained and strain hardening processes (dislocation tightening). Grain size depends on the level of accumulation of strain that is affected due to dynamic recovery and work temperature which will determine grain growth. Strain hardening causes a large increase in the amount of dislocation density so that the dislocation movement will be difficult to move. When dislocations move, there is a buildup of dislocations at the grain boundaries and lattice defects or because the dislocations move on different slip fields causing interlocking and making dislocations difficult to move [18].

The results of tensile test and elongation of Al/SiC composites, basically the tensile strength value will be inversely proportional of elongation. If the value of tensile strength increases, the results of elongation will decrease but different things can be seen at 350°C where at that temperature the tensile strength value is directly proportional to elongation seen increasing while at temperatures 450°C and 550°C the value of tensile strength and elongation decreases can be seen in Figure 3. The value of tensile strength and elongation decreased with increasing SiC because the SiC particles had three main effects, namely (1) during the tensile test, the SiC particles acting independently of the dislocation movement resulted in increased strength, (2) the presence of SiC particles in the aluminum matrix resulted in more dislocations to gather in their area, because the particles cannot be changed and the matrix undergoes plastic deformation so that the strain mismatch is produced between the particles and the matrix. Strain mismatch causes dislocations to be stored near the particle-matrix interface [9] (3) SiC tensile strength is about 27 times greater than aluminum, therefore, dislocation density on the matrix near the interface between aluminum particles and SiC increases [14]. When the SiC phase in the matrix increases the dislocation movement will be obstructed which causes a decrease in elongation value. If the SiC particles in a brittle state cause matrix with particle constraints that can change the local state of stress.

4. Conlussion

Based on the research that has been done, it can be concluded that:

1. Hardness decreases as the preheating temperature increases because the higher of preheating temperature produces the larger of grain size.
2. Tensile strength increases by addition of SiC particles. Elongation value cannot be used as a reference because the data produced varies. Eelongation value due to uneven spread of grain.
3. The pre-heating of temperature produced the larger the resulting bonding interface diamètres. This is evidenced by the increased bonding interface diameter of 36.304 μm

References

[1] Marklines A 2017 *A. I. P. Indonesia - Flash report*, Sales volume
[2] Brian N 2007 *Handbook of Valves and Actuators: Valves Manual International 1st Edition*
[3] Rezayat M, Akbarzadeh A, and Owhadi A 2012 *The Minerals, Metals & Materials Society and ASM International* **43A (6)** 2085-2093
[4] Zrnik J, Dobatkin S, and Mamuzic I 2008 *Metalurgija* **47 (3)** 211-216
[5] Agus P, Lauri K, Kaspar K, Renno V and Jaana-Kateriina G 2014 *Key Engineering Materials* **604** 273-276
[6] Agus P, Lauri K, and Renno V 2016 Proceedings of the Estonian Academy of Sciences **65 2** 132–137
[7] Montazeri and Pour M Materials Science and Engineering A**639**. 705-716 (2015)
[8] Alizadeh M and Paydar M H 2010 *Journal of Alloys and Compounds* **492 (1–2)** 231–235
[9] Majid N, Amir H, and Moch T 2015 *Ceramics International*
[10] Sajjad A 2011 *Materials Science and Engineering A* **528 (13–14)** 4462–4467
[11] Sajjad A 2016 *Metallurgical and Materials Transactions A: Physical Metallurgy and Materials Science* **47 (10)** 5136–5145
[12] Satish D R, Feyissa F, and Kumar D R 2018 *International Journal of Materials, Mechanics and Manufacturing* **6 (2)** 123–126
[13] Gheorghe G and Carmela G 2012 *Metallurgy and Materials Science* **2** 10–19
[14] Wagih A 2018 *Journal of Alloys and Compounds* **752** 137–147
[15] Chang C P, Sun P L, and Kao P W 2000 *Acta Materialia* **48 (13)** 3377–3385
[16] Agus P, Lauri K, and Renno V 2015 *Advanced Materials Research* **1123** 114-118
[17] Agus P, Lauri K, and Renno V 2015 *Procedia Chemistry* **16** 473-479
[18] Saito Y, Utsunomiya H, Tsuji N, and Sakai T 1999 *Acta Metallurgica* **47** 579 – 583

The comparison of the alkali-treated and acid-treated naturally mined Philippine zeolite for adsorption of heavy metals in highly polluted waters

E C G Dignos, K E A Gabejan, E M Olegario-Sanchez, and H D Mendoza

Department of Mining, Metallurgy, and Materials Engineering, University of the Philippines, Diliman, 1101 Philippines

Email: eleanor.olegario_sanchez@upd.edu.ph

Abstract. Natural zeolite can be modified by single or combined treatment of heating and chemical modification using acids, bases, and inorganic salts. In this study, the Philippine natural zeolite (PNZ) was modified with alkali and acidic solution. The alkali-treatment and acid treatment used sodium hydroxide (NaOH) and acetic acid (CH_3COOH) as chemical modifier respectively. Various concentration of the solution (1M and 2M) and temperature (25°C and 50°C) was used during modification of the PNZ. The XRD of Na-PNZ and Ac-PNZ shows mineralogy of clinoptilolite and mordenite phases of the natural zeolite. The SEM images shows porous and plate-like morphologies of both modified materials. These property shows surfaces for adsorption of metals from wastewater solutions. The Na-PNZ and Ac-PNZ was used as adsorbents for the treatment of synthesized copper solution. The copper uptake of Na-PNZ from an aqueous solution is higher compared to the Ac-PNZ at 99.58%. The adsorption capacity sequence of the modified materials are as follows, 1M Na-PNZ > 2M Na-PNZ > 1M Ac-PNZ > 2M Ac-PNZ. The temperature used in the modification appears to have a significant effect on the adsorption capacity of the modified PNZ samples, the increase in temperatures resulted to a decrease in the uptake of copper.

1. Introduction

Natural zeolite can be modified by single or combined treatment of heating and chemical modification using acids, bases, and inorganic salts. Chemical and thermal treatment of zeolite may result in cation migration and thus affect the cation location and pore opening. Sorbent properties are manipulated during zeolite modification by the application of "pore engineering", which is a popular term for the methods used in modifying zeolites [1]. The adsorption characteristics of any zeolite are dependent upon the detailed chemical and structural properties of the adsorbent. The composition, the Si/Al ratio, cation type, number and location also influence the adsorption. These properties can be further modified by chemical treatments to improve the separation efficiency of the unmodified, raw natural zeolite. Treatments by acids and bases, and surfactant impregnation by ion exchange are among the commonly employed methods to modify and change the hydrophilic/hydrophobic properties of the zeolites [1].

Alkali treatment of zeolites causes desilication, resulting to the lowering of the Si/Al ratio (SAR) and an increase in the relative concentration of cations in the framework. Due to a higher negative charge within the framework, the metal adsorption capacity of the zeolite thus improves. A research by Ouki et al. [4], was conducted to investigate the effect of alkali treatment to the exchange capacity of clinoptilolite and chabazite, two known natural zeolites. The modified zeolites were used to treat samples containing Pb^{2+} and Cd^{2+} ions. Results show an increase in the exchange capacity of both zeolites as compared to their natural counterpart [3].

Different treatments however result to different effects on the exchange capacity of the zeolites. Based on the same study by Ouki [4] the zeolites which undergone alkaline treatment has an inferior or nearly equivalent exchange capacity as compared to zeolites treated with salt solution at ambient temperature. A research by Panayotava, studying the uptake of Cu^{2+} of the alkaline and salt treated clinoptilolite at ambient temperatures reported the same result. The Cu^{2+}exchange capacity of the alkaline and salt treated zeolite was said to be equivalent. [3-4] The modification using the combination of salt and alkali solution resulted also to mixed results depending on the type of zeolite, the metal ion type, and the metal ion concentration. [3].

Acid treatment on natural zeolites results to the removal of impurities blocking the pores of the zeolites, thus resulting to further elimination of cations to change into H-form and finally dealuminate the structure. Acid wash is said to have great influence on the effective pore volume of the zeolite as well to its surface area. However, zeolite treatment using acid has a detrimental effect on the minerals' metal adsorption capacity. Acid treatment causes an increase in the Si/Al ratio in zeolite thus decreasing its net negative charge. Also, in the event of a natural zeolite having K^+ ions, acid treatment increases the relative concentration of K^+thus reducing the easily exchangeable Na^+ ions because of the decrease of the Na^+/K^+ ratio. The research conducted by Panayotova [5] reported a lower exchange capacity of acid treated clinoptilolite as compared to untreated clinoptilolite. [1-2].

To summarize, researches show a similar trend on the effect of the different modification on the exchange capacity and porosity of the zeolites. The combination of salt and alkaline treatments, at ambient temperature, shows the collaboration in increasing the metal adsorption capacity of the zeolite. Natural zeolites pre-treated with salt and subsequently treated with alkali solution reported a superior exchange capacity as compared to zeolites treated only with salt or only with alkali solution. However, treatment using strong acid decreases the metal adsorption capacity of the zeolite. Acid treated zeolites has an inferior exchange capacity to that of untreated or natural zeolites. [1][2][3]

The resulting trend then shows an exchange capacity or metal adsorption capacity in the following order: salt treated > alkali treated > untreated/natural zeolite > acid treated, at ambient temperatures [3]

This study will explore the applicability of naturally mined Philippine zeolite (PNZ) modified by alkali (NaOH) and acid treatment (CH3COOH) to the treatment of contaminated wastewater with from industrial effluents.

2. Experimental

The material was acquired from a zeolite mine, SAILE Industries, Inc., located in Mangatarem, Pangasinan, Philippines. The zeolite sample was dried and grinded to achieve 80% passing of 0.5mm particle size using a 0.5 mm laboratory grade test sieve. It was then washed and dried at 110°C for 12h. The chemicals used were technical grade, supplied by Alyson's Chemical Enterprises, Inc. such as NaOH pellets, $CuSO_4$, and CH_3COOH. All solutions were prepared using a deionized water.

2.1 *Modification of the PNZ*

The PNZ was modified into two treatments – acid and alkaline modification using acetic acid (1M and 2M CH_3COOH) and sodium hydroxide (1M and 2M NaOH) respectively. Thirty (30) grams of PNZ was mixed in a 200 mL of the modifying solution at 200 rpm at RT for 4h. The resulting modified materials are named as Na-PNZ and Ac-PNZ for alkali treated and acid treated materials respectively.

2.2 *Characterization of modified PNZ*

The surface morphology of the modified materials was analysed using Hitachi S3400N scanning electron microscope. The mineralogy of the samples was determined using X-ray diffractometer (Shimadzu Maxima XRD-7000).

2.3 *Batch Adsorption Studies*
The modified PNZ materials were assessed to determine its adsorption efficiency for copper. Five (5) grams of the Na-PNZ and Ac-PNZ samples were mixed separately with synthesized 50mL of synthesized copper solution (50 ppm). The solutions were mixed at 200 rpm for one hour at ambient temperature using a Corning hot plate. All the eluate was tested for copper concentration using Shimadzu AA6300 Atomic Absorption Spectrometer.

3. Results and discussion

3.1 Morphology of the modified PNZ
The SEM images of Na-PNZ and Ac-PNZ using a scanning electron microscope are shown in Figure 1. The Na-PNZ exhibits a platy morphology. The Ac-PNZ on the other hand exhibits a very smooth globular morphology. From these morphologies, the adsorption of each sample can be deduced. Since Na-PNZ exhibits a rough surface, the adsorption sites are expected to be higher, as compared to the Ac-PNZ with a smoother surface. This translates to the results that the Na-PNZ have high uptake as compared to the Ac-PNZ shown in Figure 3.

Figure 1. SEM images of (a) Na-PNZ and (b) Ac-PNZ at 10,000x magnification.

3.2 X-Ray diffraction patterns
Shown in Fig. 2, the major crystalline mineral phases of the Na-PNZ and Ac-PNZ are clinoptilolite (Na,K,Ca)2-$3Al_3(Al,Si)_2Si_{13}O_{36}\cdot 12H_2O$ and mordenite $(Ca, Na_2, K_2)Al_2Si_{10}O_{24}$.

There are four distinct peaks observed in the X-ray diffraction patterns of Na-PNZ and Ac-PNZ. These peaks are similarly observed at 10, 22.2 and 27.8 and 35 2θ values. Guide lines superimposed on these peaks are shown in the Figure 5. There are shifts of some peaks attributed to the PNZ modification processes which resulted from the changes in the SAR of the modified materials. Acidic and alkaline modifications altered the SARof the PNZ detected in the XRD analyses. Acid treatment causes an increase in the SAR and decreases its net negative charge [1-2], while alkaline treatment decreases the SAR of the PNZ.

3.3 The effect of acid and base treatment on the adsorption capacity and metal uptake
Fig. 3 shows the effect of the modification of PNZ with acidic and alkaline treatments of varying concentrations of acetic acid (CHCOOH) and sodium hydroxide (NaOH), and modification temperatures. Results show high copper recovery using the NaOH modified zeolite, while a significantly lower copper uptake was observed for PNZ modified using acetic acid.

The Na-PNZ recorded an uptake of more than 99% copper uptake. Under alkaline conditions, the treatment causes desilication for the zeolite sample, which in turn results to the lowering of the silicon-to-aluminum ratio, SAR, and an increase in the relative concentration of cations in the framework. Due to a higher negative charge within the framework, the metal adsorption capacity of the zeolite thus improves. According to Ouki [4], different treatment however results to different exchange capacity of the zeolite. Moreover, if the contact time is not enough, the desired performance in recovery is not achieved and not exhibited.

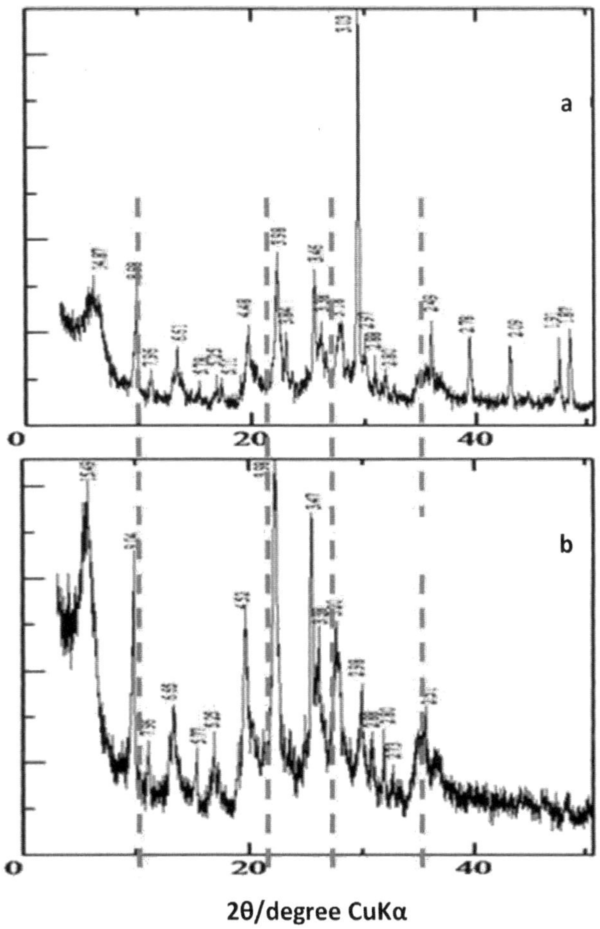

Figure 2. XRD patterns of (a) Na-PNZ and (b) Ac-PNZ at 10,000x magnification.

Acid treatment on natural zeolites improves the pore volume as well to its surface area of the material [2]. Soaking with acid results to the removal of impurities blocking the pores of the zeolites, thus resulting to further elimination of cations to change into H-form and finally dealuminate the structure. Acid treatment causes an increase in the silicon-to-aluminum ratio, SAR, thus decreasing its net negative charge. In the event of a natural zeolite having K+ ions, acid treatment increases the relative concentration of K+, reducing the easily exchangeable Na+ ions because of the decrease of the Na+/K+ ratio. As shown in Fig. 3, as the concentration of the modifying solution of acetic acid increases, the copper uptake decreases. The acid treatment exhibited detrimental effect on the adsorption capacity of the Ac-PNZ.

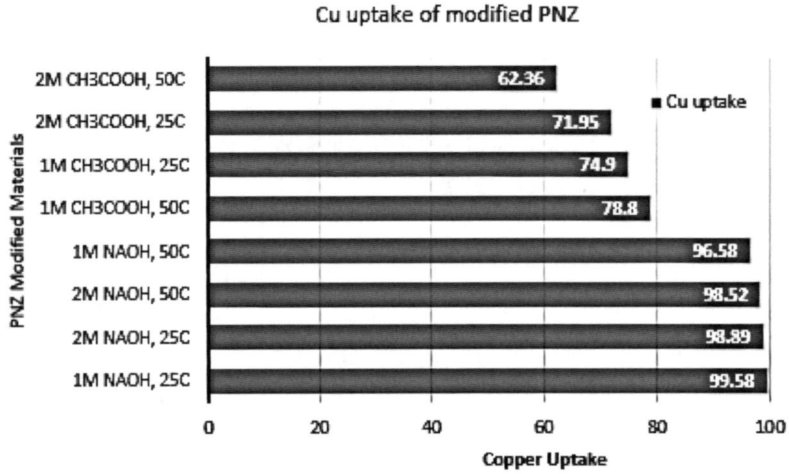

Figure 3. Copper uptake of different types of modified PNZ samples (Na-PNZ and Ac-PNZ) at varying concentrations and temperatures.

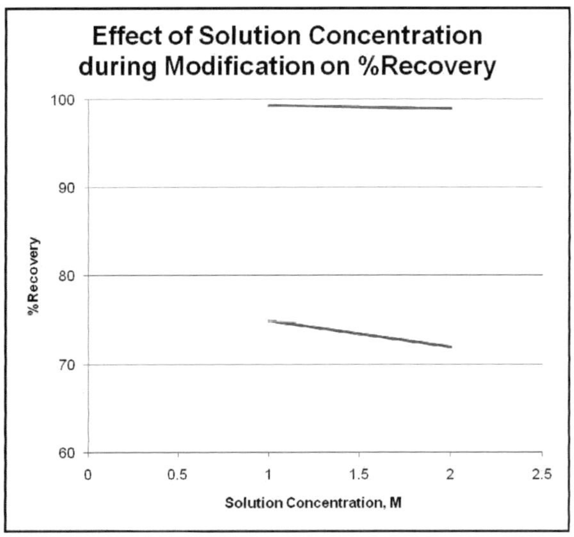

Figure 4. Effect of solution concentration used in modification to the %recovery of Cu of modified zeolite. Red – Ac-PNZ (CH3COOH), Blue – Na-PNZ (NaOH).

It is also important to note that increasing the concentration of the solution, either acidic or basic, results to a decrease in the metal recovery of the zeolite samples. Shown in Figure 4, the Na-PNZ (1M NaOH) held at ambient temperature has an uptake of 99.57% which decreased to 98.89% as the concentration of NaOH is doubled. Similar with the Ac-PNZ samples, the recovery of Ac-PNZ (1M CHCOOH) at ambient temperature is 74.90% but decreased to 71.94% at an increase of acid concentration to 2M at the same temperature. This is attributed to the structural changes occurring during the modification process, changing the SAR, and consequently the net charge of the material resulting to an improved or decreased heavy metal recovery from the test solution [1-2].

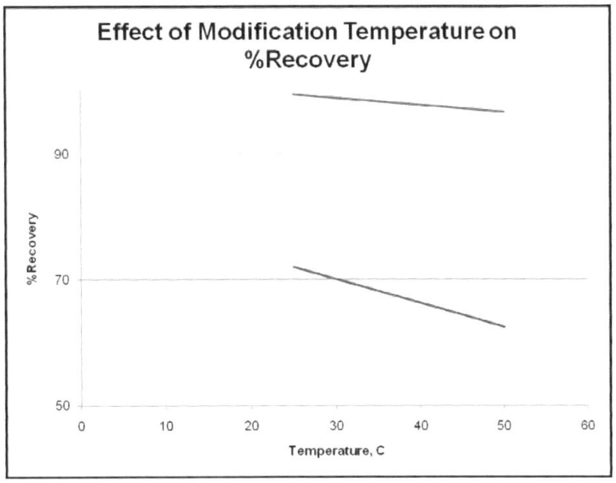

Figure 5. Effect of temperature on the recovery of copper of modified zeolites. Red – Ac-PNZ (CH_3COOH), Blue – Na-PNZ (NaOH)

In Figure 5, the temperature used in the modification technique appears to have a significant effect on the adsorption capacity of the modified PNZ samples but is not pronounced enough to be observed. For 1M NaOH, 2M NaOH and 2M CHCOOH - modified PNZ, the increase in temperature at constant acid/base concentrations resulted to a decrease in the uptake of copper from the solution.

The order of the performance of modified PNZ samples according to the copper uptake (50 ppm copper concentration) are arranged as follows, Na-PNZ (1M) >>> Ac-PNZ (1M). This order agrees to the current literature available that acid treated zeolite exhibits inferior adsorption capacity to that of alkali-treated natural zeolites [1].

4. Conclusion

The PNZ was successfully modified using acid-treatment with CH_3COOH (Ac-PNZ) and alkali-treatment with NaOH (Na-PNZ). It was observed that modification by acid-treatment decreases the metal uptake of the adsorbent, PNZ, due to the change in the crystal framework (shift of XRD peaks) of the PNZ brought about by the acid modification. The surface morphology of Na-PNZ and Ac-PNZ indicates a rough and smooth surface respectively. The Na-PNZ modified with 1M NaOH, 25°C has a higher metal uptake compared with the Ac-PNZ.

Acknowledgment

The researchers would like to acknowledge SAILE Industries Inc., for providing us the necessary zeolite samples for the project. The faculty and staff of the Department of Mining, Metallurgical and Materials of the University of the Philippines in Diliman, Quezon City for the facility use in the material modification, characterization and adsorption studies of the modified PNZ adsorbent materials.

References

[1] Kini G C 2006 *Characterization of Synthetic Zeolites Optimized for Heavy Metal Uptake from Wastewater*
[2] Dubey, Abha, Goyal, Deepti 2013 *Zeolites in Wastewater Treatment* Royal Society of Chemistry. Knovel
[3] Margeta K, Logar N K, Šiljeg M, and Farkas A 2013 *Natural Zeolites in Water Treatment – How Effective is Their Use, Water Treatment Walid Elshorbagy and Rezaul Kabir Chowdhury* Intech Open

[4] Ouki S K and Kavannagh M 1999 *Water Science Technology* **39 (10-11)** 115-122
[5] Panayotova M and Velikov B 2002 *Journal of Environmental Science and Health Part A* **37 (2)** 139-147

IOP Conf. Series: Materials Science and Engineering **478** (2019) 012031 doi:10.1088/1757-899X/478/1/012031

Effect of post-heat treatment on the creep behaviour at 550°C of NiCr+Cr_3C_2 coating produced by HVOF on the SUS 304 substrate

E Martides[1], A Rijaluddin[2], D H Prajitno[3], E Junianto[1], and B Prawara[1]

[1]Research Center for Electrical Power and Mechatronics, LIPI
[2]Metallurgical Engineering, UNJANI
[3]Science Center and Nuclear Technology Application, BATAN

Email: erie001@lipi.go.id

Abstract. Austenitic Stainless Steel Grade 304 which is used for power plants boiler pipes often occurs creep phenomenon that caused a decrease in performance due to work at elevated temperatures. Surface coating of SUS 304 with Metal Matrix Composites Coating feedstock material 60%NiCr + 40%Cr_3C_2 is deposited using high-velocity oxygen fuel (HVOF) technique. The post-annealing process at 600°C at an argon atmosphere, holding time for 3 hours, and cooling in the furnace. Creep resistance test was carried out at a temperature of 550°C with a load of 54 kg for 10080 hours on three specimens, i.e non-spray, as-spray, as-annealed, then followed by microstructure hardness testing and porosity observation. As-annealed specimens were assessed that the lowest creep resistance value, increased hardness in the coating and substrate and decreased the percentage of porosity in the coating.

1. Introduction

Austenitic stainless steel grade 304 is a non-magnetic metal widely used for super-heaters boiler tube in power plants applications, because it has properties that are good formability and weldability without annealing treatment before, and can be used for cryogenic and elevated temperatures applications [1]. SUS 304 has single phase and face-centered cubic (FCC) crystals as the influence of the dominant elements namely Cr and Ni which are intentionally obtained on very large amounts of steel for corrosion resistance and high temperatures stability [2]. However, the presence of high Cr and C elements in SUS 304 is a trigger for sensitization phenomenon by carbides forming, and subsequent stress corrosion cracking in the super-heaters boiler in some power plants [3].

Power plant systems in Indonesia have 4 types based on temperature and pressure operational, i.e subcritical, supercritical, and ultra-supercritical, with operational temperatures between 540°C-760°C and operational operations between 170-350 bars [4]. The application of SUS 304 material at elevated temperatures with a long duration, besides causing sensitization and corrosion phenomena, it also will increase in length depending on the time (creep). Failure on the boiler tube because of localized prolonged heating will occur wall thinning, reducing thickness, fish-mouth opening as visually. As the impact, it will very dangerous for safety and also economic losses because the power plant will be shut down to take corrective action [5].

To avoid sensitization, corrosion and the creep rate of SUS 304, special treatment is needed for SUS 304 materials to be resistant to creep and high-temperature corrosion, so that the material has a longer service life. A fairly effective treatment can be done by adding coatings on the SUS 304 surface with Metal Matrix Composites powder coating using High-Velocity Oxygen Fuel (HVOF) method and continued with post-heat treatment process.

Metal Matrix Composite coating is often used as a protective coating because its properties are suitable for stainless steel part in elevated temperature applications. Metal that used as a matrix usually has high ductility and corrosion resistance, while the reinforcement materials have high hardness and melting point temperature. HVOF sprayed NiCr-Cr_3C_2 coatings are one of the most popular candidates for protection of material from high-temperature erosion and have been successfully used to protect parts with an erosive and corrosive environment at service temperature up to 800°C. Utilization of carbide, oxide and cermets as a reinforced materials will increase not only hardness value of coating, but also wear and corrosion resistance value which is ideally suited as protective coating in corrosive and erosive environments [6].

Heat treatment after HVOF thermal spray coating process will greatly affects creep properties, because of residual stresses related to the coating process and due to mismatch of thermal expansion and elastic modulus between substrate and coating [7]. Furthermore, the HVOF coating results also will form the porosity, the fraction of melted and partially melted particles which are relatively difficult to avoid, which can degrade the quality of the coating and cause weakening of cohesion bonds on intersplats that are susceptible to corrosion attacks. As a anticipation, post-heat treatment can be carried out to improve microstructure, increase the adhesion of coatings, densification of layers, release residual stress, reduce porosity, increase coating uniformity, reduce oxide in layers, increase carbide precipitates, modify chemical composition.

To sum up, deposition metal matrix composite into 304 stainless steel surface will increase mechanical properties in particular hardness value and creep resistance especially when 304 stainless steel parts are used in aggressive environment. The important thing which determines quality of coatings are coating material selection and coating deposition process.

2. Experimental

Surface preparation of boiler pipes in accordance with SUS 304 specifications as a substrate that is cutting specimens adjusted to the specimen creep resistance test and giving a radius at the sharp end to avoid stress concentration. The blasting process uses 24 mesh of white aluminum oxide powder to clean the surface of the substrate from the impurity and to increase the adhesion of the coating (adhesion).

Ball mill process with 272 rpm rotation speed and sieving process carried out on NiCr powder as a matrix to obtain 45 μm powder sizes. The reinforcing powder of Cr_3C_2 was added according to the predetermined ratio, NiCr (60%): Cr_3C_2 (40%). then mixed in a V-type Blending machine for 8 hours with a rotation speed of 64 rpm. Blending powder is heated in furnace at 100°C for 15 hours to remove the moisture content in the powder. Substrate coating using the method of High Velocity Oxy Fuel (HVOF) to deposit each variation of powder by coating gradually with the target bond coat 50-75μm and top coat 125-150 μm with constant parameters namely; spraying distance of 200 mm, gun angle and substrate of 90°, and pre-heat on the substrate surface at a maximum of 100°C.

In this study, the post heat treatment experiment was carried out in a tube furnace in an argon gas atmosphere with a pressure of 1.25 kgf / cm^2 with a rate of 0.6 ° C / s at a temperature of 600°C. After reaching the desired temperature, detention was carried out for 3 hours as shown at figure 2.

Creep resistance testing was performed to determine the durability of 3 specimens, i.e non-spray, as-sprayed and as-annealed. The parameters of the creep resistance test on the three specimens were constantly, ie using a static load of 54 kg with a test temperature of 550 °C and long dimension changes was calculated and observed every 15 minutes. Coating characterization was conducted by micro vickers method to determine the hardness distribution in coating with 3 times indentation in the cross-section with 50 gram load. Microstructure of coating and porosity observation are observed in cross section using SEM method, with surface preparation by grinding and polishing uses abrasive paper.

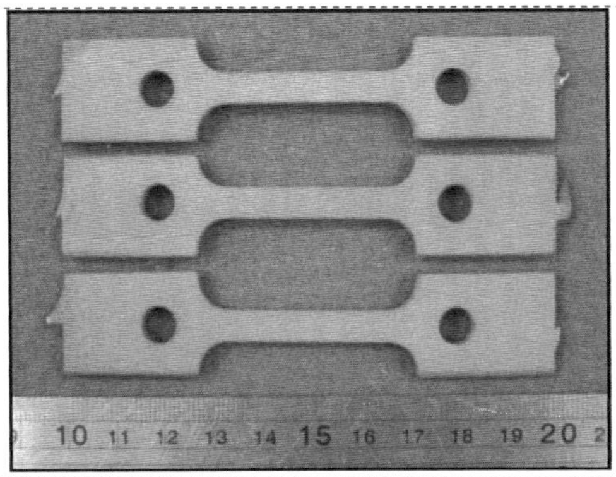

Figure 1. creep resistance test specimens after HVOF process

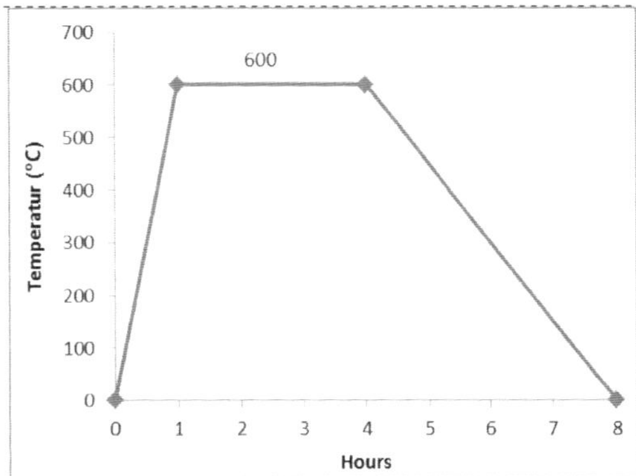

Figure 2. Schematic of post-annealing process after HVOF process

The quality of MMC coating that are already attached on 304 stainless steel surface can be seen from the characterization results using microhardness tester, SEM, and creep resistance machine methods. In this experiment, creep value is the most important mechanical properties of coating that must be observed to prove quality coating, apart from hardness value. The hardness value of MMC coating will increase on average up to 59% of the initial hardness value of stainless steel, if there is no defect or porosity in large quantities [8]. To determine the presence of defects or porosity in the coating can be seen by using optical microscopy which is strongly determined by the parameters that used to deposition coating material into stainless steel surface. Creep value will be related to hardness value, and presence of porosities

3. Result and Discussion

The results of creep resistance test with a temperature of 550 ° C, a load of 54 kg can been seen in strain diagram of the time shown in figure 3 and 4. The minimum value of creep rate (steady state creep) on the specimens from the test creep at a temperature of 550°C and a load of 54 kg is determined by obtaining the slope of the curve (secondary creep). Where at this stage, the creep rate will decrease over time until it reaches an almost stable state, so that the rate of creep undergoes a small change in time.

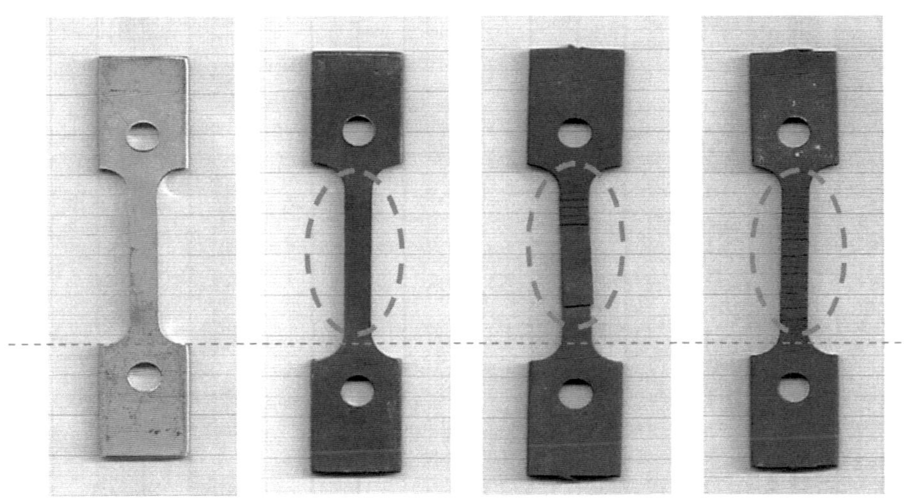

(a) (b) (c) (d)

Figure 3. Creep resistance specimens (a) initial (b) *non-spraying* (c) *as-spraying* (d) *as-annealed*

As-annealed specimen has the lowest value of creep resistance which is the lowest elongation and strain, the smallest percent reduction in cross-sectional area and the smallest creep rate value. The creep resistance of the as-annealed specimen is affected by the post-heat treatment which improves the adhesion that can be seen from the visual evidence in figure 3. Post-heat treatment process makes the coating more dense, which increases adhesion automatically due to external energy in the form of heat energy that pushes the atoms to move inward. The crack direction in the necking area occurs following the stress that occurs in the specimen, namely the pulling force.

After creep resistance test, coating on as-sprayed specimen falls out and peels from the substrate that occurs in the necking area. It happens because the specimen is in high pressure and temperature conditions, and finally undergoes coating loss, but in the as-annealed specimen has undergone post-annealing process there is no loss as in the second sample, only in this third sample cracking occurs in the coating.

Figure 4 shows elongation after creep resistance test. Elongation in non-spray specimen is 10.20 mm with strain (ε) of 0.10111023. The strain value is obtained from the calculation of the change in length (Δl) divided by the initial length of the specimen which is 100.88 mm. After the stretching test process has been calculated, the reduction in cross-sectional area that occurs in the first sample is obtained by a percentage reduction of 8.05%. While from the calculation of the creep rate obtained the value of the creep rate at steady state creep is 0.0058 mm / hour which means that the non-spray specimens experience a creep of 0.0058 per hour. Using the same calculation, the as-sprayed specimen was 9.86 mm long with a stretch value of 0.098501499 which had an initial length of 100.10 mm. The reduction in cross-sectional area on the second specimen is 4.59%, of course, has a reduction in percent reduction than the first specimen. The creep rate on this as-sprayed specimen is 0.0033 mm / hour. Whereas the as-annealed specimen experienced a 9.51 mm long stretch with a strain (ε) 0.0938 which had an initial length of 101.30 mm. Reduction in cross-sectional area in as-annealed specimens has better creep resistance than before, which is 2.44. The creep rate value in the third sample is also lower than the two previous samples with the creep rate of 0.0029 mm / hour.

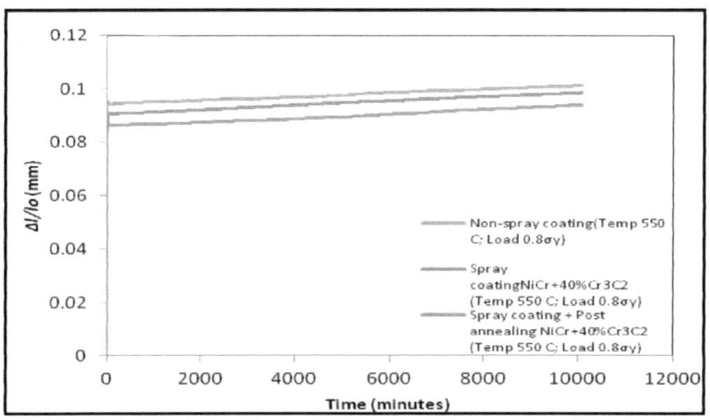

Figure 4. Strain vs time diagrams of creep specimens test

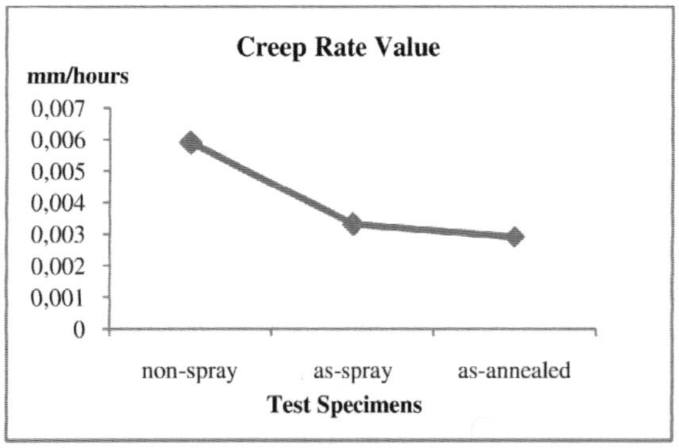

Figure 5. Creep rate value of MMC coating on 304 Stainless Steel

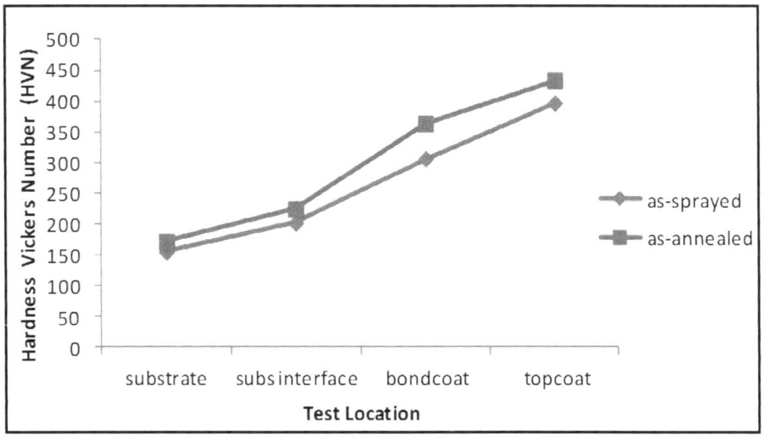

Figure 6. Hardness Vickers value of MMC Coating

From Vickers hardness test results, it was found that the increasing of hardness value on the post-annealing results with hardness on the topcoat was 432.97 HVN while the unheated one had a hardness of 396.68 HVN, in bond coat as-spray had a hardness of 306.25 HVN, whereas in bond coat the annealed axles have a hardness of 363.64 HVN. From figure 6, it can be seen the comparison of the hardness between the as-sprayed and the as-annealed specimens. Likewise, the order of hardness from the substrate test point until the topcoat experienced an increase in violence. So, the curve above shows a linear increase in violence. Increasing of hardness value to the coating caused by post-annealing reduces the residual stress in the specimen [9]. This residual stress is in the form of high pressure from the result of spraying powder melt with high speed. Other causes are due to porosity decreasing in the coating will increases violence. The influence of post-heat treatment also affects the formation of fine crystalline as to produce grain uniformity [10].

The study of porosity weight and porosity sizes was characterized using the Image-J program from SEM. The results of this characterization resulted in a porosity decreased at as-annealed specimen. The initial porosity (as-sprayed) is 1.24% with an average porosity size of 12.68 μm, while the porosity of the as-annealed sample is 0.59% with an average porosity size of 6.19 μm. The results of visual observations are also seen from the decrease in porosity in the SEM results which are shown in figures 7 and 8.

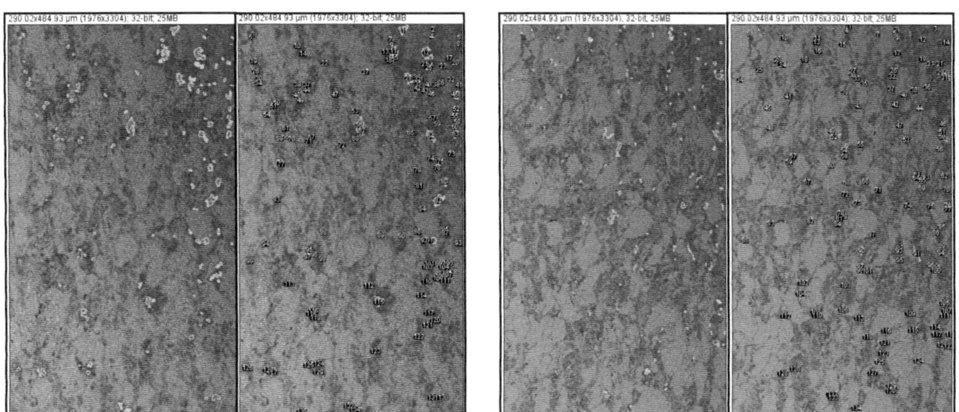

(a) (b)

Figure 7. Porosity measurements using imageJ (a) as-sprayed specimen ; (b) as-annealed specimen

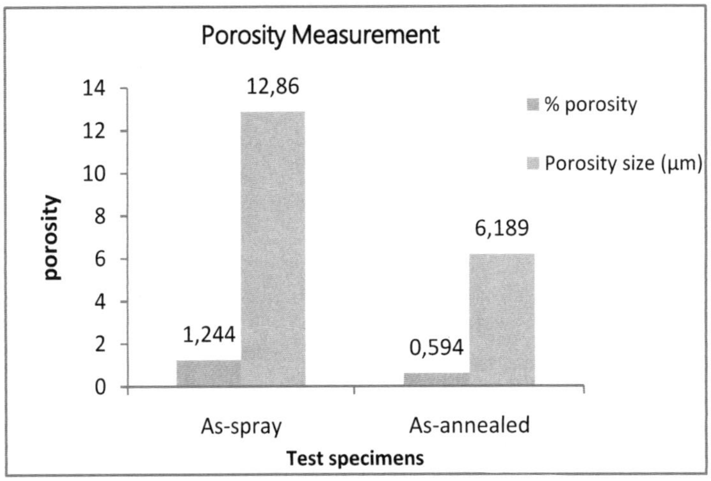

Figure 8. Coating porosity measurement diagram

In the thermal spray coating processes, porosity occurs due to the presence of oxygen gas that trapped in the coating, caused by distance of spraying too far or the presence of contaminants such as oil, grease, air. Besides that, porosity is caused by inadequate pressure on every powder coating to the substrate apart from distance between gun and substrate and powder feed rate, so that the splats shape not uniform and unable to fill the empty space underneath [11]. Basically, the effect of post-annealing on decreasing porosity is due to the presence of heat in the form of energy which pushes the layer more tightly so as to eliminate the porosity in the layer [12,13].

4. Conclusions

Post-heat treatment process has influences to mechanical properties of 304 stainless steel. As-annealed specimen has the smallest creep rate of 0.0029 mm/hour with elongation of 9.51 mm and the highest hardness value of 363.64 HVN (bondcoat) ,432.97 HVN (topcoat). Post-heat treatment process has an effect on the decrease in percent of porosity, where the as-spray sample has 1.24% porosity with an average size of porosity of 12.68 μm, while the porosity of the as-annealed sample is 0.59% with an average size of porosity of 6.19 μm. Reduction of cross-sectional area with post-annealing results produces the smallest value of 2.44%.

Acknowledgments

The authors would like to thank to all researcher and technicians at Research Centre for Electrical Power and Mechatronics, All of lectures and students in Metllurgical Engineering, UNJANI, and all of researcher in BATAN. This work is a part of research that have been supporting by LIPI Priority and Advanced Research Projects, sub program development of material, energy and manufacturing engineering

References

[1] Atlas Steel 2013 *Stainless Steel Grade Datashee* Atlas Steel Technical Department Australia
[2] Aalco 2017*Stainless Steel: Alloys in Stainless Steel Elements* Aalco Metals Ltd Cobham-UK
[3] Saad A A, Mahallawi I E I, Abdel-Karim R, Rashad R 2009 *Engineering Failure Analysis* **16** 433-448
[4] Di Gianfrancesco A 2016 *Materials for Ultra-Supercritical and Advanced Ultra-Supercritical Power Plants* **Woodhead Publishing** Cambridge-UK
[5] Parit A N, Tadamalle A P, Ramaswamy 2014 *Thermal stress and Creep Analysis of Failed Tube of Secondary Super Heater* (5 th International & 26th All India Manufacturing Technology, Design and Research Conference (AIMTDR 2014) December 12th–14thIIT Guwahati, Assam, India)
[6] Bhatia R, Singh H, and Sidhu B S 2012 *Asia Journal of Engineering and applied Technology* **1(2)**

[7] Sudhansu B 2007 *Chapter 6-Oxidation and Corrosion Resistant Coatings* (High Temperature Coatings) 71-154
[8] Tao K, Zhou X, Cui H, and Zhang J 2008 *Material Transactions* **49(9)** 2159 - 2162
[9] Hung-Hua, Ting-Yi H, Tzu-Te L, and Ming-Der G *Int. J. Electrochem. Sci.* **13** 9399 – 9415
[10] Xiaomei C, Jiangang Q, and Xiaotian H 2017 Coating **7(9)** 146
[11] Gustavo B S 2015 *Soldagem & Inspeção* 238-252
[12] Jianguo Z and Kang M 2014 *Theoretical & Applied Mechanics Letters* **4** 021008
[13] Öz A and Samur R 2013 *Metalurgija* **52** 368-370

Effect of repeated TIG welding cycles on microstructure evolution and pitting corrosion behavior of UNS 32760 super duplex stainless steel

Alfian, R Riastuti, and Winarto

Department of Metallurgy and Materials Engineering, Faculty of Engineering, University of Indonesia

Email: alfianzero@gmail.com

Abstract. Microstructure evolution and pitting corrosion behavior in repeated TIG welding cycles of UNS32760 super duplex stainless steel were studied. The microstructural evolution was observed through optical electron microscope. Meanwhile, pitting corrosion was investigated by means of potentiodynamic polarization and gravimetric test. The specimens were evaluated to simulate repeated welding cycles of original weld, Repair-1 (R1), Repair-2 (R2) and Repair-3 (R3) respectively.The result show that the chromium nitride precipitates starts appear in heat-affected zone R-2 welding cycles which followed by the growth of metastable pits.

1. Introduction

Super Duplex Stainless Steel (SDSS) is a material that is formed by a unique combination of ferrite and austenite microstructure that ideally has the same large volume fraction that offers an interesting combination of mechanical properties and corrosion resistance [1,3]. These excellent properties depend on its chemical composition (Cr, Mo, N and Ni) and the right balance between ferrite (α) and austenite phases (γ) [4]. Therefore, SDSS is widely used in the oil, chemical, petrochemical, nuclear and marine industries because of its attractive combination of mechanical strength and higher corrosion resistance in a variety of aggressive environments [2,5].GTAW is the most common type of welding used in DSS and SDSS materials in various industries [1].Rapid heating and cooling cycles in the welding process can interfere with the α / γ phase balance and result in precipitation of intermetallic phases such as sigma, chi, secondary austenite and chromium nitrite. These can reduce mechanical properties and corrosion resistance of super duplex stainless steel materials [1,2,4].

Many studies have been carried out related to changes in microstructure due to the welding process in SDSS materials which have an impact on mechanical properties and corrosion resistance, but studies and references in repeated welding cycles of SDSS materials are very rare. Some information obtained from several standards / codes such as the NORSOK M-601 clause 7 only allows for one weld repair on SDS material. DNV-OS-F101 Table C-7 Appendix C, prohibits repetition of welding repairs to SDSS material

[6]. In fact, because of the difficulty in obtaining quality welds of SDSS material, repaired welding are often carried out. The present study investigated the effect of heat cycles on microstructure evolution and pitting corrosion behavior of UNS 32760 repeated welds. The TIG welding process was executed on 2" sch. 160 pipes with 6G welding position and carried out with appropriate welding parameters.

2. Methodology

2.1. Material and welding process

The base material employed is a commercial UNS 32760 SDSS produced by Tubacex Tubos Inoxidables. The alloy has been solution annealed and quenched by water into 2" pipe of about 8.74 mm thickness. The chemical compositions of base metal and filler are shown in Table 1. The pipes were butt-welded with TIG welding. The welding parameters are listed in table 2. Welding is employed by GTAW with DC-EN polarity and 6G position. The backing and shielding gas used are Ar + 2% N. Test coupons are prepared as shown in Figure 1 and given a datum mark or line 60 mm from the edge of the bevel area as a reference line in cutting and preparing samples for the repair process.In the R-1 sample preparation, the OW sample was cut and machined to form a bevel on the same fusion line area as OW with reference to the datum line to keep the heat-affected zone the same. The beveling process for R-2 and R-3 sample applied the same.

Table 1. Chemical compositions of base metal and filler

Element (%)	C	Si	Mn	P	S	Cr	Mo	Ni	Cu	W	N	Al	PREN
Base Metal	0.02	0.5	0.7	0.02	0.001	25.7	3.7	7.00	0.7	0.6	0.27	-	42.02
Filler	0.01	0.5	0.5	0.02	0.001	25.1	3.9	9.2	0.1	0	0.27	0.01	42.1

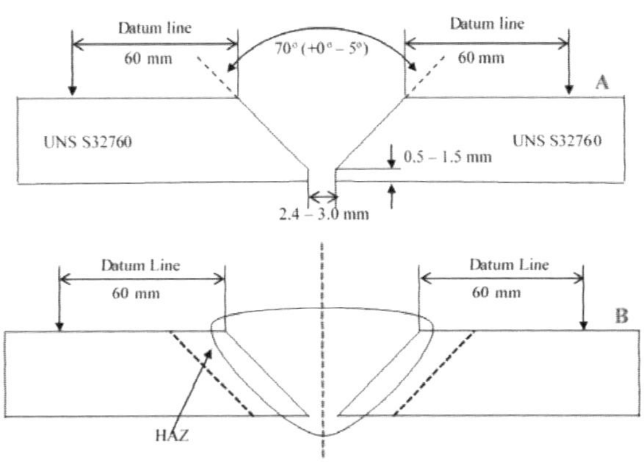

Figure 1. Joint Design: (A) Preparation and fit up of welding test coupons, (B) Cutting scheme of repaired test coupons.

Table 2. Welding Parameter

Joint Design	Welding current (A)	Welding voltage (V)	Welding speed (mm/min)	Shielding gas flow: Ar (L/min)	Backing gas flow: Ar (L/min)
Single V Groove	60 - 78	10	21.52 - 83.12	10 - 20	5 - 15

2.2. Microstructure characterization

To observe the optical micrographs of the WM, HAZ, and BM, the specimens were polished with silicon carbide up to 1000 grit followed by a cloth polishing with 0.05μm alumina powder and etched in 30% NaOH electrolyte [7]. The volume fraction of the ferriteand austenite phases was evaluated carefully by means of a BX53M Olympus optical microscope.

2.3. Corrosion and Electrochemical Measurements

Gravimetric test were conducted at 22°C for 72 h in the ferric chloride solution according to ASTM G48 Method A. The specimen dimension is 25 by 50 mm (1 by 2 in.) with 10mm Thickness. The solution was prepared by dissolving 100 gram of FeCl3.6H2O in 900 mL of deionized water with constant pH. After the soaking time, the specimens were rinsed with water and brushed with nylon to remove corrosion products then dipped in acetone and left to dry at room temperature. The weight losses of the specimens were recorded [8, 9].Electrochemical Measurement were carried out with a GAMRY Reference 600+ in a three-electrode cell held in a water bath. The counter electrode was a platinum foil, and the saturated calomel electrodes (SCE) were used as the reference electrode. All potentials were given against the SCE. The test solution, 1 mol/L NaCl, was made by dissolving 58.45 g of NaCl in water until the solution mixture reaches a volume of 1L. The working electrodes were mounted in epoxy resin. The specimens were passivated with 20% nitric acid (HNO3) at ambient temperature for 1 h to avoid crevice corrosion in specimen/resin interfaces then abraded with 600 grit paper.The minimum test area is 1 cm2. The potentio-dynamic polarization tests were conducted at a scanning rate of 1 mV/s from -0.7 V to a very anodic potential at 23oC ± 2oC. The potential at which the current density exceeded 100 μA/cm2 is defined as pitting potential.Before corrosion and electrochemical tests were conducted, the specimens have been polished with silicon carbide up to 1000 grit [10-13].

3. Result and discussion

3.1. Microstructure Evolution

Fig. 2a shows microstructure of the base metal. The light area is austenite while the dark is ferrite. Ferrite and austenite phases extend and assemble along the direction of the material plane with the same volume fraction. There are no precipitates in both ferrite and austenite. This indicates the volume balance of the ferrite and austenite phases. The appearance of microstructure at the base material is the same for all weld repairs.

Fig. 2b exhibits how the heat cycle in the welding process destroys the balance of ferrite and austenite phases. In this zone there are several types of austenite: Grain boundary austenite (GBA), Widmänsttenaustenite (WA), intra-granular austenite (IGA)and secondary austenite (γ_2). During the welding process, GBA is formed on the ferrite grain boundary because there is a large amount of free energy at this location. WA continues to form and grow around GBA. At the same time IGA and secondary austenite are formed in this zone because there are many nickel elements as austenite stabilizers carried by filler metals [4]. It can also be seen that GBA is coarser compared to WA and IGA. This indicates instability in the ferrite and austenite interfaces [2].

In Fig. 2c, the GBA is clearly visible with a thinner and smoother structure with a small amount of IGA. In this zone the ferrite volume fraction is far more compared to austenite. This is due to the high temperature of welding [2].Fig. 2d has almost the same structure as fig. 2c with GBA is more coarse and thicker and relatively more austenite volume fractions characterized by more IGA.

Fig. 2e is the HAZ R-2 microstructure. R-2 HAZ is relatively narrow compared to OW and R-1 so that the structure is not clearly visible. The narrow HAZ as a transition area is strongly influenced by both BM and WM structures with a higher volume of austenite fraction compared to ferrite. In this condition, the chromium nitride (Cr_2N) precipitate which is marked by black spot on the ferrite grains appeared.

Fig. 2f shows the HAZ which is slightly wider than fig. 2e more ferrite volume fractions. GBA looks clearer with a combination of fine/thin and thick/rough types and there are more chromium nitride precipitates.

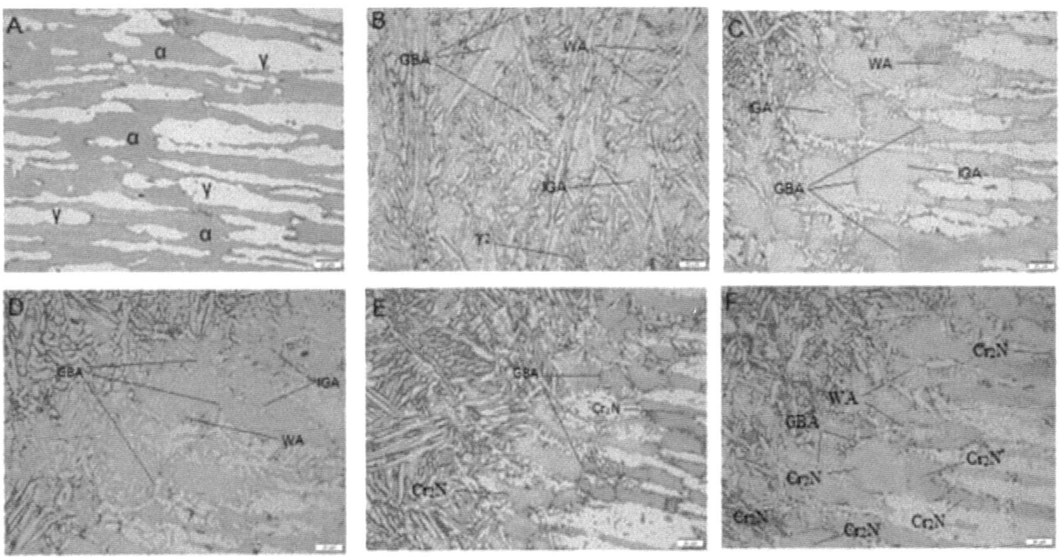

Figure 2. Optical microstructure of BM, HAZ and WM of repeated TIG welding: (a) base metal, (b) weld metal, (c) HAZ for OW, (d) HAZ for R-1, (e) HAZ for R-2, (f) HAZ for R-3.

3.2. Pitting Corrosion Resistance

Gravimetric Test and Potentiodynamic anodic polarization technique were used to evaluate the effect of repeated TIG welding cycles on the pitting corrosion resistance of the test specimens.The results of the gravimetric tests are listed in table 3. It is obvious that R-3 welding cycles has highest corrosion rate during the test followed by R-2, OW and R-1.Theanodic polarization curves are presented in Fig. 3. The polarization pattern on all specimens is relatively the same where the specimen is passivated before the occurrence of pitting.The R-3 has the largest passive current density followed by R-2, OW and R-1.The appearance of the current peaks triggered under Epitt in R-2 showed that the specimen was attacked by metastable pits [2] as shown in figure 4.The current density magnitude of the specimens showed that the R-3 has the highest corrosion rate during the measurement. The results of the polarization measurements are listed in table 4. The open circuit potential (OCP) value starts with negative value, continuously increase to positive and decrease to more negative. Thermodynamically, the more negative the OCP, the higher tendency the corrosion takes place [4, 12]. As well as OCP, Ecorr starts with negative values, increase to positive and back to negative. The value of Epitt showed significantly decrease by increasing the number of repeated welding cycles. Taking both corrosion rate and anodic polarization into consideration, it can be concluded that R-3 has the poorest pitting corrosion resistance followed by R-2.

Table 3. Gravimetric Test Result of repeated TIG welding

No	Specimen	Area (cm2)	Weight before test (g)	Weight after test (g)	Weight loss (g/cm2)	Corrosion Rate (mdd)
1	OW	33.7408	49.0500	49.0483	0.00005	1,68
2	R1	35.8517	53.2932	53.2914	0.00005	1,67
3	R2	32.9367	48.3204	48.3165	0.00012	3.95
4	R3	34.6789	48.4726	48.4454	0.00078	26,08

Table 4. Electrochemical parameters of therepeated TIG welding

No	Specimen	OCP (V vs SCE)	Ecorr (V vs SCE)	Epitt (V vs SCE)
1	OW	-0.222	-0.4559	0.7135
2	R1	0.1899	0.01792	0.7276
3	R2	0.009132	-0.3206	0.6869
4	R3	-0.2403	-0.394	0.6835

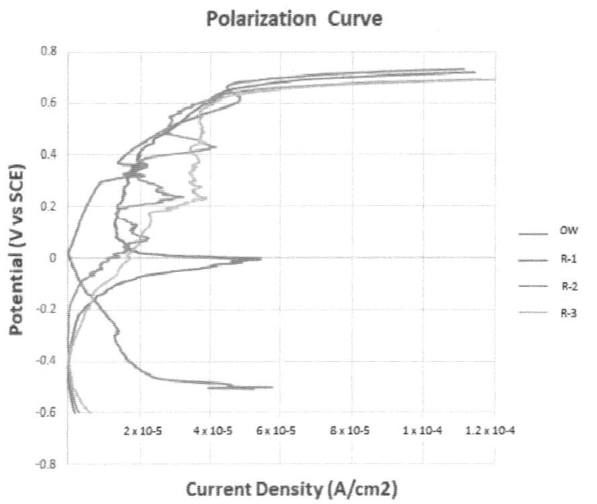

Figure 3. Polarization curves of repeated TIG welding

IOP Conf. Series: Materials Science and Engineering **478** (2019) 012032 doi:10.1088/1757-899X/478/1/012032

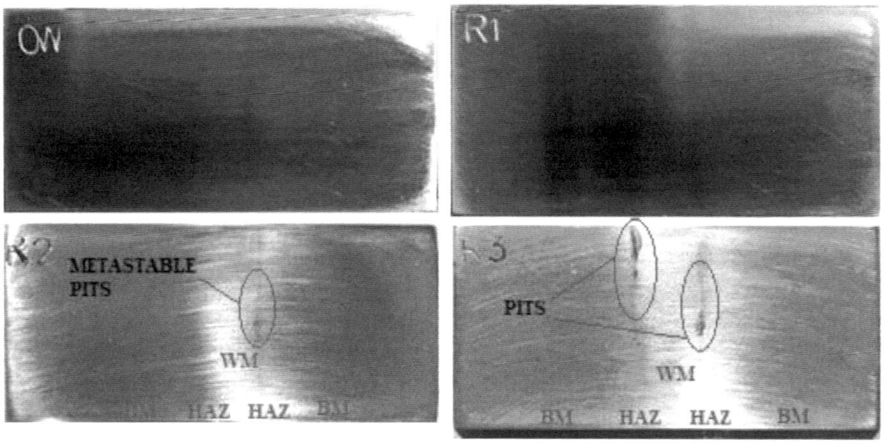

Figure 4. Specimens of repeated TIG welding after gravimetric tests

4. Conclusion

The effect of repeated TIG welding cycles on microstructure evolution and pitting corrosion behavior of UNS 32760 super-duplex stainless-steel were investigated by morphological observation and electrochemical measurements. The more repetition of welding cause the coarser grain growth, the more ferrite volume fraction and the more chromium nitride precipitates in heat-affected zone. The chromium nitride precipitates and the metastable pits start appearing in R-2 welding cycles and make the pitting corrosion resistance start to decrease.

References

[1] Prabhu P and Rajnish G 2016 *Engineering Science and Technology, an International Journal* **19** 1076–1083
[2] Ziying Z, Hui Z, Huizhen Z, Jun H, and Jiarui J 2017 *Corrosion Science* **121** 22–31
[3] Hua T, Yiming J, Bo D, Tao S, Juliang X, Jin L 2009 *Material Characterization* **60** 1049-1054
[4] Shaoning G, Junsheng S, Lingyu G, Hongquan W 2015 *Journal of Manufacturing Processes* **19** 32–37
[5] Bo D, Zhiyu W, Yiming J, Hao W, Juan G, Jin L 2009 Electrochimica Acta **54** 2790–2794
[6] Danny S M and Turnad L G 2016 ARPN Journal of Engineering and Applied Sciences **11 (14)**
[7] ASM International 2014 *Standard Practice for Microetching Metals and Alloys (ASTM E407)*
[8] ASM International 2011 *Standard Test Methods for Pitting and Crevice Corrosion Resistance of Stainless Steels and Related Alloys by use of Ferric Chloride Solution (ASTM G48)*
[9] Jyoti B, Samantha L, Rouzbeh A, Vikram G, Roberto O, Denis L, and Faisal K 2017 *Journal of Loss Prevention in the Process Industries* **47** 10-21
[10] ASM International 2004 *Standard Reference Test Method for Making Potentiostatic and Potentiodynamic Anodic Polarization Measurements (ASTM G5)*
[11] Alonso-Falleiros N, Hakim A, and Wolynec S 1999 *Corrosion* **5(5)** 443-448
[12] Luo H, Dong C F, Cheng X Q, Xiao K, and Li X G 2011 *Journal of Materials Engineering and Performance*
[13] Sun Y T, Wang J M, Jiang Y M, and Li J 2018 *Materials and Corrosion* **69** 44-52

IOP Conf. Series: Materials Science and Engineering **478** (2019) 012033 doi:10.1088/1757-899X/478/1/012033

Root cause failure analysis of reducer weld-joint leakage for liquid outlet of slug catcher

R A T Gurning, D Priadi, D Ferdian, and Suryadi

Department of Metallurgy and Materials Engineering, Faculty of Engineering, University of Indonesia

Email: rolly.gurning@yahoo.com

ABSTRACT – This paper presents a failure analysis of reducer weld-joint leakage for liquid outlet of slug catcher. The leak point was found on the area of weld joint with position relative to clock position from 9 to 3 o'clock. In order to determine morphology and mechanisms of crack were observed through laboratory tests including chemical composition test, weld macro analysis, microstructure analysis, and corrosion product analysis. Failure analysis result was found that corrosion process consumed the root weld material and cause leakage. The corrosion is occurred due to combination effect of galvanic cell, root weld microstructure, turbulence and corrosive fluids. Scaling was found based on scale analysis. Therefore the pH was indicated high enough and showed the small chance for corrosion due to carbon dioxide (CO_2). Detailed investigation also revealed that the inconsistent of weld root which resulting excessive root penetration over 3 mm. An excessive weld root about 5.6 mm is observed which was as result of improper instant repair. This profile of weld root will disrupt the flow and cause the weld root to be more anodic and vulnerable to corrosion.

1. Introduction

Investigation of piping failure is very common to be used in the oil and gas industry. Understanding the failures of piping system may help the industry to design and operate their piping system carefully. There must be several different aspects to be considered as the probability of failure of piping systems at various weld joints, branches, nozzles and piping itself. Nondestructive Testing (NDT) inspection is one of inspection method which may control and determine the condition of internal piping systems. However in some cases, piping failure due to leakage may not be detected using NDT inspection. Failure analysis and fracture investigations of surface flaw which might have undetected using NDT inspection, may not become through thickness during the components life time [1].

There are some cases which may lead the piping leakage on weld joint area. The residual stress of the elbow pipe and the heat effect of the welding process are the main problem of piping leak and contribute the cracks which may be caused by stress corrosion cracking [2]. Failure due to corrosion attack in combination with misalignment of the pipe, may lead to an accelerate corrosion rate in carbon steel API 5L grade B [3]. The premature failure on the longitudinal weld of ERW pipe also occurred due to the sediment deposits and the elevated Cl^- concentration [4].

In this case study, piping leakage was found on the area of weld joint position on spool line of slug catcher. This spool line was used to deliver two phases fluid. Previously there were also found leakage on that spool line. However, the repair had been conducted using welding technique. Therefore, in order to prevent another failure in the reducer weld joint and tee spool weld joint, the failure analysis may be conducted on the leakage point of reducer weld joint material, ASTM A 105 and ASTM A 106 Grade B, and tee spool weld joint material, ASTM A 105 and ASTM A 234.

IOP Conf. Series: Materials Science and Engineering **478** (2019) 012033 doi:10.1088/1757-899X/478/1/012033

2. Methodology

The failure analysis of reducer weld-joint leakage was investigated by testing the spool leakage in the laboratory. The laboratory tests that was carried out on the spool leakage material including chemical composition test using Optical Emission Spectrometry (OES), weld macro analysis using Optical Macroscopy, microstructure analysis using Optical Microscopy, and corrosion product analysis using X-Ray Diffraction (XRD).

2.1. Visual and Macroscopy Examination

Macroscopy examination was conducted on weld joint leakage. This examination was the next step after conducting visual examination in any metallurgical failure investigation. Both of this examination may help to identify the damage surface of material and also possible visible surface defects. Basically, both of these examination were categorized as non-destructive examination. However in some cases, macroscopy examination required polished and etched material to examine the material sample.

2.2. Chemical Composition Test

The damage or failure pipe were already determined the analyzed location for chemical composition sample. The chemical composition test was conducted in accordance with ASTM A751 and ASTM E 415. The chemical analysis was conducted at the facilities of CMPFA Company using WAS AG-Foundry Master with 8 metal base detectors including Fe, Al, Ni, Pb, Cu, Mg, Zn, Ti.

2.3. Microstructure Analysis

Scanning Electron Microscopy (SEM) was used for the metal characterization on the metal structures, as well as in fracture surfaces analyzing. Field Emission SEM was used to analysis the failure with magnification up to 500.000x of multi specimens. Utilizing of SEM may revealed the surface morphology of defects characteristics due to high resolution and field depth of SEM. Basically, using SEM was also followed by XRD analysis to detect the corrosion product.

2.4. Corrosion Product Analysis

As mention above on microstructure analysis, XRD analysis is usually used to determine corrosion product of failure sample. The XRD analysis using Philips, PW 1430 Generator to determine any corrosion products in this weld joint leakage pipe. There was a thing to note that the x-ray peak for any different elements might overlap.

3. Results and Discussion

3.1. Visual Examination Result

Visual examination was performed on the spool material after sectioning. The sectioning of spool material is made in order to observe the internal surface of the spool material. Examination is made by unaided eyes. The visual examination of the reducer 4"x2" may be shown Figure 1, 2, and 3, respectively.

Based on internal view of the spool reducer R.W #1 and R.W#2, they were suggested that the corrosion product found on the top surface area. The color was nearly dark brown, very crumble and also easily wiped away. However, generally the layer of scale and corrosion product was mainly found on the bottom surface of reducer. The combination of color was dark brown and earth yellow. The corrosion product was very hard and sticks onto the steel surface. The erosion marks are not observable in the internal of reducer and on the both flange mouth, inlet and outlet mouth.

Visual examination also performed on Tee-Weld 4" spool. The internal view of tee-weld 4" may be shown in Figure 4. The appearance of corrosion product and layer of scale were found on the bottom surface of tee. The characteristic and combination color were similar with layer found in the reducer

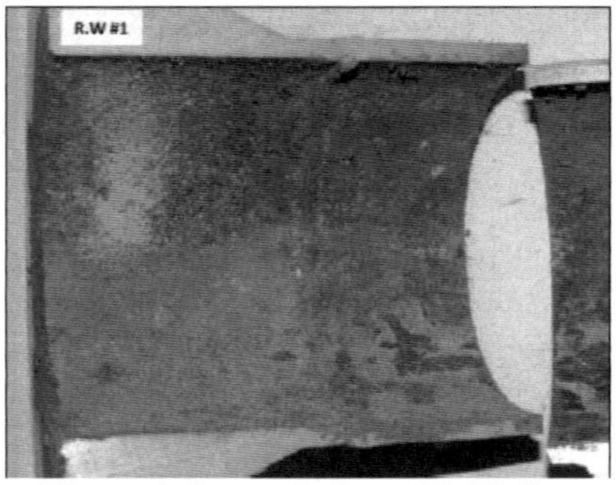

Figure 1. Internal view of R.W #1

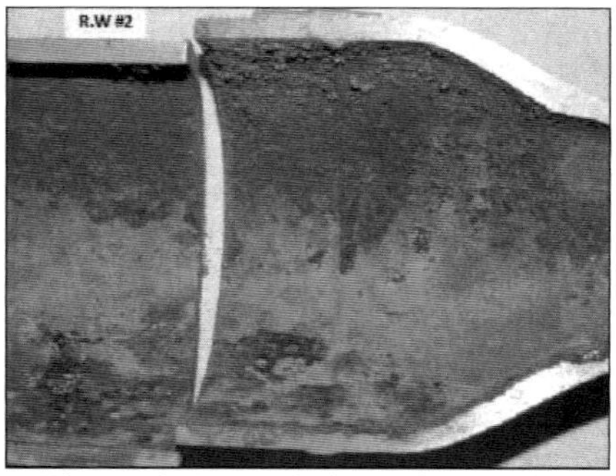

Figure 2. Internal view of R.W #2

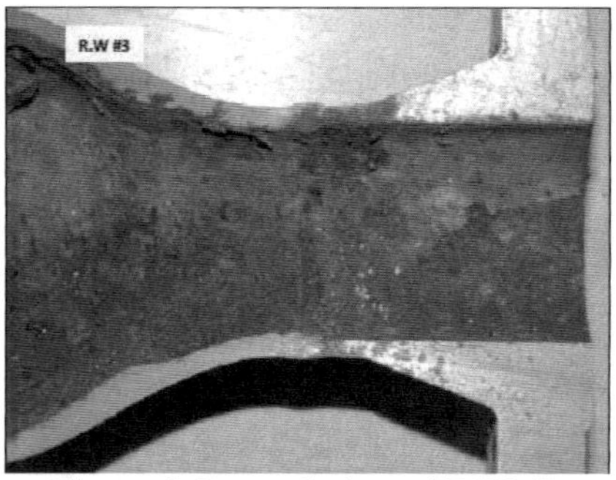

Figure 3. Internal view of R.W #3

Figure 5 and 6 shows that there were weld root degradation indications on weld joint of T.W#4 and T.W#5. The weld root degradation may be resulted either from improper of welding technique or improper instant repair. This weld root degradation may cause the turbulent flow of fluid in the weld-joint pipe and also may cause the root weld more anodic and vulnerable to corrosion. However, the corrosion product were found on the 2" straight pipe with dark brown of its color. Based on its fluid service and the visual result, the corrosion product may be siderite ($FeCO_3$). XRD analysis may show the dominant corrosion product in the straight pipe.

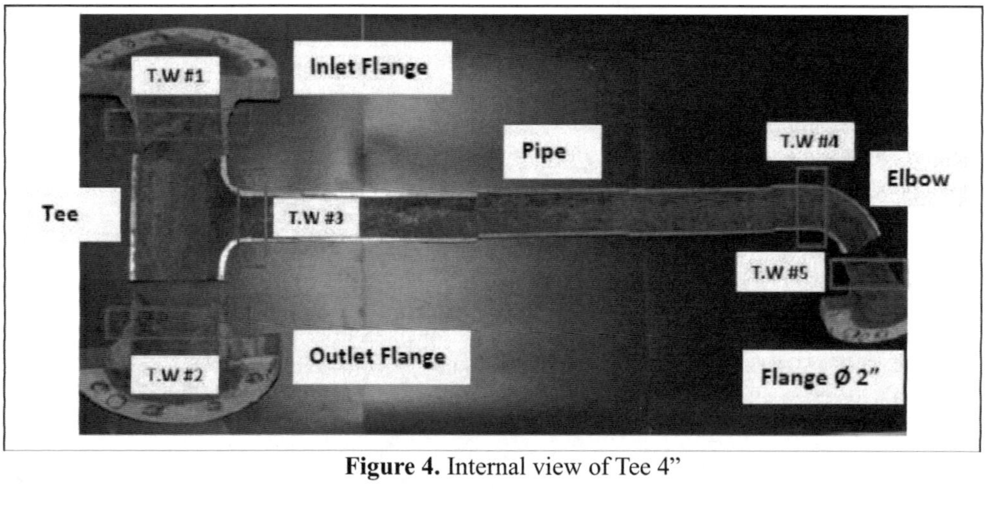

Figure 4. Internal view of Tee 4"

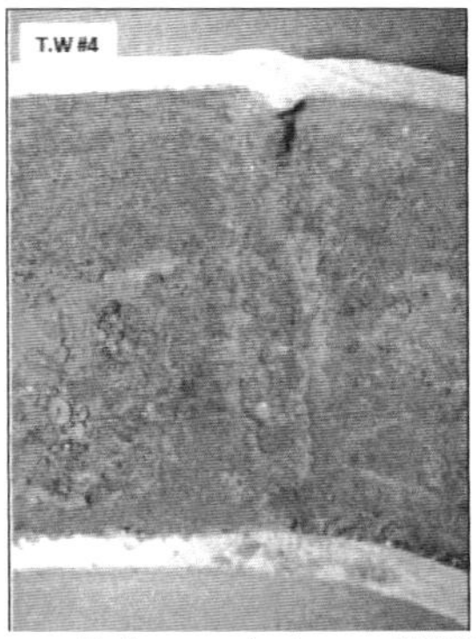

Figure 5. Internal view of T.W#4

Figure 6. Internal view of T.W#5

3.2. Chemical Composition Result

The chemical composition test was already conducted on the pipe spool, weld metal spool, inlet and outlet flange. The chemical composition of the flanges and reducer body conformed to the material specification as shown in Table 1. Each of joint weld metal chemical composition was shown in Table 2. There was no significant difference in the chemical composition, therefore the chemical composition of reducer and reducer weld joint are still normal.

The chemical composition test results of tee spool and tee spool weld joint may also be shown in Table 3 and Table 4. The results also found that there was no significant difference in the chemical composition, therefore the chemical composition of tee spool and tee spool weld joint are still normal.

Table 1. Chemical composition test results of reducer

Part	C (%)	Mn (%)	Si (%)	P (%)	S (%)
Inlet Flange Ø 4"	0.18	1.15	0.26	0.009	0.007
Outlet Flange Ø 2"	0.19	1.17	0.24	0.007	0.004
ASTM A 105	**0.35 max**	**0.60-1.05**	**0.10-0.35**	**0.035 max**	**0.040 max**
Reducer Body Ø 4"	0.20	0.64	0.21	0.015	0.009
ASTM A106 Grade B	**0.30 max**	**0.29-1.06**	**0.10 min**	**0.035 max**	**0.035 max**

Table 2. Chemical composition test results of reducer weld joint

Part	C (%)	Mn (%)	Si (%)	P (%)	S (%)
R W1	0.086	1.44	0.72	0.010	0.006
R W2	0.087	1.46	0.75	0.009	0.006
R W3	0.100	1.47	0.71	0.009	0.006

Table 3. Chemical composition test results of tee spool

Part	C (%)	Mn (%)	Si (%)	P (%)	S (%)
Inlet Flange Ø 4"	0.19	1.20	0.24	0.007	0.003
Outlet Flange Ø 4"	0.16	1.23	0.19	0.008	0.006
Outlet Flange Ø 2"	0.18	1.12	0.16	<0.003	0.011
ASTM A 105	**0.35 max**	**0.60-1.05**	**0.10-0.35**	**0.035 max**	**0.040 max**
Elbow Ø 2"	0.16	0.78	0.20	0.012	0.005
Tee Ø 4"	0.20	0.66	0.21	0.023	0.009
ASTM A 234	**0.30 max**	**0.29-1.06**	**0.10 min**	**0.05 max**	**0.058 max**
Pipe Ø 2"	0.19	1.03	0.17	0.012	0.009

ASTM A106 Grade B	**0.30 max**	**0.29-1.06**	**0.10 min**	**0.035 max**	**0.035 max**

Table 4. Chemical composition test results of tee spool weld joint

Part	C (%)	Mn (%)	Si (%)	P (%)	S (%)
TS W1	0.094	1.40	0.66	0.010	0.007
TS W2	0.087	1.30	0.59	0.009	0.006
TS W3	0.089	1.37	0.65	0.009	0.005
TS W4	0.112	1.32	0.59	0.010	0.006
TS W5	0.095	1.22	0.59	0.010	0.008

3.3. Weld Macro Result

Macrophotography test was conducted on the weld joint spool. Samples are cut from spool material. Its cross section is prepared with the following steps such as grinding, polishing and etching. Figure 7 shows macrophotograph result of weld joints on reducer 4"x2". Based on Figure 7, there is no abnormal weld root on weld joint reducer 4"x2". The weld root height was found about 2.5 mm on each sample. However, Figure 8 shows the excessive weld root on T.W#2. The excessive root of T.W#2 may be caused by the improper instant repair. This excessive weld root may disrupt the fluid flow inside the pipe and increasing the preferential weld corrosion occurred. The preferential weld corrosion may be influenced by several parameters including flow conditions, scaling effects, the environment, steel composition and also welding procedure [5].

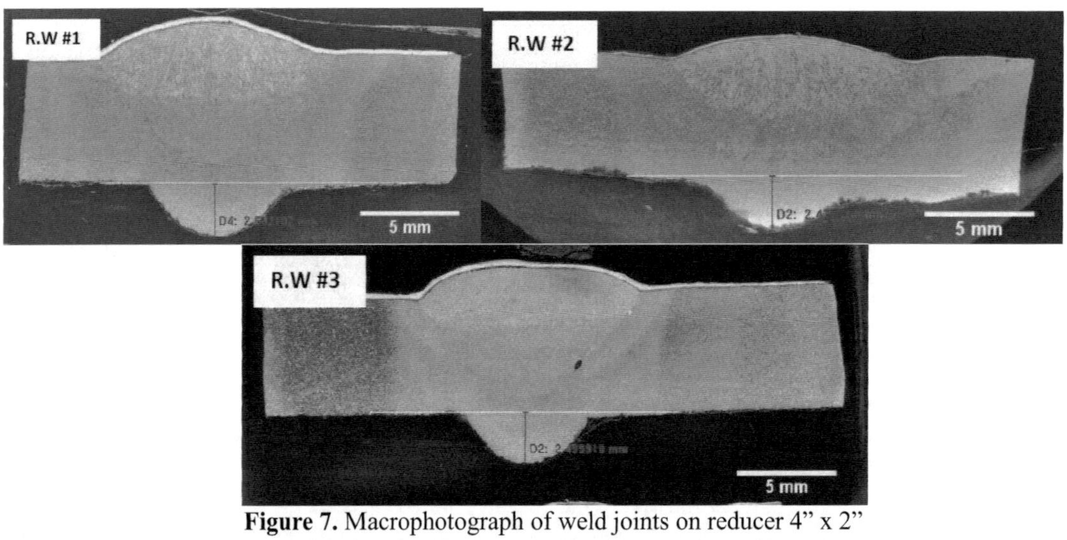

Figure 7. Macrophotograph of weld joints on reducer 4" x 2"

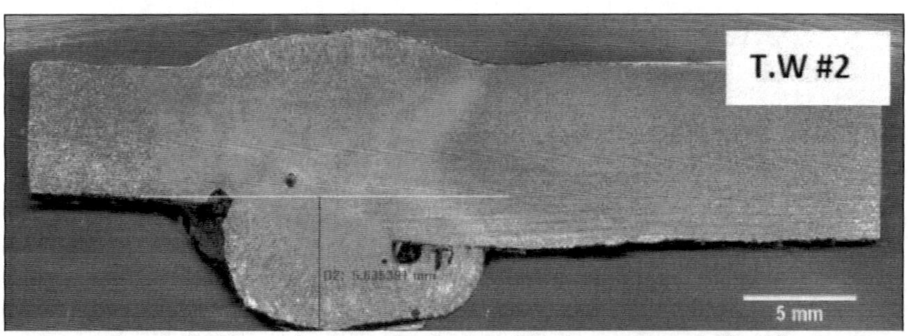

Figure 8. Macrophotograph of tee weld joints on spool 4" x 2"

3.4. Microstructure Result

The analysis of microstructure was performed on the sample after weld macro analysis has been conducted. Microstructure analysis was conducted using optical microscope with magnification of 500x. Detail observation and photography was made on specific area on the cross section of sample. The microstructure of each weld joints was found normal. There was also no detrimental microstructure on the weld joint such as martensite.

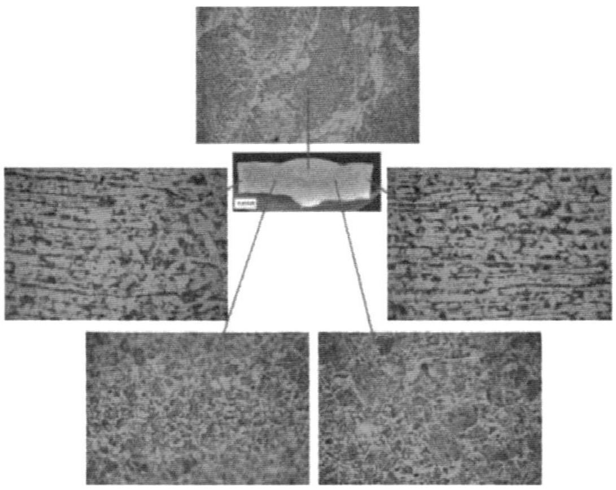

Figure 9. Microphotograph of weld joint R.W#2

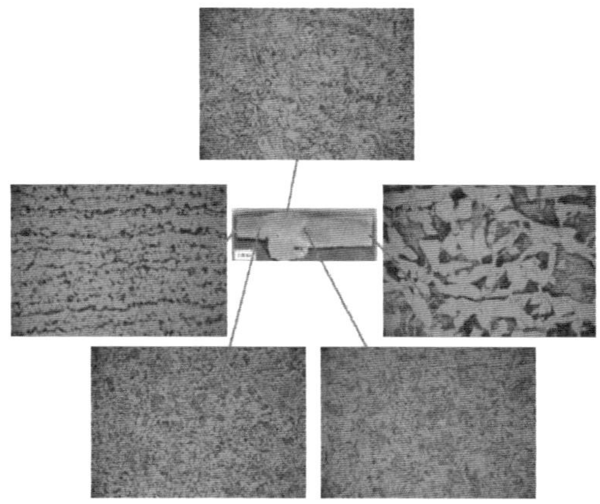

Figure 10. Microphotograph of weld joint T.W#2

Figure 9 shows the macrophotograph of weld joint R.W#2. The microstructure analysis was taken on HAZ area. There was no abnormal microstructure on R.W#2. The combination of ferrite and pearlite was found normal on each sampling analysis. The excessive root sampling of T.W#2 was also observed on Figure 10. However, the microstructure was also found normal and there was no detrimental microstructure. Therefore, failure due to microstructure may be negligible.

3.5. Corrosion Product Result

In order to determine the compound of corrosion product in the internal surface of both reducer 4" x 2" and tee 4", X-Ray Diffraction (XRD) was used to analyze it. There are four sampling point that was taken in the internal surface of pipe spool. Table 5 shows the description of the sample point for XRD analysis.

Table 5. Code description of corrosion product analysis

Code	Description
Bottom Corrosion Product Reducer (BCPR)	Sample taken from layer of scale and corrosion products with combination color of dark brown and earth yellow in reducer
Top Corrosion Product Reducer (TCPR)	Sample taken from dark brown and crumble corrosion product in reducer
Bottom Corrosion Product Tee (BCPT)	Sample taken from layer of scale and corrosion products with combination color of dark brown and earth yellow in tee
Top Corrosion Product Tee (TCPT)	Sample taken from dark brown and crumble corrosion product in tee

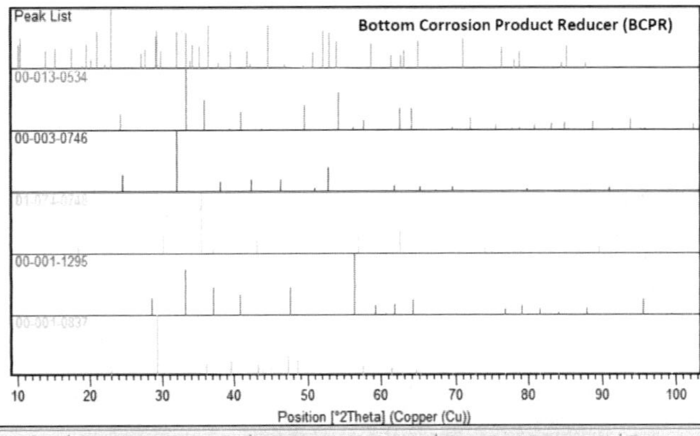

	Visible	Ref. Code	Compound Name	Chemical Formula	Score
1	✔	00-013-0534	Hematite, syn	Fe2 O3	2
2	✔	00-003-0746	Siderite	Fe C O3	4
3	✔	01-074-0748	Iron diiron(III) oxide	Fe3 O4	1
4	✔	00-001-1295	Pyrite	Fe S2	1
5	✔	00-001-0837	Calcite	Ca C O3	5

Figure 11. XRD of BCPR

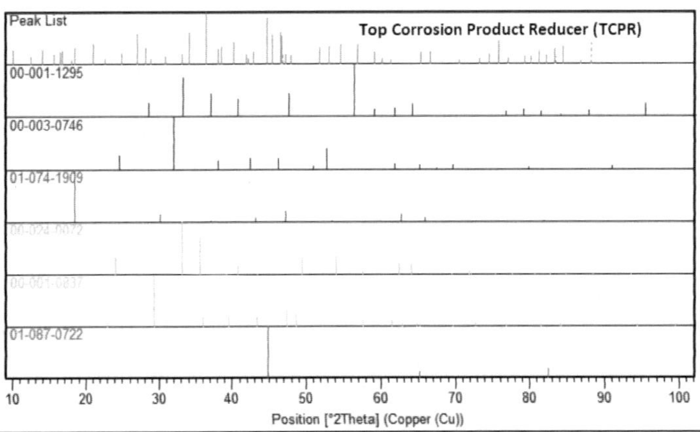

	Visible	Ref. Code	Compound Name	Chemical Formula	Score
1	✔	00-001-1295	Pyrite	Fe S2	Unmatched Strong
2	✔	00-003-0746	Siderite	Fe C O3	1
3	✔	01-074-1909	Iron diiron(III) oxide	Fe3 O4	5
4	✔	00-024-0072	Hematite	Fe2 O3	1
5	✔	00-001-0837	Calcite	Ca C O3	1
6	✔	01-087-0722	Iron	Fe	8

Figure 12. XRD of TCPR

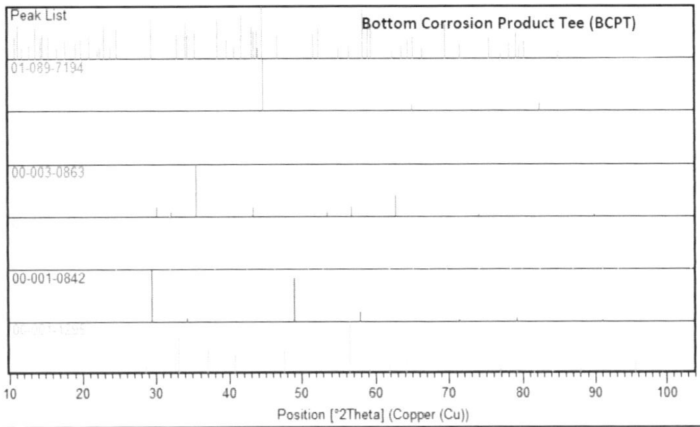

	Visible	Ref. Code	Compound Name	Chemical Formula	Score
1	✔	01-089-7194	Iron	Fe	12
2	✔	00-001-0837	Calcite	Ca C O3	3
3	✔	00-003-0863	Magnetite	Fe3 O4	2
4	✔	00-003-0746	Siderite	Fe C O3	1
5	✔	00-001-0842	Chalcopyrite	Cu Fe S2	3
6	✔	00-001-1295	Pyrite	Fe S2	Unmatched Strong

Figure 13. XRD of BCPT

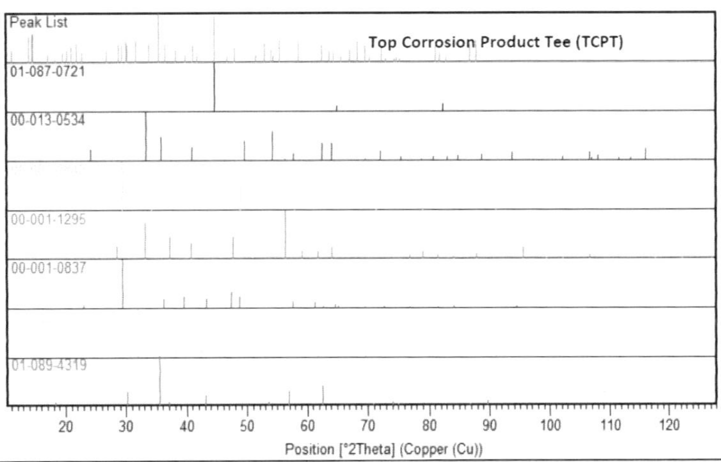

	Visible	Ref. Code	Compound Name	Chemical Formula	Score
1	✔	01-087-0721	Iron - $-alpha	Fe	18
2	✔	00-013-0534	Hematite, syn	Fe2 O3	1
3	✔	00-001-0842	Chalcopyrite	Cu Fe S2	4
4	✔	00-001-1295	Pyrite	Fe S2	Unmatched Strong
5	✔	00-001-0837	Calcite	Ca C O3	2
6	✔	00-003-0746	Siderite	Fe C O3	2
7	✔	01-089-4319	Iron diiron(III) oxide	Fe3 O4	9

Figure 14. XRD of TCPT

Table 6. Summary of XRD test

Code	Corrosion Product Findings	Indications
BCPR	Fe_2O_3, $FeCO_3$, Fe_3O_4, FeS_2, $CaCO_3$	Corrosion of steel by water, Carbonate scale, Bacterial corrosion, Sweet corrosion
TCPR	$FeCO_3$, Fe_3O_4, Fe_2O_3, $CaCO_3$, Fe	Corrosion of steel by water, Carbonate scale, Sweet corrosion
BCPR	Fe, $CaCO_3$, Fe_3O_4, $FeCO_3$, $CuFeS_2$	Corrosion of steel by water, Carbonate scale, Sweet corrosion

TCPR	Fe, Fe_2O_3, $CuFeS_2$, $CaCO_3$, $FeCO_3$, Fe_3O_4	Corrosion of steel by water, Carbonate scale, Sweet corrosion

Corrosion products may be shown from Figure 11 to Figure 14. Table 6 shows summary of XRD test of BCPR, TCPR, BCPR, and TCPR. Generally, corrosion products that occurred based on XRD tests may be caused by carbonate scale, bacterial corrosion, sweet corrosion and corrosion of steel by water. There was also presence of pyrite (FeS_2) in the corrosion product. However, it may not be considered due to the reading from XRD spectrum showed "unmatched strong". Therefore, FeS_2 may not exist in the corrosion product.

3.6. Discussions
Visual inspection revealed that there was scale formation in the internal spool pipe and reducer. Based on scale analysis, it contains calcite ($CaCO_3$) and siderite ($FeCO_3$) as majoring compounds. Calcite is the scaling product which may be predicted by the Langelier Saturation Index (LSI). Failure of the calcite may occur when the calcite layer develop the internal stress larger than the adhesive bonds [6]. According to XRD analysis, the environment is supporting to form a protective scale in the form of calcite. However, the detection of siderite in the scale is revealed that the environment is also corrosive. Siderite is the commonly product of sweet corrosion due to dissolve CO_2 gas into the water. Table 6 also shows that there are another compounds in the scale collected. There are two corrosion products from the result of steel corroded by oxygen which are magnetite (Fe_3O_4) and hematite (Fe_2O_3). Both corrosion products may be dominated by oxygen reaction due to the pH was probably high (pH=8). The suspect of microbial induced corrosion may not be proved since no evidence in the corrosion product after XRD test. The presence of FeS_2 is unmatched with the peak database of compound.

The preferential attack occurred and cause the leakage due to the weld root is more vulnerable to deterioration. The combination of flow, corrosion and weldment are the potential source of this problem since the leakages was found at the flow pattern change area. The flow pattern may change from laminar into turbulent since the protruding weld root occurred in the internal surface of spool pipe. The key influence of corrosion under turbulent flow regimes was the surface conditions formed during pre-corrosion [7]. Then, the corrosion attacks the exposed metal since the turbulent flow break the film layer. According to previous failure history, both of spool pipe and reducer had a leakage history on the 9 to 3 o'clock position. However, it has been repaired by welding technique. Therefore, the evidence of leakage may not be retrieved. It was only reported that a preferential weld attack occurred on the root weld since there was combination of flow and corrosion. The unstable condition of film layer was selectively disrupted on the weld metal then may cause the preferential weld corrosion in the pipe surface [8].

According to the weldment quality, the cross sectional of a weld is generally proportional to the amount of heat input. In multi-pass welding, the portion of previous weld pass is refined since the heat from each pass tempers the weld bead. A decreasing level of microstructure refinement of the root by the subsequent passes is the one of some factors which may increase preferential weld metal corrosion attack at weldment[9]. The root is observed to be the first pass that has undergone grain refinement due to multiple pass heating. The combination of fine root grains, turbulence of flow-induced, and the fluids, may increase the corrosion process on the weldment. The scale, which may be produced by the service fluids, at the root weld are easily damaged and corroded since the influence of high shear stress, flow regimes and fine grains in the weldment area.

Figure 11 and Figure 12 also shows that some weld roots are still in acceptable height for excessive root penetration according to BS EN ISO 5817. The average height are still below 3 mm on each sample of reducer and tee spool. However, T.W #2 is the only one sample which is measured about 5.6 mm. The T.W #2 sample is suspected as product of repair process which made on tee spool. This type of excessive weld root will abruptly change the flow pattern inside the spool tee and may lead to erosion or corrosion of the weld joint.

4. Conclusions and Recommendations
The corrosion process consumed the root weld material and may cause leakage. The corrosion process may be occurred since combination effect of galvanic cell, root weld microstructure, turbulence and corrosive fluids. According to scale analysis, presence of scale shows the high pH and gives the small probability of CO_2 corrosion damage to occur. In a range condition of pH 5.5-6.5 at temperature of 55 °C, there will be no corrosion product of CO_2 corrosion formed on the steel surface [10]. In other hand, according to microstructure aspect, the root weld has fine grains. It was the result of multiple pass with high input. The evidence of high heat input was shown by wide bead of the weldment. An excessive weld root about 5.6 mm

has been observed and resulting the improper of instant repair process. The weld root profile will disrupt the flow and may cause the weld root to be more anodic and vulnerable to corrosion.

In order to minimize the inconsistency of weld joint quality and the improper of welding process, it is recommended to check the remaining weld joint and replace any joint with improper weld profile which may affect the flow pattern and threats such as flow accelerated corrosion.

References

[1] Khan I 2011 *International Journal of Pressure Vessels and Piping* **88** 223-230

[2] Jiang S, Zheng Y and Duan D 2013 *Engineering Failure Analysis* **27** 203-212

[3] Ul-Hamid A and Tawancy H M 2009 *Engineering Failure Analysis* **16(3)** 825-832

[4] Tavares S, Pardal J, Mainier F, Igreja H D, Barbosa E and Rodrigues C 2005 *Engineering Failure Analysis* **61** 100-107

[5] Lopez D, Perez T, and Simison S 2003 *Material and Design* **24**

[6] Keysar S, Hasson D, Semiat R, and Bramson D 1997 *Ind. Eng. Chem. Res.* **36** 2903-2909

[7] Barmatov E and Hughes T L 2017 *Corrosion Science* **123**170-181

[8] Alawadhi K and Robinson M 2011*Corrosion Engineering* **46(4)** 318-329

[9] Lee C M, Bond S, and Woolin P 2005 *Prefential weld corrosion: Effects of weldment microstructure and composition* in *NACE 2005 Houston*, Texas

[10] Nazari M H, Allahkaram S, and Kermani M 2010 *Materials and Design* **31(7)** 3559-3563

Molecular Dynamics Simulation of Hydrogen Adsorption on Silica

J F Fatriansyah, D Dhaneswara, M H Abdurrahman, F R Kuskendrianto, and M B Yusuf

Metallurgy and Materials Engineering Department, Faculty of Engineering, Universitas Indonesia

E-mail: jakafajarf@gmail.com

Abstract. Hydrogen is one of the future energy because it is enviromentally friendly. However, there still some problem in the storage method of hydrogen. In its development, there are many candidates that are still in research as a hydrogen storage medium. Also in several studies, it was found that Silicon based material is a promosing candidate. In this study we conduct the effect of various pressure to the adsorption of hydrogen on Silica with molecular dynamics simulation using Lennard-Jones potential. The simulations indicate that Silica has a good hydrogen storage capability where pressure and time affect the amount of hydrogen adsorbed.

1. Introduction

In recent years, there has been many research carried out with the concept of green energy because this is a way to save our environment. In addition, fossilbased energies diminish their existence, so that an alternative energy is needed to replace these kind of resources. One of the renewable energy that get attention by researchers is hydrogen. Hydrogen characteristic is friendly for the environment. The hydrogen reaction does not produce CO_2 as a side product [1]. Hydrogen also has the highest energy density of any hydrogenbased fuel[2]. With this character, currently hydrogen has been developed for fuel cell applications in vehicles, mobile and other portable applications [3,4]. But there are still some challenges in storing hydrogen. As we know, hydrogen can explode when get a contact with air and the way to store it is very difficult[2].

In its development, there are many ways to store this hydrogen. Adsorption becomes an interesting method and continues to be developed when compared to other methods such as physical storage as compressed gas and cryogenic liquid hydrogen[5-8]. The advantage of using the adsorption method were it take cheaper cost, the use of operating pressure lower and the storage system design is not complex[9]. Beside the method, the materials that use as an storage medium must be considered. They must have a high surface areas and usually is porous materials. So far, hydrogen strorage medium were carbon nanotubes, graphite, graphene, metal organic frameworks and many more [10-12]. Beside that,silicon based material usually used as a storage material such as silicon surface, silicon nanotube, and zeolites [13-15].

Silica is also a silicon based material with porous and can be a promising candidate as a hydorgen storage media. Silica has a good thermal stability, harmless and inexpensive. The properties like thermal stability is needed because in some hydrogen strorage mediumit is constrained in its applications as an automotive applications whereas at room temperature and moderate pressure can not store the amount of hydrogen is needed [16]. Therefore silica can be a solution for that problem. Beside that silica has high surface area. In this work, we want to study the effect of various pressure in adsorption of hydrogen on silica by using molecular dynamics.

2. Experimental Methods

In this study, we perfomed simulations using LAMMPS. The hydrogen molecules were modeled in round particles and amorphous silica was build by the inorganic builders in the VMD. All interactions that occur in this process were defined using the lennard jones equation. The LJ parameters used in this study are shown in Table 1. The placement of amorphous silica particles is carried out regularly in all parts of the cube wall. Whereas for the placement of the hydrogen is done randomly but remains close togther. The results obtained will be in the form of weight percent. The simulation results that carried out on LAMMPS, it can be visualized to see how many atoms are adsorbed on silica

.**Table 1.** Lennard-Jones Interaction Parameter [17].

Atom	σ(Å)	ε(kJ/mol)
H	0	0
Si	3.385	2.4522
O	3.17	0.6503

$$U = 4\varepsilon\left\{\left(\frac{\sigma}{r}\right)^{12} - \left(\frac{\sigma}{r}\right)^{6}\right\} \quad (1)$$

Cross parameter that used to identfy the pair coefficient of different atom, which each particle is termed as i and j is know with Lorentz-Berthelot mixing function.

$$\varepsilon_{ij} = \sqrt{\varepsilon_{ii}\varepsilon_{jj}} \quad (2)$$

$$\sigma_{ij} = \frac{(\sigma_{ii} + \sigma_{jj}}{2} \quad (3)$$

In this study, the simulation conditions was set in fix temperature and some variations in pressure. The temperature was set in 273 K with the variation of the pressure is 1 atm, 2 atm, 4 atm, 6 atm, 8 atm, and 10 atm. In this study we just set one kind of temperature because above 273 K, the atom motion will randomized. So it still difficult for us to capture the interaction. The hydrogen storage capacity in silica can be calculated with equation 4.

$$\rho_W = \frac{N_H.m_H}{N_{Si}.m_{Si} + .N_O.m_O + N_H.m_H}. \quad (4)$$

N are the number of atoms or molecule while m are atomic relative masses. In this simulation, the pressure variations that we use are calculated by the modification of ideal gas equation. So that the amount of hydrogen needed at the specific pressure can be known.

$$(P + a\left(\frac{n}{V_m}\right)^2)(V_m - n.b) = nRT \quad (5)$$

a and b in this equation represent van der Waals coefficient for hydrogen, while V_m is volume of a hydrogen molecule and R is universal gas constant [18].

3. Result and Discussion

This simulation made in different pressure conditions with fixed temperature at 273 K. Amount of pressure that we obserbed in this hydrogen adsorption was 1 atm, 2 atm, 4 atm, 6 atm, 8 atm and 10 atm. The amount of hydrogen in the initial conditions of each pressure varies according to equation 5. From equation 5, we can calculated the amount of hydrogen for initial condition where the result was in table 2. The hydrogen concentration were calculated in different pressure at each 10000 time step. The total running time in each simulation is 200000. The capacity of silica in adsorbing hydrogen in this simulation is shown in table 3. We also take an value in each pressure at the same time step to know the trend at each pressure that wre given in table 4. The results were also plotted in figure 2 and 3.

Table 2. Initial hydrogen molecules amount in various pressure.

Pressure	1 atm	2 atm	4 atm	6 atm	8 atm	10 atm
Amount of Molecules	7	15	30	44	59	74

Table 3. Hydrogen Concentration at 273 K.

Time Step	Pressure (atm)					
	1	**2**	**4**	**6**	**8**	**10**
10000	0	0	0	0	0	0
20000	0	0	0	0	0	0
30000	0	0	0	0	0	0
40000	0	0	0	0	0	0
50000	0.000777932	0.00155567	0	0	0	0
60000	0.00155586	0.00155567	0.0031106	0.0015549	0.000777303	0
70000	0.00155586	0.00155567	0.0031106	0.0023324	0.00310921	0.003108
80000	0.00155586	0.0023335	0.004665	0.006219	0.00621842	0.004662
90000	0.00155586	0.0038891	0.007776	0.009329	0.0108822	0.007771
100000	0.00155586	0.0046670	0.009331	0.010884	0.0124368	0.013988
110000	0.00155586	0.0062226	0.009331	0.012439	0.0155461	0.016319
120000	0.00155586	0.0046670	0.009331	0.015549	0.017878	0.020205
130000	0.00311173	0.0046670	0.009331	0.017104	0.0202099	0.025645
140000	0.00311173	0.0046670	0.010887	0.018659	0.0248737	0.031084
150000	0.00311173	0.0046670	0.010887	0.018659	0.0248737	0.032639
160000	0.00311173	0.0046670	0.013220	0.023324	0.0279829	0.034193
170000	0.00311173	0.0046670	0.013997	0.021769	0.0295375	0.034193
180000	0.00155586	0.0046670	0.013997	0.021769	0.0295375	0.034193
190000	0.00155586	0.0046670	0.013997	0.021769	0.0295375	0.034193
200000	0.00155586	0.0046670	0.013997	0.021769	0.0295375	0.034193

Table 4. Hydrogen Concentration at 273 in 170000 Timestep.

Pressure	1 atm	2 atm	4 atm	6 atm	8 atm	10 atm
Hydrogen Concentration (Wt %)	0.00311173	0.0046670	0.0139978	0.0217696	0.0295375	0.0341933

In Table 2 we can see that the amount of initial hydrogen molecules will increase by the increasing of pressure. This is in accordance with theory of ideal gas equation. The value in table 3. Indicates that at 10 atm the hydrogen concentration become higher.

IOP Conf. Series: Materials Science and Engineering **478** (2019) 012034 doi:10.1088/1757-899X/478/1/012034

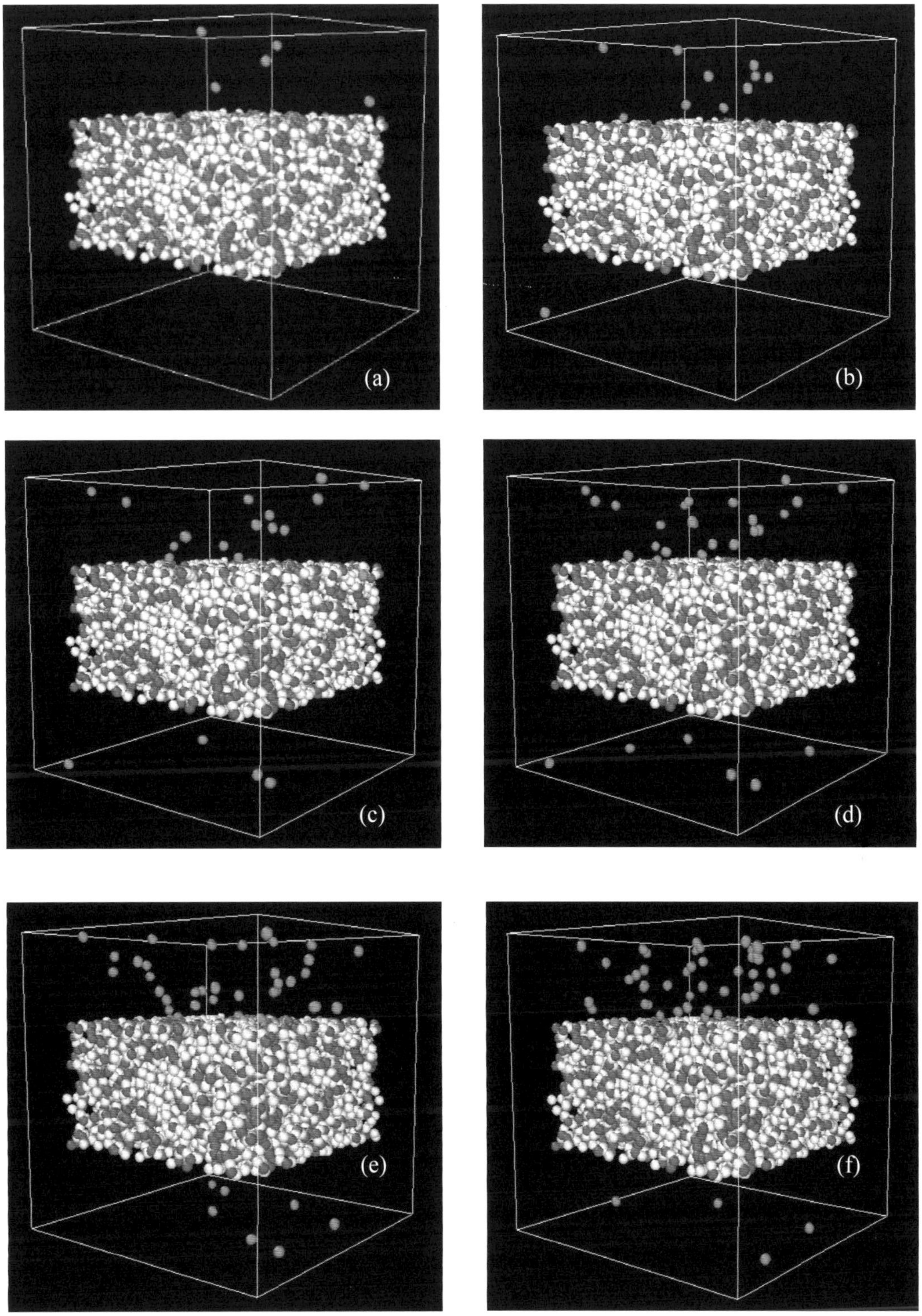

Figure 1. The hydrogen adsorption in170000 time step at (a) 1 atm, (b) 2 atm, (c) 4 atm, (d) 6 atm, (e) 8 atm, (f) 10 atm.

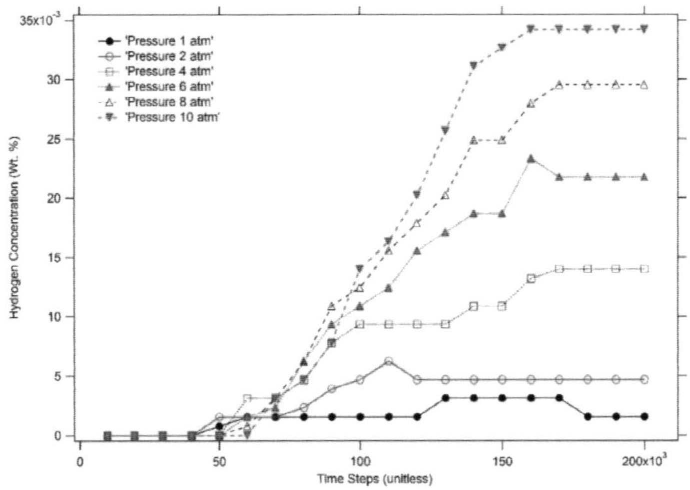

Figure 2. The hydrogen molecules adsoption at 273 K.

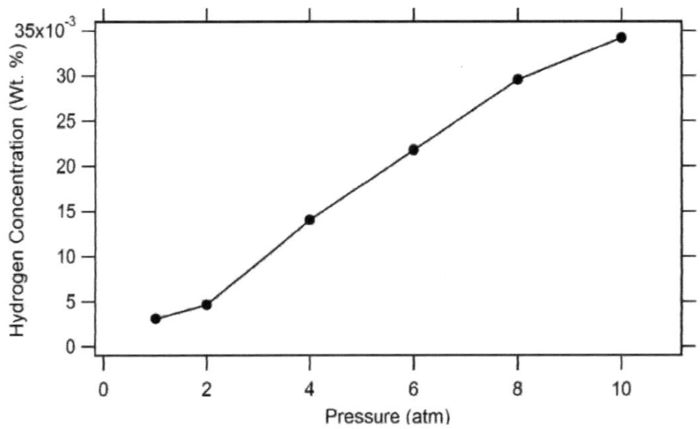

Figure 3. The hydrogen concentration adsoption at 273 K in 170000 time step.

We can see the figure 1. show the mechanism of hydrogen adsoprtion in 170000 time step at various pressure figure 1. The red atom represented hydrogen. Silica molecules were shown by the blue and white atoms, where silicon was blue and the white atom was oxygen. From figure 2, it is shown that the hydrogen adsorption is pressure dependent. As the pressure increases, the hydrogen adsorption will also increases. This is in accordance with another literature that prefers adsorption at high pressure [19, 20]. It also can be seen that with increasing of the running time, the hydrogen concentration that can be adsorbed increases. This graph indicates the amount of adsorption hydrogen in silica will be stable in some running time. The adsorption become saturated in some point because the material has maxiumum capacity to adsorb gas or fluid depending on the interaction between them. In this simulation at all pressure the hydrogen concentration will stable at 180000 time step. From Figure 3 we can know the optimum hydrogen concentration that can be adsorbed with various pressure occurs at the 1700000 running time. We can see that from table 4 that the greatest hydrogen concentration in silica was in 10 atm with 0.0341933 Wt %. This results shows that this simulation can describe the real phenomena of hydrogen adsorption. This simulation can also demonstrated the BET (Brunauer, Emmett and Teller) adsorption measurement.

4. Conclusion

We had succesfully obtained the value of hydrogen concentration in hydrogen adsoption on silica using LAMMPS, VMD as builder and Ovito as visualizer. From the result we can say that silica is on of promising candidate for hydrogen storage. We observed that pressure has a good effect in hydrogen adsorption. When pressure increase, the hydrogen concentration adsorption becomes higher. Our simulation also visualize the real mechanism of hydrogen adsorption by demonstrating time dependence of adsorption, saturation and the pressure dependence of adsorption.

Acknowledgments

This writing is supported by DRPM (Universitas Indonesia Research Directorate) under PITTA2018 grant, Grant No. 2420/UN2.R3.1/HKP.05.00/2018 There is no conflict of interest between the authors and the institution regarding thepublication of this article.

References

[1] Graetz J 2009 *Chemical Society Reviews* **38(1)** 73–82
[2] Durbin D J and Malardier-Jugroot C 2013 *International Journal of Hydrogen Energy* **38** 14595-14617
[3] Kundu PP and Dutta K 2016 *Compendium of Hydrogen Energy* **Volume 4: Hydrogen Use, Safety and the Hydrogen Economy** Woodhead Publishing Series in Energy 2016, 111-131
[4] Ahluwalia R K 2007 *International Journal of Hydrogen Energy* **32(9)** 1251–1261
[5] Zheng J, Liu X, Xu P, Liu P, Zhao Y, and Yang J 2012 *International Journal of Hydrogen Energy* **37(1)** 1048–1057
[6] Aceves S M, Petitpas G, Espinosa-Loza F, Matthews M J, and Ledesma-Orozco E 2013 *International Journal of Hydrogen Energy* **38(5)** 2480–2489
[7] Lototskyy M V, Yartys V A, Pollet B G and Bowman R C 2014 *International Journal of Hydrogen Energy* **39(11)** 5818–5851
[8] Züttel A 2003 *Materials Today* **6(9)** 24–33
[9] Zhang F, Zhao P, Niu M, and Maddy J 2016 *International Journal of Hydrogen Energy* **41(33)** 14535–14552
[10] Froudakis G E 2011 *Materials Today* **14(7–8)** 324–328
[11] Ströbel R, Garche J, Moseley P T, Jörissen L, and Wolf G 2006 *Journal of Power Sources* **159(2)** 781–801
[12] Murray L J, Dinc M, and Long J R 2009 *Chemical Society Reviews* **38(5)** 1294–1314
[13] Dürr M and Höfer U 2006 *Surface Science Reports* **61(12)** 465–526
[14] Lan J, Cheng D, Cao D, and Wang W 2008 *Journal of Physical Chemistry C* **112(14)** 5598–5604
[15] Du X M, Huang Y, and Wu E D 2011 *Recent Trends in Materials and Mechanical Engineering Materials, Mechatronics and Automation* **Pts 1-3 55–57** 1518–1522
[16] Dhaneswara D, Utami S, Adriyani A, Putranto D A, and Delayori F 2018 *IOP Conf. Series: Journal of Physics: Conf. Series* **1011** 012017
[17] Mashayak S Y and Aluru N R 2012 *International Journal Chemistry Theory Computation* **8** 1828-1840
[18] Bratu P, Brenig W, Gro S A, Hartmann M, Höfer U, Kratzer P, and Russ R 1996 *Phys. Rev. B* **54(8)** 5978–5991
[19] Moellmer J, Moeller A, Dreisbach F, Glaeser R, and Staudt R 2011 *Microporous and Mesoporous Materials* **138(1–3)** 140–148
[20] Wu C D, Fang T H, and Lo J Y 2012 *International Journal of Hydrogen Energy* **37(19)** 14211–14216

High-strength aluminum alloy of ultrafine grained by consolidation-ECAP

A Pramono[1], L Kollo[2], L Kommel[2] and R Veinthal[2]

[1]Department Metallurgy, Universitas Sultan Ageng Tirtayasa, Banten, Cilegon, Banten 42435, Indonesia
[2]Department Materials Engineering, Tallinn University of Technology. Ehitaja tee 5, 19086 Tallinn Estonia.

Email: agus.pramono@untirta.ac.id

Abstract. Technology of Severe Plastic Deformation (SPD) has emerged as a widely known procedure for the fabrication of ultrafine grained metals and alloys. Equal Channel Angular Pressing (ECAP) are one of methods of SPD has been developed for many applications; military devices, aerospace and automotive components. This paper examines recent developments related to the use powder of Aluminum alloys processed by consolidation-ECAP, for grain refinement; Al-5Mg, were compared by another type of Aluminum, which has also been researched. Characterization by mechanical, physical and microstructured. In the current study powder of material wrapped in copper capsule to be compressed and heated at a temperature of 400°C in hot press under the pressure of 450-500 MPa. Afterward the powder in solid condition was cooled in the air. The sample results of Aluminum by ECAP process then followed heat treatment with type Annealed, for heating at a temperature of 346°C for 1,5 continuing 415°C during 2 hours followed by heating at a temperature of 205-230°C for 4 hours. This paper describes the effect of heat treatment on the characteristics of the powder material on Al-5Mg are based on changes in the mechanical properties and microstructures.

1. Introduction

The latest development of metal processing technology is Severe Plastic Deformation (SPD), It has updating conventional metal processing technology. For conventional metal forming processes, the imposed plastic strain is generally less than about 2.0. When the pressing forces are carried out up to a plastic strain of greater than 2.0, the thickness and the diameter become very thin and are not suitable to be used for structural parts. In order to impose an extremely large strain on the bulk metal without changing the shape, then SPD are compatible for that processes [1-2]. There are many SPD processes already developed. Various SPD processes such as; Equal Channel Angular Pressing (ECAP) [3], High Pressure Torsion (HPT) [4], Accumulative Roll Bonding (ARB) [5], are a development for the manufacture of metallic materials to achieve superior nature. At this moment, ECAP has emerged as a widely recognized procedure for the manufacture of ultrafine-grained metals and alloys. To grain refinement including modifying of conventional-ECAP in order of improve process in efficiency and techniques on increasing the scale of procedures so that very difficult processing can be smoothly to pressed. All of basic and experimental parameters with alternate functions in microstructure during operations of pressing processes. Significant progress has been made in ECAP development in recent years, so it shows there are excellent prospects for future successful merge of the process. Among SPD technologies, ECAP has advantages for several reasons; First, it can be applied to billets large enough so that there is potential to produce materials that can be used in a wide range of structural applications. Secondly, it is a relatively simple procedure that is easy to do on various alloys and, except for development of dies only, processing by ECAP uses the equipment available in most laboratories. Thirdly, ECAP can be developed and applied to materials with different crystal structures and materials for many ranges from alloy tensile hardening to intermetallics and metallic-matrix composites. Fourth, reasonable homogeneity is achieved through most of the as-pressed billets providing the pressing that keeps the strain high enough. Fifthly, the process may scale-up to suppress relatively large samples and, there is potential for developing ECAP for use in commercial metal-processing procedures [1,3,6]. ECAP is one kind of SPD processes by combining pressing and tensile stress in the mold. ECAP is manufacturing process for the metal that has ability to introducing significant grains refinement into large bulk samples, can produce fine grain structures and combined with oxide inclusion from high- strength particle surface [7-9]. The processing scheme for ECAP Consolidation is shown in Figure 1.

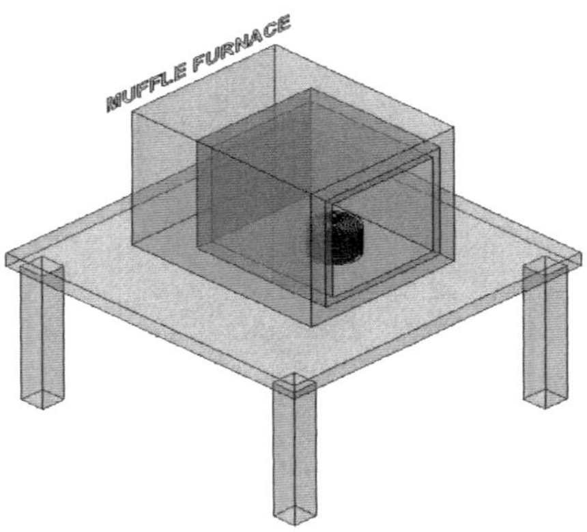

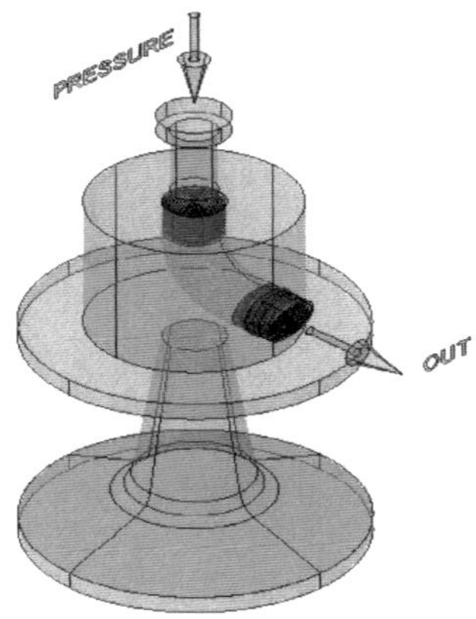

(a) (b)

Figure 1. Schematic processes of Consolidation ECAP: (a) Pre-heating Al-5Mg on the muffle furnace; (b) Pressing Al-5Mg inside in the dies on ECAP Consolidation.

Aluminum Alloys type Al-Mg has good mechanical properties, high corrosion resistance and weldability, it was different processing Al-Mg subject to dynamic loading such as impact or collision with other objects. Therefore, the mechanical properties and constitutive behavior of these alloys at both low and high levels of deformation need to be investigated properly [10]. ECAP combines compressive and tensile stress in the mold to generate products with severe deformation. It has the ability to introducing significant grain refinement into large bulk samples. Typically, it reduces the grain size into the sub-micrometer level, The resulting grain reaching 1-10 μm, on the size of a grain that was ultrafine grained (UFG) and thereby produces materials which are capable to give unusual physical and mechanical properties [3,9]. Although the mechanical and physical properties of all crystalline materials are determined by several factors, the average grain size of the material generally plays a very significant, and often a dominant, role. Thus, the strength of all polycrystalline materials is related to the grain size, d, through the Hall–Petch equation which states that the yield stress, σy, is given by :

$$\sigma y = \sigma o + ky.d^{-1/2} \quad (1)$$

where σo is termed the friction stress and ky is a constant of yielding. It follows from equation (1) that the strength increases with a reduction in the grain size and this has led to an ever-increasing interest in fabricating materials with extremely small grain sizes [3].

This study investigates the relationship microstructure on mechanical properties of the ECAP Consolidation process with heat treatment of Al-5Mg and characterization of hardness and compressive strength to get the relationship between the mechanical properties to microstructures.

2. Experimental Methods

Starting material by Al-5Mg powder, average particle size: ±70μm. The Materials come from Aluminum Powder Company Ltd, ALPOCO. The powder was wrapped in the copper sheet and heated in a resistance furnace at 400-500°C with holding time of 1 hour. The samples were transported to the non-heated ECAP die during 3 – 4 seconds and pressed at the rate of 5 mm/s through the angle of 90° for one pass. For each sample front stopper as back-pressure was used. The pressure for compacting 16×16 mm square specimens was in the range of 300-550 MPa. For improve the properties of Al-5Mg, heat treatment was performed after ECAP Consolidation. Annealing was conducted by heating the sample up to 345°C and holding for 1,5 hours. ECAP consolidation process series of Aluminum Alloys

and heat treatment as shown in Fig. 2

Figure 2. The preparation process of aluminum-based nano powder ECAP Consolidation

3. Characterization of Aluminum Alloys (Al-Mg)

Characterization of mechanical properties was traced by Vickers hardness (HV10) and compressive strength test. All of samples measured on specimens with diameter ± 5mm and height of 8mm. Scanning Electron Microscope (SEM) JEOL JSM-6510LA was used to determine the grain morphology and precipitate distribution. To obtain the results achieved in Al-5Mg, then pure Aluminum is used as a comparison.

4. Result Discussion

4.1. Micros-structured

The microstructure shown fine grains in the matrix phase Al has been spread evenly among Mg_5Al as black color a precipitates, which is randomly dispersed as flat shape precipitates (Figure 3(a)). After the annealed process occurs grain boundary grain growth followed by reduction of aluminum as a matrix size, but Mg_5Al precipitate grain size has magnification as shown in figure 3(b). Annealing at high temperatures, the precipitated particles are embedded and greatly interact with the remaining grain boundaries. As long as the annealing density of the dislocation is reduced due to the effects of the recovery by increasing the ductility [7].

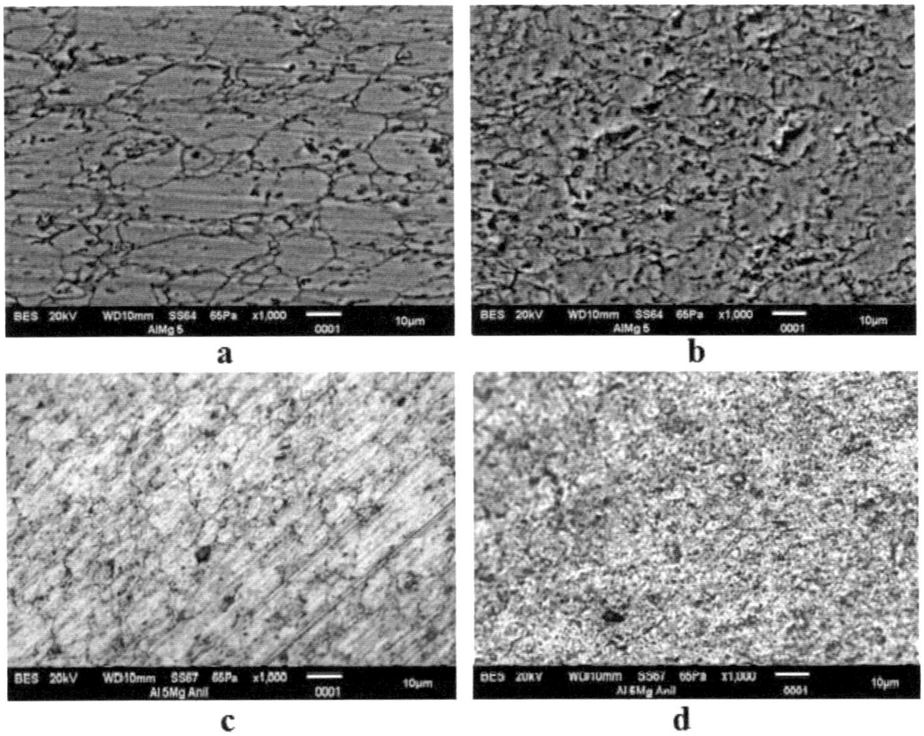

Figure 3. The microstructure of Al-5Mg by ECAP consolidates a). Al-5Mg as received ECAP-Consolidation b). Al-5Mg anneal process c). Calculation of Image-J Al-5Mg as received d). Calculation of image-J Al-5Mg result of anneal

The grain calculations used Image-J, result an average grain size of 9.6μm on the consolidated ECAP shown figure 3(c). After annealing process, the grain changes become 9.2μm, this grain change is not to significant. In figure 3, seen in the annealing process there are Mg5Al that visible precipitates evenly distributed between the aluminum grain boundaries. This condition cause of uniformly distributed hardness and increases mechanical properties. Morphological grains of Al-$_5$Mg by ECAP becomes oval grains with a grain morphological pressure line. The oval grains of ECAP resulted are due to grain shifts from the high force by ECAP pressure process, between the compressive and tensile stresses which make the grains shift into ovals, after receiving an annealing treatment (figure 3(d)) Mg as precipitates spread over several aluminum grain boundaries, the aluminum grain becomes large while the precipitate is spread evenly.

4.2. Mechanical Properties Characterization

Mechanical properties measurements by hardness vicker (HV10) and compressive strength (MPa). Measurement of mecahnical properties to get information of homogeneity of strength Al-5Mg as received by ECAP consolidation and anneling process. UFG capable to increase the strength with great reinforcement on ECAP consolidation, associated with considerable improvements during intensive plastic deformation [8-9]. The Result of Al-5Mg by ECAP consolidated samples was presented in tables 1. Hardness values of Al-5Mg are 148 HV, after annealing, the hardness decreases down to 102 HV. The decline in hardness on the annealing process caused by grain growth in the heating stages, that resulting in coarse grains. This will be followed by an increase of ductility. Increase in ductility cause by the grains growth of equiaxed shaped, resulted in the loss of the dislocation network as well.

Compressive strength in figure 4 shown directly proportional of strength value Al-5Mg by ECAP-Consolidation 450 MPa, higher than compressive strength of annealed process 200 MPa. The compressive strength value will be proportional to the hardness, but after heating of annealing type, then the value of compressive strength and hardness will decrease, this decrease is due to the Al-5Mg as precipitate which is less evenly distributed in the aluminum matrix, so the hardness and strength will tend to decrease, but the annealing will cause the grain coarsening that increases on ductility.

Table 1. Mechanical Properties of Al-5Mg by ECAP-Consolidation.

Materials	Hardness (HV10)	Compressive Strength (MPa)	Ductility (%)
Al-5Mg as it is	148	450	15
Al-5Mg Anneal	102	200	35
AA6061	141	250	30
AA7075	121	395	2.5

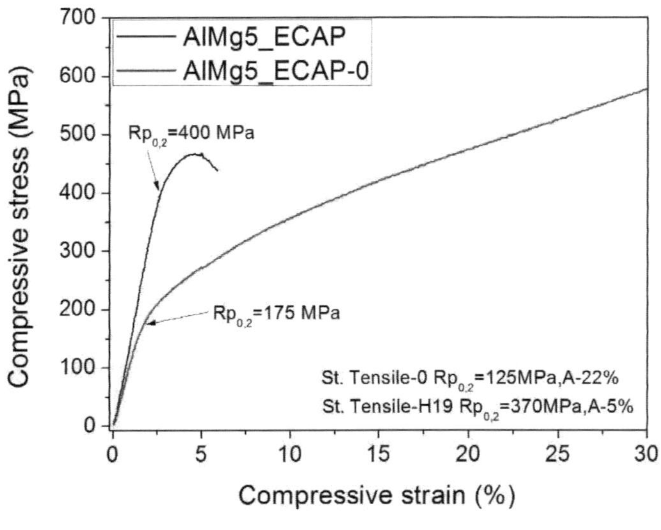

Figure 4. Compressive strength of Al-5Mg by ECAP-Consolidation.

4.3. Comparative results of the aluminum process by ECAP Consolidation.

Aluminum alloys have elements that act as precipitates, Mg_2Si on AA6061 series acts as precipitates [7], $MgZn_2$ inside AA7075 acts as precipitate [8]. Al-5Mg compounds act as precipitates as well. High-strength aluminum alloys containing Mg_2Si, $MgZn_2$ and Al-5Mg, which are easily reached, but for formability processes is not sufficient to be applied. Therefore, much research has focused on increasing the formability of these alloys through heat treatment. High strength aluminum alloy, such as alloy AA7075, which combines the power of high-density ratio with excellent mechanical properties, is widely used for aviation applications, superplastic forming into high-volume fabrication of components in the automotive. Consumer products industry is currently limited because of the relatively low strain rate. AA6061 are some of the alloys materials, mostly used currently covering the whole range of industry. Applications of AA6061 easier to process than AA7075 because of properties the material elasticity, AA7075 tend to form internal cracks as a result of micro-segregation. That way heat treatment is required to improve the properties of the material when it is processed by ECAP [11]. At Al-5Mg the most superior character is compressive strength, compared with: AA6061 and AA7075. To increase ductility applied heat treatment of annealing type. The heating process is needed as a step for the application of component manufacturing processes. But compared to other kind of aluminum, AA7075 was lower on formability, this is seen in low ductility. In figure 5 it has been proven that research with AA7075 will tend to form cracks during the process. Workability of AA7075 usually constrained by the microstructure of the material, deposit of precipitates such as $MgZn_2$ element typically forming a micro-segregation, which lead to cracking when processed materials [12]. Based on plastic strains of structure of the alloys was determined, Fig 5 represents the results of the failure of sample formation process due to limited ductility AA7075 due to micro-segregation of produce cracks [11-12].

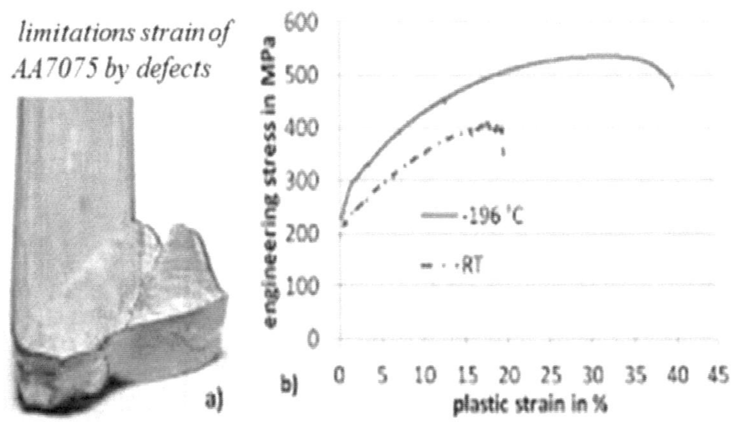

Figure 5. Limitations due to the formation of cracks AA7075 by ECAP process [12].

4. Conclusions

The microstructure shown fine grains in the matrix phase Al has been spread evenly among Mg5Al as black colour a precipitates, which is randomly dispersed as flat shape precipitates. After the Anil process occurs grain boundary grain growth followed by reduction of aluminum as a matrix size. Annealing at high temperatures, the precipitated particles are embedded and greatly interact with the remaining grain boundaries.

The decline in hardness on the annealing process caused by grain growth in the heating stages, that resulting in coarse grains. This will be followed by an increase of ductility. Increase in ductility cause by the grains growth of equaxed shaped

Compressive strength directly proportional of strength value Al-5Mg by ECAP-Consolidation 450 MPa, higher than compressive strength of annealed process 200 MPa. The compressive strength value will be proportional to the hardness, but after heating of annealing type, then the value of compressive strength and hardness will decrease, this decrease is due to the Al-5Mg as precipitate which is less evenly distributed in the aluminum matrix, so the hardness and strength will tend to decrease, but the annealing will cause the grain coarsening that increases on ductility.

Al-5Mg the most superior character is compressive strength, compared with: AA6061 and AA7075. To increase ductility applied heat treatment of annealing type. The heating process is needed as a step for the application of component manufacturing processes

References

[1] Vladimir M and Segal 1999 *Materials Science and Engineering* **A271** 322–333
[2] Azushima A, Kopp R, Korhonen A, Yang D Y, Micari F, Lahoti G D, Groche P, Yanagimoto Y, Tsujii N, Rosochowski A, Yanagida A 2008 *CIRP Annals - Manufacturing Technology* **57** 716-735
[3] Valiev R Z and Langdon T G 2006 *Progress in Materials Science* **51** 881-98
[4] Valiev R Z, Krasilnikov N A, and Tsenev N K 1991 *Materials Science and Engineering A* **197** 35–40
[5] Saito Y, Utsunomiya H, Tsuji N, and Sakai T 1999 Acta Metallurgica **47** 579-583
[6] Valiev R Z and Langdon T G 2011 *Metallurgical and Materials Transactions A* **42 (10)** 2942-2951
[7] Agus P, Lauri K, Kaspar K, Renno V, and Jaana-Kateriina G 2015 *Applied Mechanics and Materials* **771** 252-256
[8] Agus P, Lauri K, Kaspar K, Renno V, and Jaana-Kateriina G 2014 *Key Engineering Materials* **604** 273-276
[9] Agus P, Lauri K, Kaspar K, Renno V, and Jaana-Kateriina G 2013 *Proceeding Toward Sustainability And Quality of Life* **1** 66-72
[10] He L, Zhang H, and Cui J 2009 *Journal of Wuhan University of Technology-Mater Sci Ed* · 198 - 201
[11] Agus P, Anistasia M, Kurnia N, and Moch F 2017 *Vanos-Journal of Mechanical Engineering Education* **2 (2)** 117-126
[12] Fritsch S, scholze M, and Wagner M F X 2012 *Materialwissenschaft und Werkstofftechnik (Materials Science and Engineering Technology)* 43 (7) 561–566

IOP Conf. Series: Materials Science and Engineering **478** (2019) 012036 doi:10.1088/1757-899X/478/1/012036

Policy on mineral resource management in the good governance perspective

T Hidayat and Y Pramadi

Research Unit for Mineral Technology – Indonesian Institute of Sciences
Jl. Ir. Sutami Km 15 Tanjung Bintang Lampung Selatan

Abstract. God lavishes the country with abundant mineral resources, so we proudly say "what do we not have?". But on the other hand, there are still many residents of this country who are still far below the poverty line. Of course, a follow-up question "what's wrong?". The research method used is qualitative method used through descriptive explorative approach. Explorative descriptive approach is done by document study. After the research concludes that there has been a "mismanagement" of mineral resources management, every mineral resource management policy should be based on the principles of good governance which are the principles of good governance, namely: 1) Participation, 2) Rule of law, 3) Transparency, 4) Responsive, 5) Consensus Orientation, 6) Equity, 7) Effectiveness and efficiency, 8) Accountability, 9) Strategic vision.

1. Introduction

Perhaps God created Indonesia in a "happy" state. Why is that ?, Because Indonesia is overflowing with abundant mineral resources, so proudly we as residents can say "what the hell do not we have?".

Based on the 1945 Constitution, the mineral resources are controlled by the state. The right to control the state shall be authorized to regulate, manage and supervise its management and / or control. The right is accompanied also with the obligation to use it for the greatest prosperity of the people. Thus, the use of mineral resources should benefit the people. The pattern of mineral resource management identical to environmental crime, crimes against humanity and state financial crime must be stopped.

Then, open conflict between government and corporate on the one hand with the people of stakeholders on the other hand, is still being held in many locations of mineral resource exploitation in Indonesia and still has not shown any signs of subsiding. This is due to the philosophy of development adopted not so much as to the fate of the people

Many encounters rejection of exploration activities and exploitation of mineral resources. Not to mention the environmental damage caused by the exploration and exploitation activities. For example, PT. Freeport has disposed of tailings by the category of B3 waste (Hazardous Toxic Substances) through the Ajkwa River. This waste has reached the coast of the Arafura sea. The tailings that Freeport dumped into the Ajkwa River exceeded the total suspended solid (TSS) standard allowable under Indonesian law. Freeport's tailings waste has also polluted the waters of the Ajkwa River estuary and contaminated large numbers of living things and threatened the waters with large amounts of acidic mine water. From the environmental audit conducted by Parametrix, it was revealed that Freeport's exhausted tailings were a material capable of producing harmful acidic liquids for aquatic life [1]

Furthermore, the results of research from Rumzi Samin and friends [2] with the title of Impact of Mining Against Socio-Economic Conditions of People of Tanjung Pinang Sub-district, Kepulauan Riau Province, indicated by the mining there was no significant influence on formal and informal education because most were financed the community itself, while from the entrepreneur there is no Corporate Social Responsibility (CSR) it. From the point of view of employment: only a small proportion of the population are employed in mining on the grounds that they do not have adequate skills, most of whom work as fishermen. From the point of income of the people: the income of the people tends to decrease as the catchment area and the catch of some fishermen are reduced due to the mining.

It should be, rich countries can menyeterahkan people. However, there are still many residents of this country who are still far below the poverty line. Of course, the question arises "what's wrong?"

The question of "what is wrong?", Will further elaborate the mineral aspects of policies made in mineral resources management, especially in the perspective of good governance.

2. Conceptual Framework

The policy referred to dmineral this paper is public policy / public policy. The earliest definition of public policy was proposed by Harold Laswell and Abraham, Kaplan Howlett and Ramesh [3] which defines public policy as "a program projected with specific objectives, values, and practices (a projected of goals, values, and practices) ".

The definition of public policy also includes mineral Attachment 1 of the Regulation of the Minister of Administrative Reform Number PER/04/M.PAN/4/2007 on General Guidelines for Formulation, Implementation, Performance Evaluation, and Revision of Public Policy in Central and Regional Government Institutions. This Ministerial Regulation, public policy is "a decision made by government or government agencies to overcome certain problems, to perform certain activities or to achieve certain goals related to the interests and benefits of the people". Mineral Ministerial Regulation, public policy has 2 (two) forms of codified codified formally and legally, and public statement publicly.

Based on both definitions, public policy has the following concepts: a. Public policy contains its goals, values, and practices; b. These public policies are made by government agencies, not private organizations; c. The public policy concerns the choice made or not by the government.

The word "Management" can be likened to "management", which also means setting or managing. Many people define management as setting, managing and administering, and that's what popular today is. Management is defined as a series of work or business undertaken by a group of people to perform a series of mineral work to reach a particular goal [4].

Understanding of management has been discussed by many experts who among others complement each other. Stoner was quoted by Handoko as saying that management is the process of planning, organizing, directing, and supervising, the efforts of organization members and users of other organizational resources to achieve organizational goals that have been established. Stoner emphasizes that management is focused on processes and systems. Therefore, if the system dmineral and planning process, organizing, directing, budgeting, and monitoring system are not good, the overall management process is not smooth so that the process of achieving the objectives will be disrupted or failure [5].

The State Administration Institute [6] defines good governance as a solid and responsible state administration, efficient and effective by maintaining a "synergistic" constructive interaction between state domains, the private sector and society. At this level, good governance is oriented to 2 (two) main points, namely: First, the ideal orientation of the country directed at the achievement of national goals. At this level, good governance refers to the democratization of the dmineral of state life with its constituent elements, such as legitimacy, accountability, scuring of human rights, autonomy and devolution of power and assurance of civilian control; Second, an ideal functioning government that effectively and efficiently mineral make efforts to achieve national goals. In this context, good governance depends on the extent to which structures and political and administrative mechanisms function effectively and efficiently.

From some sense of good governance above, it can be identified the principles contained dimineralnya. UNDP [7], proposes the characteristics of good governance, as follows: 1) Participation; 2) Rule of law; 3) Transparency; 4) Responsive; 5) Consensus Orientation; 6) Equity; 7) Effectiveness and effeciency; 8) Accountability; 9) Strategic vision.

Thus, the conceptual framework of the research undertaken is the actions undertaken by the government with a view to achieving the objectives that have been formulated dmineral conduct activities to process data or data collection conducted by the organization of the state well and regularly on the management of MINERAL RESOURCES mineral perspective good governanc**e.**

3. Research Methods

The method used mineral this research is qualitative research method [8]. Qualitative method used by explorative descriptive approach. The descriptive explorative approach is done by document and field study. Document and field studies are conducted to find information on the MINERAL RESOURCES MANAGEMENT Policy in Indonesia. Document studies conducted at related institutions. After obtained the data will be analyzed descriptively explorative.

4. Results and Discussion

In the preparation, implementation and evaluation of any public policy pertaining to the management of mineral resources in Indonesia, it should be based on the principles contained by good governance.

As previously outlined the principles are:

Participation; Every citizen has a decision-making voice, either directly or in an intermediate institution of legitimacy representing his or her interests. Such participation is built on the basis of the legitimacy of association and of speaking and participating constructively. Public policy formation is done through a process often called public policy formulation. This process begins the input (input) in the form of demands and support from the community related to community life. The inputs are grouped or identified one by one into proposals. Proposed or recommended inputs are discussed together by policy makers such as the government, senat/local senat, community leaders, religious leaders, and academics. The discussion results in a joint decision called policy or output (output). The output or output is then applied and evaluated. The importance of public participation in the formulation of public policy is the result of cooperation of various actors, both government, society, experts, and social institutions. The forms of public participation of the mineral to formulate public policy can be done by: 1) The community can form opinions (thoughts) through the mass media that the community desperately need the welfare, mineral every policy of mineral resources management. 2) Communities provide input to the problems facing the community by sending information to the government through various by showing facts in the field. 3) Delivering their aspirations when members of senat visit villages/villages that are villagers/villages. 4) To criticize public policies issued by the government that are not impartial to the interests of the community. 5) Participation can also be demonstrated by providing moral support to the policy makers. Dmineral this stage of society must play an active role to control whether the input from the community is accommodated or not.

Rule of law; The legal framework should be fair and undertaken indiscriminately, especially the law for human rights. In the implementation of public policy is brought back through the laws and regulations that are born to accommodate its purpose. Law as social control as well as social construction tools must necessarily capture the complexity between dialectical relationships of various factors. Thus, the ministerial policy on the management of MINERAL RESOURCES in Indonesia must be fair and have a positive impact on the community.

Transparency; Transparency builds on the validity of information flows. Processes, institutions and information that are directly acceptable to those in need. Good governance demands the government to ensure the open access of information to the public to public policy of the government, both the context of the public policy process, the allocated budget for policy implementation and the evaluation and control of the policy practice undertaken. The openness of community information access here becomes important so that the community can oversee the process of implementing government policies so that the community can ensure that the policy is actually implemented for the benefit of the people's welfare. Furthermore, information on governance administration has the benefit to anticipate corruption practices on the implementation of government activities in the form of budget leak, collusion, which makes the practice of public policy implementation is not optimal.

Responsive; Institutions and processes should try to serve every stakeholder. In this case, any policy adopted as a management framework for MINERAL RESOURCES shall be "serve" the community as a stakeholder, not the other way around.

Consensus Orientation; Good governance mediates the different interests to get the best choice for the wider interests, both the dmineral policies and the procedures. Good governance will act as a mediator for different interests in order to achieve the best possible consensus or opportunity for the interests of each party, and where it is possible to apply to policies and procedures established by the government and to the broader interests of society.

Equity; every society has equal opportunity to gain prosperity and justice. A policy of oriented smoothing is a policy whose outcome is fairly distributed.

Effectiveness and effeciency; Processes and institutions produce according to what has been outlined using the best available resources. Every process of activities and institutions is directed to produce something that really suits the needs through the best utilization of the various resources available and the management of public resources is done efficiently (efficient) and effective (effective).

Accountability; mineral government, private sector and civil society decision makers are responsible to the public and stakeholder institutions. This accountability depends on the organization and nature of the decision made, whether the decision is for the internal or external interests of the organization. Mineral decision makers of public, private, and civil society organizations have public accountability for each activity of the activities undertaken.

Strategic vision; Leaders and the public should have a broad and far-reaching perspective of good governance and development in line with what is required for such development. Leaders and communities have broad and long-term perspectives on good governance and human development, along with the need for development.

Until late 2014, the government's policy on applying the principles of good governance in terms of management of MINERAL RESOURCES is still far from satisfactory results. The governance of MINERAL RESOURCES which has not been based on the principles of good governance has resulted in increasing distance between communities and surrounding minerals. In addition, the uneven distribution of and utilization of MINERAL RESOURCES also causes many communities, including indigenous and tribal peoples, to be only dmineral viewers of the utilization of their surrounding mineral resources.

Implementation of good governace conducted by the government so far there are still many gaps that must be improved mineral terms of management of MINERAL RESOURCES as an improvement of the national economy. One of them aadalah, where the management that happened during this is the number of profits gained by foreign rather than the profits derived by the state itself. Furthermore, profit-sharing is still very concerning, where the benefits are shared by more central government runs than to local governments. This is what makes many protests against the management of MINERAL RESOURCES in Indonesia. The management of MINERAL RESOURCES is not yet maximized, it is still beneficial to foreigners and most foreign companies do not pay attention to the surrounding environment, the community economy, and the principle of sustainable development. Where the principle of sustainable development is meeting the needs of the present without sacrificing the needs of future generations. But what happens next generation needs will soon be taken up by foreigners if there are no restrictions and no firm rules governing mineral resources management.

5. Conclusions and Suggestions

The principles contained mineral good governance should be the "life" dmineral every preparation, implementation and evaluation of public policy management of MINERAL RESOURCES in

Indonesia. It is imperative that all of this mineral wealth be utilized to the greatest possible prosperity of the people as mandated by the 1945 Constitution

Acknowledgment
The authors acknowledged financial and facility supports from Research Unit of Mineral Technology, Indonesian Institute of Sciences.

Reference

[1] WALHI 2006 *Laporan WALHI Tentang Dampak pencemaran Lingkungan Hidup OperasiFreeport-Rio Tinto di Papua*, Jakarta.
[2] Rumzi Samin dan kawan-kawan 2008 *Dampak Penambangan Terhadap Kondisi sosial ekonomi Masyarakat Kecamatan Tanjungpinang Kota Provinsi Kepulauan Riau*, Jurnal Perbatasan FISIP UMRAH, Riau.
[3] Howlett, Michael, and Ramesh M 1995 *Policy Sciences and Political Sciences Mineral Studying Public Policy: Policy Cycle and Policy Subsystem* Toronto: Oxford University Press.
[4] Arikunto S 1993 *Prosedur Penelitian Suatu Pendekatan Praktek*. Jakarta: Rineka Cipta
[5] Qalyubi S 2007 *Dasar-Dasar Ilmu Perpustakaan dan Informasi. Yogyakarta: Jurusan Ilmu Perpustakaan dan Informasi (IPI)* Fakultas Adab UIN Sunan Kalijaga
[6] Peraturan Menteri Pendayagunaan Aparatur Negara Nomor PER/04/M.PAN/4/2007 tentang Pedoman Umum Formulasi, *Implementasi, Evaluasi Kinerja, dan Revisi Kebijakan Publik di Lingkungan Lembaga Pemerintah Pusat dan Daerah*
[7] Bhattacharya 1972 *Administrative Organization for Development, International Institute of Administrative Science* Brussel
[8] Bogdan, Robert C, and Steven J T 1992 *Introduction to Qualitative Research Methods: A Phenomenological Approach in the Social Sciences* John Wiley and Sons

The study of low carbon steel pack carburizing using cow bone and coconut shell

Miswanto[1], T O Rajaguguk[1], and S Sumardi[2]

[1]Department of Mechanical Engineering, Malahayati University, Lampung 35144, Indonesia
[2]Research Unit for Mineral Technology, Indonesian Institute of Sciences, Lampung 35361, Indonesia

Email: slametsumardi99@gmail.com

Abstract. Despite of the ability of cow bone and coconut shell to be able to be processed into a charcoal, so far these have been only used as fuel or left aside becoming waste, in order to improve the economy value of cow bone and coconut shell waste, it requires processing into a charcoal. The cow bone and coconut shell can be obtained from copra farmers and cow slaughtering house farmers. In this research cow bone and coconut shell charcoal were used as carbon sources to improve low carbon steel hardness. The objective of this research was to find out the influences of temperature and charcoal composition to low carbon steel hardness. A process to use in this research was pack carburizing by using cow bone and coconut shell charcoal and applied temperatures of 900°C, 950°C, and 1000°C during 3 hours of retention. After pack carburizing, the %C content was improved between 0.710% and 1.20%. This improvement was followed by hardness value up to 64.04 kgf/mm^2 at 900^0 C, 64.20 kgf/mm^2 at 950°C, and 64.30 kgf/mm^2 at 1000°C.

1. Introduction

Damage is often found in metal material, and some of its causes are low carbon percentage and incompatibility between the metal characteristics and desired strength [1,2]. This may cause that the casting industry reduces carbon level at steel casting, so that it influences hardness score and steel classification [3]. Hence, the objectives of this research were to improve metal carbon level and hardness.

This research used pack carburizing method by using cow bone [4] and coconut charcoal media. In this method, carbon was added to the low carbon material by using cow bone and coconut shell charcoal media. Cow bone and coconut charcoal were selected because of its abundant availability. Data from National Statistical Bureau in 2015 showed that the cow meat productions in Java, Sumatera, Sulawesi, Kalimantan, Bali/Nusa Tenggara, Maluku/ Papua islands were 297,063 kg/year; 99,251 kg/year; 34,063 kg/year; 31,334 kg/year; 32,735 kg/year; and 10,373 kg/year respectively. The highest and the lowest cow productions were in Java Papua islands respectively. The high cow production in Indonesia affects the environment because of large amount of cow bone waste.

Culinary business in Indonesia has been using cow been so far, and some industries also use cow bone as adsorbent. However this cow bone usage does not influence significantly the amount of cow bone waste in Indonesia, because the cow meat consumption increase each year in Indonesia while it is not followed by using the waste from cow slaughtering houses. Based on this problem, to help government in overcoming waste in Indonesia, the author used cow bone and coconut shell charcoal as the carbon source in the pack carburizing. Cow bone has sufficient carbon element to use as media in pack carburizing.

2. Experimental Procedures

2.1. Charcoal Making

Charcoal making was done in a controlled air supply system, where the burning process was controlled by controlling air supply to the burning chamber. In the part where charcoal has been burnt up into charcoal, the air supply ventilation hole was closed and the holes in upper line were opened so that the burning process would only go to the opened ventilation holes. This process was repeated until the last ventilation hole on the upper line was closed. In this process the material used for carbon addition was inserted into the can or container which has been ready with the flames, and then it was piled up again with charcoal as burner medium and then the container was closed tightly until the desired temperature and retaining time, or until all of the material has became charcoal.

In this research the author used cow bone and coconut shell charcoal media. To make a coconut shell, the coconut shells were first cracked and then were burnt into charcoal. The same process was applied for the cow bone. The cow bone was collected, cleaned, and dried under sunrays at 30°C temperature for 16 hours. After drying, the cow bone was crashed to facilitate burning process and charcoal making. The flow chart of making charcoal can be seen in Figure 1 and Figure 2.

Figure 1. Cow bone Charcoal making

Figure 2. Coconut shell Charcoal

2.2. Carbon content analysis of the charcoal
The following analyses are required to calculate the total carbon content of the charcoal.

a) Analysis of total water content (SNI 13-376-1994)

$$m_{ar}\,(\%) = \frac{m2-m3}{m2-m1}\,x\;100$$

m_1 = empty dish weight (gram)
m_2 = dish + sample weight (gram)
m_3= dish + sample weight after heated (gram)
mar = total water content in the sample

b) Analysis of moisture in the air (SNI 13-3477-1994)

$$mad\;(\%) = \frac{m2-m3}{m2-m1}\,x\;100$$

c) Analysis of volatile matter: (SNI 13-3999-1995)

$$volatile\;(\%) = \frac{m2-m3}{m2-m1}\,x\;100 - mad$$

d) Analysis of ash content (SNI 3478: 2010)

$$ash\;content\;(\%) = \frac{m3-m1}{m2-m1}\,x\;100$$

e) Analysis of fixed carbon: (SNI 13-3479-1994)

$$Fc\;(\%) = 100 - (mad + ash\;content + volatile)$$

2.3. Filtering charcoal particle size
In this stage, charcoal was sieved to obtain desired particle size according to the need for *pack carburizing* [4,5].

2.4. Carburizing specimen making
Steps in making specimen are follows. Specimens were cut in a striped plate of 70 mm x 20 mm x 3 mm by using cutting. After specimens were cut into desired dimension, specimens were then polished by sanding process with polishing machine.

2.5. Container making
In the solid material carbonizing (*packcarburizing*), a box or container was required to place the specimen and charcoal. The container box was made of 2 mm thick steel plate which was cut and welded in to the dimension of 100 mm long x 90 mm wide x 50 mm height. The container box cover was made precisely according to the container box dimension to ease opening while the container box was still hot.

2.6. Specimen cleaning
The specimen was cleaned impurities attached to the specimen, such as impurities that might be attached at polishing process to remove rust and to get flat surface. The objective of this stage was to have good process of diffusion in the carbonizing process and to get good testing results. The sample cleaning processes were as follow. The physical cleaning process was done by sanding with grinding machine to make smooth the uneven-surface and to remove scratch and chips on the specimen surface.

3. Results and discussion

3.1. Charcoal Making Result
Table 1 show that the time required to burn 9 kg of cow bone into charcoal was 330 minutes to produce 2.5 kg charcoal. The yield from cow bone burning was 27.78%. It took 279 minutes to burn 8 kg coconut shell into 1.16 kg charcoal. The yield was 29.12%.

Table 1. The yield result of charcoal making

Burning	Initial weight (kg)	Charcoal product (kg)	Time (t)	Yield (%)
Cow bone	9	2,5	330	27,78
Coconut shell	8	1,16	279	29,12

Some analyses on the coconut shell to use as pack carburizing media were done. The result can be seen in Table 2. The average score of total of water content is 6.5%, moisture is 0.33%, volatile matter is 14%, and ash content is 3%. The total carbon in fixed carbon analysis is 82.67% C.

The same analyses were done to the cow bone charcoal. Table 3 shows that the average score of total of water content is 1.55%, moisture is 0.35%, volatile matter is 16.50%, and ash content is 80.00%. The total carbon in fixed carbon analysis is 3.15% C.

Total carbon of the cow bone charcoal shows the lowest score of %C, but cow bone contains of high content of calcium phosphate and calcium carbonate. Both calcium phosphate and calcium carbonate are used as *energizer* in the *pack carburizing* process (Pramuko and Purboputro, 2006). Table 2 presents chemical composition of cow bone.

Table 2. Chemical composition of the charcoal

Chemical composition	Cow bone charcoal	Coconut shell charcoal
Total of Water content	1,55%	6,50%
moisture	0,35%	0,33%
Volatile matter	16,50%	14,00%
Asch content	80,00%	3,00%
total of carbon	3,15%	82,67%

3.2. Chemical composition test

Table 3 shows that the higher carburizing temperature, then the higher the carbon (C) addition is to the specimen. This can be seen from the increase of % C in each respective temperature variation. It increases up to 0.89% at 900°C form 60% coconut shell charcoal + 40% cow bone charcoal. 1.12% C increase also occurs at 950°C temperature with 100% cow bone charcoal. It is followed 1.12% C increase at 1000°C with 100% coconut shell charcoal. This carbon increase is influenced by the higher temperature during carbonizing process, where carbonizing media at higher temperature is degraded into CO and then degraded into active carbon is able to diffuse into specimen, so that the carbon content in the specimen increases [6].

Table 3. The metal content of specimen before and after *pack carburizing* process.

Temperature (°C)	Time (Hour)	Charcoal Composition	Metal Element (%)						
			Fe	C	Si	Mn	P	S	Cr
RM			98.3	0.106	0.157	0.769	0.021	0.016	0.112
900	3	100% AB	97.6	0.776	0.148	0.695	0.069	0.076	0.096
		80% AB + 20% AT	97.6	0.710	0.157	0.762	0.012	0.051	0.108
		60% AB + 40% AT	97.7	0.897	0.146	0.664	0.018	0.026	0.092
		40% AB + 60% AT	97.6	0.886	0.153	0.690	0.018	0.020	0.099
		100% AT	97.6	0.846	0.139	0.701	0.090	0.057	0.106
950	3	100% AB	97.5	1.00	0.148	0.683	0.013	0.016	0.106
		80% AB + 20% AT	97.5	1.05	0.153	0.656	0.013	0.015	0.100
		60% AB + 40% AT	97.4	1.08	0.151	0.672	0.014	0.017	0.112
		40% AB + 60% AT	97.4	1.09	0.147	0.666	0.016	0.017	0.094
		100% AT	97.4	1.12	0.159	0.686	0.016	0.024	0.105
1000	3	100% AB	97.4	1.20	0.152	0.669	0.011	0.019	0.097
		80% AB + 20% AT	97.3	1.15	0.145	0.639	0.026	0.017	0.091
		60% AB + 40% AT	97.4	1.12	0.154	0.667	0.017	0.020	0.102
		40% AB + 60% AT	97.4	1.14	0.153	0.685	0.011	0.016	0.102

100% AT	97.3	1.15	0.152	0.730	0.018	0.016	0.113

Noted: RM=Raw Material, AB=Ccoconut shell charcoal, AT= Cow bone charcoal

3.3. Rockwell hardness test result

Rockwell hardness test was conducted at the base material before the same test was conducted to the pack carburizing specimen. The Rockwell hardness test at the base material was done in the spherical steel ball with initial load of 10 kg and total load of 150 kg with Rockwell B hardness test (HRB). The result is shown in Table 4.

Table 4. Rockwell Hardness Test (HRB) to the *Raw Material*

A	B	C	D	E	Average
57.2	58.5	57	57	55	
56.5	59	56.2	58	55.4	
56.7	60.8	58	57	55	
56.8	**59,43**	**57.06**	**57.33**	**55.13**	**= 57.15**

The Rockwell hardness test (HRB) at the base material, which is shown in the Table 5, obtained the average score of hardness of 57.15 kgf/mm^2.

Table 5. Rockwell Hardness Test (HRC) by using varying charcoal composition

Temperature (^{o}C)	Time (Hour)	Charcoal Composition	Rockwell Hardness Test (HRC)
		100% AB	60.63
		80% AB + 20% AT	61.28
900	3	60% AB + 40% AT	62.24
		40% AB + 60% AT	63.62
		100% AT	64.04
		100% AB	64.20
		80% AB + 20% AT	63.15
950	3	60% AB + 40% AT	62.72
		40% AB + 60% AT	62.03
		100% AT	61.43
		100% AB	64.30
		80% AB + 20% AT	63.83
1000	3	60% AB + 40% AT	63.21
		40% AB + 60% AT	62.63
		100% AT	62.08

*Note: AB = coconut shell charcoal, AT = cow bone charcoal

Data of Rockwell hardness test (HRB) shows that the base material hardness score is 57.15 kgf/mm^2. Data in the Rockwell hardness test (HRC) in the material undergoing pack carburizing shows hardness improvement. It could be seen in the indenters used in the Rockwell hardness test. Steel ball indenter was used in the Rockwell hardness test (HRB) of base material, while diamond cone indenter of the material undergoing carburizing was used in the Rockwell hardness test (HRC).

Table 5 shows that the hardness score at 900°C temperature shows significant hardness to the material (64.04 kgf/mm^2) at charcoal composition of 100% AT (cow bone charcoal). The 950°C temperature with 100% AB (coconut shell charcoal) show highest hardness improvement at 64.20 kgf/mm^2, while at 1000^{0}C highest improvement with 100% AB is 64.3 kgf/mm$^{2.}$

4. Conclusion

The method used in this research was the controlled air supply system. Coconut shell charcoal making took 279 minutes at 450°C, while cow bone charcoal making took 330 minutes at 746^{0}C. The higher the temperature, the higher was the score of %C between 0.710 %C to 1.20 %C. At 900°C, the charcoal composition of 60% AB (coconut shell charcoal) + 40% AT (cow bone charcoal) produced the best hardness carbon improvement of 0.897 %C. 100% AT produced best carbon-improvement of 1.12 %C at 950°C. While at 1000°C, 100% AB produced the best carbon improvement of 1.20 %C. Therefore, the higher the temperature was the higher the score of %C. The initial *low carbon steel* hardness score was 57.15 kgf/mm^2. After the carburizing process, it increased up to 64.04 kgf/mm^2, 64.20 kgf/mm^2, 64.30 kgf/mm^2 at 900°C, 950°C, and 1000°C respectively. Therefore, the higher temperature was the greater the hardness at both composition of 100%AT (cow bone charcoal) and 100%AB (coconut shell charcoal).

Acknowledgment

The authors would like to thank the Research unit for Mineral Technology and Malahayati University that has facilitate for this research.

References

[1] Smallman R E 1991 *Metalurgi Fisik Modern* PT Gramedia Pustaka Utama, Jakarta
[2] Surdia T 1996 *Teknik Pengecoran Logam* Pradnya Paramitha, Jakarta
[3] Callister 2009 *Materials Science and Engineering: an Introduction* Eight Editian Wiley, USA
[4] Ihom A P 2013 *African Journal of Engineering* **Vol. 1** 97-101
[5] Pramuko I dan Purboputro 2006 *Jurnal Media Mesin* **7(1)** 9-16
[6] Mujiyono dan Arianto L S 2008 *Jurnal Teknik Mesin* **10(1)** 8-14

IOP Conf. Series: Materials Science and Engineering **478** (2019) 012038 doi:10.1088/1757-899X/478/1/012038

Addition of feldspart minerals as a substitution material on ceramics clay manufacture

M Amin[1], Y I Supriyatna[1], Y Hendronursito[1], D C Birawidha[1], Maulida[2], Latifah[2], and P Karokaro[2]

[1]Research Unit for Mineral Technology – Indonesian Institute of Sciences, Jl. Ir. Sutami Km 15, Tanjung Bintang, Lampung Selatan
[2]Physics Department, Faculty of Mathematics and Natural Science – Lampung University

Email: muha047@lipi.go.id

Abstract. The use of clay minerals as the main raw material for pottery and sand additives has many weaknesses, among others, easy to crack, have high water absorption and low compressive strength. Therefore it is necessary to develop a better method of making traditional ceramics by providing additional materials such as mineral feldspar. Mineral feldspart can be used as a ceramic raw material because it is functional as filler and glass forming phases. Clay and mineral feldspart raw materials were analyzed using XRF method, followed by mixing with the composition of clay soil mixture as much as 65% compared to feldspart by 35%, printed with the shape of cube size 50x50x50mm. Drying in oven with temperature 110°C for 8 hours. Combustion in muflle furnace at temperature variation is 800°C, 900°C, 950°C and 1000°C and holding time is 2 hours, 3 hours, 4 hours and 5 hours. Test of compressive strength and porosity is done to see the physical properties of pottery ceramics and characterization by using XRF and XRD. The result of compressive strength test shows the temperature of 1000°C with 3 hours reaches the optimum value of compressive strength which is 9.52 Mpa, while the porosity value reaches the optimum point at heating temperature 1000°C with time of 5 hours that is equal to 16,19%. Characterization of XRF shows that ceramic pottery contains SiO_2 of 64.074%, and Al_2O_3 of 25.287% according to XRD characterization indicates the dominant phase formed by mineral quartz or silica oxside (SiO_2).

1. Introduction

Ceramic clay (earthenware) is a part of ceramics, this term in the modern context encompassing a very large inorganic material, ceramic clay containing nonmetallic and metallic elements made with various manufacturing techniques (the traditional ceramic is clay) that is burned at a certain temperature up tothe chemically has been transformed into a permanent form. The mechanical properties of ceramics clay are usually a strong material,hard but fragile and easy to broken. Furthermoreceramic claycorrosion resistance,have low density and high melting point. The quality of ceramic clay affected by raw materials used and temperature heating [1]. Main raw material of ceramic clayare clay or clay minerals. This is according to Jone et.al (2015), the quality of pottery depends on the physical properties of clay minerals as the main raw material and sand as an additional raw material [2]. In the wet conditions clay mineral have plastic properties, but in a dry conditionsbecome hard and when it burned become solid and strong [3]. The clay composed of oxide compounds such as SiO_2 and Al_2O_3 which are materials for the skeleton and form the ceramic body. Aplication of clay minerals as a main material for ceramic clay and sand as additives has many weaknesses,such as easy to crack, have high water absorption and have low compressive strength [4]. Therefore it is necessary to develop a better method of making traditional ceramics by providing additional materials such as mineral feldspart.Mineral feldspart can be used as ceramic material because it is functional as filler and glass forming [5]. Feldspart stone is the raw material needed for the production of ceramic tiles, household appliances, glazes, etc., because when it burned at temperature 1120-1300°C give effect of porosity change, bulk density and water absorption on pottery [6]. The industrial of ceramics and glass are the main consumers of feldspart. In the manufacture of glass, feldspart used as a flux agent and as a source of alumina (Al_2O_3), alkali (Na_2O and K_2O), and silica [7]. When ceramic clay mixed with feldspart minerals are molded and burnedthe microstructure will be changefrom the pottery product to mullite ($3Al_2O_3.2SiO_2$) and insoluble quartz crystal (SiO_2) that functioning into glass phase formation [8]. Based on the research of Indiani et.al (2009) the use of feldspart minerals in the manufacture of ceramics are useful as a melting agent to form glass phase at temperatures below 1100ºC [9]. Research conducted by Rivai (2016), reported that after passing temperature 600ºC physical and chemical clay will changes to be hard, solid and can't be destroyed by water [1]. This change is called ceramic change. This conclusion is reinforced by research conducted Ratri et.al (2008), at temperatures between 400-600°C chemically bonded water and other substances in clay minerals will evaporate [4]. After the temperature more than 600°C clay will change into a solid, hard and permanent mineral. In the range oftemperature 900°C the mullite phase will be formed. Mullite is a very stable compound that makes ceramic clay are hard, compact, and solid. So the ceramic clay has a high compressive strength and low absorption

2. Methode

In this research, raw material of clay used from Rejosari, Natar, Lampung Selatan, and the feldspart mineral from Nyukang Harjo, Lampung Tengah. Those raw materials were analyzed using XRF method as seen at theTable 1.

Table 1. XRF analysis results of the raw materials ceramic clay

Elements	Chemical Composition (%)	
	Clay	Feldspart
SiO_2	50.55	67.30
Al_2O_3	32.61	14.02
Fe_2O_3	2.43	2.24
K_2O	0.53	4.59
TiO_2	0.19	0.13
CaO	0.77	2.25
P_2O_5	0.14	3.20
NiO	0.12	0.020
SO_3	0.018	0.069
MnO	0.006	0.044
MgO	0.15	0.059
Cr_2O_3	0.003	0.015
SrO	0.002	0.020
PbO	0.004	0.006
CuO	0.001	0.007
Na_2O	1.05	0.005

Next, ceramic clay material is cleaned from impurities, dried in the oven with temperature 110ºC or with sunl for 3 hours and sieved with 20 pass mesh sifting.

Be pondered according to the mixing composition between clay and feldspart with ratio clay: feldspart = 65%:35%, stirring using mixer machine for 5 minutes with 10% water added from it weight.And printed with cube shape size 50x50x50mm. Drying in oven with 110ºC temperature for 8 hours. Heating in the muflle furnace at temperature variation of 800ºC, 900ºC, 950ºC and 1000ºC and the holding time 2 hours, 3 hours, 4 hours and 5 hours, It cooled and compressive strengthtested, at the Civil Engineering laboratory in University of Lampung, The porosity test and characterization with XRF and XRD methods were performed in the laboratory of BPTM – LIPI Lampung.

3. Result and Discussion

The results of ceramic clay compressive strength testing can be seen at the graphic bellow in Figure 1.

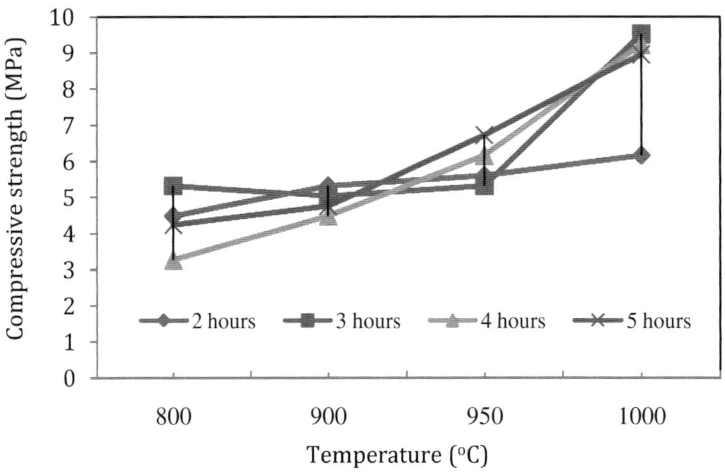

Figure 1. Graph of the comparison between compressive strength value to the temperature and heating time of ceramic clay

Testing porosity of physical properties in ceramic clay is to know the empty space in the sample. Yafie et.al (2014), reveals that shaft is an empty space or cavity formed in a compacting process that causes air to be trapped between powders during sintering [11]. The existence of a shaft inside a material will affect its mechanical properties. At 5hour variation time have porosity value respectively, at heating temperature 800°C with 5 hours is 20.85%, its value down at heating temperature 900°C with 5 hours is 20.43%, and back down at heating temperature 950°C with 5 hours is 18.89% and reached the optimum value at heating temperature of 1000°C 5 hours is 16.19%. The porosity value of an object greatly affects the value of its compressive strength. The more pores on the ceramic clay the strength of a material will decrease. The results of a study published by Gonggo et.al (2013) explain that the magnitude of the porosity value is due to the feldspart minerals that act as a fuser not yet completely fused to fill the cavities between the clay mineral granules [12]. Ceramic that have been heating at 900°C still contain minerals from feldspart. To reduce the porosity of ceramic required combustion at temperatures above 900°C, because the felders of feldspart begin to melt at a temperature of 900°C-1100°C. This is in accordance with this study, where the increasing temperature and heating time decreases the porosity percentage. Because there has been a good melting process to fill other pore spaces on ceramic, although the melting process has not been perfect.

Ceramic clay XRF characterization as performed to determine the amount of oxide contained in the sample in percent are presented in Table 2.

Table 2. X-ray fluorescence characterization results of ceramic clay with some variations of temperature and heating time.

Elements (%)	Heating Time (°C)				
	800 (3 hours)	900 (3 hours)	950 (3 hours)	1000 (3 hours)	1000 (5 hours)
SiO_2	61.169	58.218	60.066	60.162	64.074
Al_2O_3	29.138	30.661	28.918	27.901	25.287
Fe_2O_3	4.863	5.208	5.252	5.732	5.358
TiO_2	0.732	0.742	0.779	0.732	0.719
CaO	0.641	0.961	1.182	0.681	0.635
K_2O	2.695	2.971	2.958	2.532	3.114
P_2O_5	0.466	0.477	0.488	0.481	0.469

Based on the analysis in Table 2, it is also known that the ratio of the highest compound content of SiO_2 and Al_2O_3 this proves that there is no significant phase change. For at each combustion temperature with 3 hours, the percentage of both compounds is in cosmetic numbers. At 800°C temperature the amount of SiO_2 and Al_2O_3 is compounds 90.307%, at 900°C the amount of SiO_2 and Al_2O_3 is compounds 88.899%, at 950°C the amount of SiO_2 and Al_2O_3 is compounds 88.984% and at 1000°C the amount of SiO_2 and Al_2O_3 is compounds 89.361%. The high compounds of Fe_2O_3, in the range 4.863%, 5.208%, 5.252% and 5.358% are the oxide that gives red color when heatting as evidenced in the reddish sample [12]. This is also supported by the XRD characterization results in the sample.

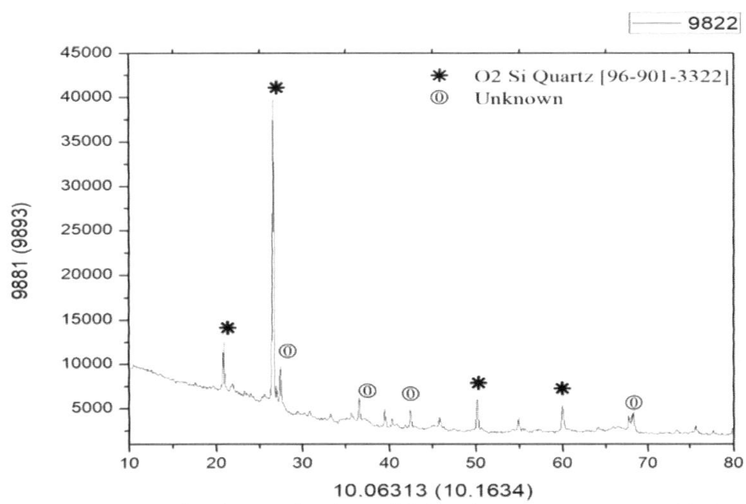

Figure 2. XRD Graph of ceramic clay with heating temperature 1000°C 3 hours

The results of XRD characterization in temperature variation 1000°C 3 hours indicating that the dominant phase formed just silica oxide (SiO_2) with the highest peak at $2\theta = 26.64°$. The research of Ratri et.al (2008) also expressed opinion that regarding the formation of a mullite phase formed above a temperature of 1000°C [4]. But in this study,ceramic clayheatingat 1000°C 3 hours only formed silica oxide. Based on the XRD graph, its known that the mullite phase ($2Al_2O_3.2SiO_2$) ceramic clay heating at 1000°C 3 hours has not been formed. The glass phase is formed more by the feldspar grains ceramic clay will decrease on 1100°C temperatures. The pores remained small the at 1150°C, because the silica granules ocurs by causing more glass phase. At 1200°C temperature, the ceramic body matures with a porosity almost toward zero. In the pseudomorphic form of the feldspar grains can be seen a large number of mullite needles formed resulting from alumina diffusing into the clay through the glass phase.

4. Conclusion
The value of the optimum compressive strength occurs at a temperature of 1000°C with 3 hours time is 9.52 MPa. This is supported by the porosity value of 17.032%. Because the higher of compressive strengthvalue is that the lower of the porosity percentage. Based on of XRD analysisresultthat the heating optimum at 1000°C with 3 hours and at 1000°C with 5 hours mullite phase ($2Al_2O_3.2SiO_2$) not formed it just inversion change from low quartz to high quartz.

Acknowledgments
The authors are greatful the Head of BPTM - LIPI who has allowed the implementation of this research and provide support in the form of licensing the use of facilities and infrastructure needed in this study.

References
[1] Rivai M and Hartono S B 2016 *Jurnal TRAKSI* **16** 2
[2] Jone Y, Utamakno L, and Cahyono Y D 2015 *Jurnal Seminar Nasional Sains dan Teknologi Terapan III*
[3] Winarno T 2016 *Jurnal Teknik* **1** 37
[4] Ratri A K, Sriatun, and Darmawan A 2008 *Jurnal Kimia Sains dan Aplikasi* **11** 3
[5] Ajayi B A and Owoeye S S 2012 *International Journal of Science and Research (IJSR)* **3** 11
[6] Radomír Sokolář and Lucie Vodová, 2014 *International Journal Of Engineering And Science* **4** 10
[7] André C S, Sílvio D C, Débora N S, Elenice M, and Schons S 2018 *Journal of Material Reasecht and Technology* **2** 5
[8] Matthew G O and Fatile B O 2014 *Research Journal of Recent Sciences* **3** 9
[9] Indiani E, Ayu N, and Umiati K 2009 *Jurnal TELKOMNIKA* **7** 2
[10] Alaa S and Wijaya D 2015 *Jurnal Fisika* **1** 1
[11] Yafie M S and Widyastuti 2014 *Jurnal Teknik Pomits* **3** 1
[12] Gonggo S T, Edyanti F, and Suherman 2013 *Jurnal Akademika Kimia* **2** 2

IOP Conf. Series: Materials Science and Engineering **478** (2019) 012039 doi:10.1088/1757-899X/478/1/012039

The effect on the zeta potential of surface modified Philippine natural zeolites (SM-PNZ) for the adsorption of anionic solutions

C E C Barola, I F C Dusaban, E M Olegario-Sanchez, and H D Mendoza

Department of Mining, Metallurgy, and Materials Engineering, University of the Philippines, Diliman, Philippines

Email: eleanor.olegario-sanchez@upd.edu.ph

Abstract. This study investigates the effect of surface modification on the surface charge of the natural zeolites (PNZ) from SAILE Industries in the Philippines by utilizing the zeta-potential technique. The SM-PNZ is prepared to be utilized in the adsorption of anionic aqueous solutions like gold cyanide [$Au(CN)^{2-}$] solutions from gold extractive companies. The experiment was designed by converting natural zeolites into its Na-form and modified using cetyl trimethylammonium bromide (CTAB), a micelles-forming surfactant, with dosages equal to 100% and 200% of ECEC (external cationic exchange capacity) of the PNZ. Further characterization of the adsorbents by SEM shows a plate-like morphology. The zeta-potential of the SM-PNZ material shows a decreased negative surface compared with the unmodified PNZ. The percentage recovery of gold cyanide to the PNZ, SM-PNZ (100) and SM-PNZ (200) was 22%, 45% and 40% respectively. Surface modification has shown to improve the adsorption of the PNZ by decreasing the negative potential of the PNZ and increasing the affinity of the SM-PNZ to anionic solutions.

1. Introduction

Naturally, zeolites have negative charges even in acidic conditions due to its substitutions of cations within the lattice, lattice imperfections, and broken Si-O-Si bonds generated from grinding. Also, zeolite is found to be permanently charged due to isomorphous substitution [1-3]. Negative charges are decreased when natural zeolite is modified in alkali pH, due to exchange of Na+ cations, which relatively has the weakest bond and nearest locality to the surface. Al and Si were observed to dissolve only in acidic solutions with strong electrolytic property at low zeolite amounts due to the strong interaction between zeolite protons and surface hydroxyl groups. Lastly, it was found out that zeolite, whether natural or modified, can be used as buffer for strong acid- weak base solution. Thus, varying the pH affects the ion exchange behavior of zeolite, aside from other factors such as temperature, cation type, and heat of solution [1].

A different study by Celik [2] determined the electrokinetic properties of natural zeolite, specifically clinoptilolite. Isoelectric properties are important factors to consider since it governs adsorption and other related processes [2]. Since surface potential cannot be measured directly, zeta potential was used to measure the potential of zeolite. It was found to be negatively charged even in very acidic conditions, thus allowing the entrance of cations and not favouring the deposition of negatively charged ions into the lattice channels to make up for the positive charge deficiency.

Two isoelectric properties were studied: isoelectric point and potential determining ion. Isoelectric point (iep) is the total interaction happening at the mineral/water interface such as hydrolytic reactions, adsorption and distribution of dissolved lattice ions while potential determining ions (pdi) are ions that alters or dictates the surface charge of the material. No isoelectric point of zeolite was observed in pH between 2-12, which means that zeolites always carries a net electrical charge whatever the pH is. Increase in pH lowers the charge due to accumulation of OH- on positively charged sites or the release of positive charges of hydroxyl groups found at the zeolite surface while decreasing the pH increases the surface charge due to adsorption of H+ onto negatively charged sites. H+ and OH- were found to preferentially dissolve due to uneven activities present in the system, thus changing the surface charge of zeolite and dictating the mechanism of adsorption. Though a negative surface charge was obtained even when H+ ions were adsorbed at lower pH, it is possible to obtain a positive zeta potential or even reverse the zeta potential through deposition of cations with higher valence. To readily adsorb anions, the zeta potential of zeolite should be reversed [2].

Though zeolite is very good adsorbent, its properties can still be altered and modified to increase its adsorption efficiency. A study by Choi [4] specified how modified zeolite can adsorb three times better than natural zeolite. However, due to its innate negative charge, only cations have the natural affinity to be adsorbed. Nevertheless, rapid advent of previous studies show how zeolites can be a versatile adsorbent can be used to recover and isolate anions including metal complexes.

A study by K. Hor [5] implemented physicochemical methods to improve the adsorption capacity and removal efficiency of zeolites. Examples of methods used and investigated were acid treatment, base treatment, combined acid-thermal, and base thermal treatment, in which the main rationale of the treatment is to improve de-alumination and de-silication processes.

Among the modification processes that were mentioned above, manipulating the electrokinetic potential of zeolites shows the greatest promise. Electrokinetic potential or zeta potential is the electric potential in the electric double layer, at the boundary between the compact layer and diffuse layer. The boundary between the compact layer and diffuse layer is also known as the Stern layer. A key in understanding the surface properties of an adsorbent is to know its surface charge. However, surface charges are impossible to measure and instead, zeta potential is widely used to determine interaction energy between particles and colloidal stability. In adsorption processes, zeta potential measurements over a wide range of pH can be used to assess the affinity of the adsorbate ions to the surface of the adsorbent [6].

Zeolites with reversed zeta potential were able to adsorb anions, but only in small amounts and the adsorption capacity for different anions can be improved with more positive zeta potentials. One way to reverse the zeta potential of zeolites is by using CTAB (Cetyl trimethylammonium bromide) to turn natural zeolites to surfactant modified natural zeolite (SMNZ), which was studied by Barczyk [7], to allow anion sorption onto zeolite surfaces. It was found out that anion sorption of zeolite is correlated with the type anion adsorbate, type of natural zeolite, and surfactant loading level.

In this study, the Philippine natural zeolite, PNZ, will be used to adsorb gold cyanide from an anionic aqueous solution. Although zeolites have been used to adsorb anions by reversing its zeta potential, no existing researches have undertaken to study if zeolite can successfully adsorb gold cyanide. The PNZ will be modified by using CTAB to alter its zeta potential and determine whether it is efficient in adsorbing gold cyanide.

2. Experimental

2.1 Materials Preparations

The zeolite (PNZ) samples from SAILE Industries in Mangatarem, Pangasinan were selected for this study. The same PNZ sample sourced and analysed by previous researchers [8-10] was determined to be chiefly clinoptilolite. The PNZ were sieved to a particle size of 1.68mm < Dp < 2.0 mm and washed with deionized water for four times, then dried in an oven at 150C for three hours. The surfactant CTAB (N-Cetyl-N, N, N trimethyl ammonium bromide) with 99% purity (sourced from Belman Laboratories) was used for the surface modification PNZ.

2.2 Sodium activation
The PNZ was soaked in 4 M NaCl solution 24 hours at a weight/volume ratio of 20% (5g dried PNZ to 25 mL NaCl solution) without agitation. After 24 hours, Na-PNZ was rinsed eight times with deionized water, and dried in an oven for 3 hours at a temperature of 150 C.

2.3 Surface Modification
CTAB was used to modify the Na-PNZ in an amount equivalent to 100% and 200% ECEC of each of the zeolites [7]. From literature [11], the computed ECEC of clinoptilolite by using the method proposed by Ming and Dixon [11] was computed to be 10 mmol/100g. Surface modification was then carried out by a ratio of five grams zeolite per 100 mL of CTAB solution with CTAB concentrations equal to both 100% and 200% ECEC of zeolites. The solutions were bottle rolled for 24 hours at 200 rpm. The suspensions were then rinsed four times with deionized water and dried in an oven for three hours at 105C.

2.4 Characterizations
The morphology of PNZ, SMNZ at 100% and 200% CEC was examined by using was examined by field emission scanning electron microscopy (Fe-SEM, S-3400N, Hitachi). To further confirm if CTAB was attached on the zeolite surface, FTIR analyses were performed on an ATR-FTIR spectrometer (IS50, Thermo Scientific Nicolet) for PNZ, SMNZs at 100% and 200% CEC. The surface charge of the materials was then measured using the Zeta Meter 4.0.

3. Results and discussion
To characterize the functionality of the modified adsorbents, the Fourier-Transform Infrared Spectroscopy (FTIR) was used to determine the presence of CTAB onto the materials was determined by analysing the spectra of the materials using the scanning electron microscopy (SEM), and Zeta Potential test were performed on the materials namely PNZ, SMNZ (100%) and SMNZ (200%).

3.1 Functionality of PNZ and SM-PNZ
Figure 6 shows the absorbance vs wavenumber curve derived from the FTIR results. The figure shows that CTAB modification was achieved at dosages equal to 100% ECEC and 200% ECEC of the natural zeolite [12]. The presence of the CTAB modifier was observed in the spectra peaks at wavenumber 2000-2500 cm_{-1}. The observed spectra correspond to the carbon tails present in the CTAB. Also, the peak at 3300-3600 cm_{-1} observed in the PNZ material was not seen in the spectra of SM-PNZ (100% and 200% ECEC) samples. Originally zeolites are hydrophilic due to its tetrahedral aluminosilicate network, but became hydrophobic at the presence of CTAB, which is a known surfactant. Lastly, peaks observed at 500-1300 cm_{-1} were all observed in all FTIR spectra, which signifies that the Si-O-Si and Si-O-Al bonds were not changed upon modification.

3.2 Morphology of the PNZ and modified SM-PNZ
The images obtained by using SEM shows that the PNZ has an irregular platy-like surface morphology. Void spaces represent pores and cavities on the PNZ surface. In comparison with the SEM images of PNZ and SMNZ, faded parts are seen on some particles which signify that an organic layer has formed aggregates over the plate like morphologies of PNZ. The organic layer is assumed to be CTAB accumulated onto the PNZ external surface. More aggregates of the organic material are observed in SM-PNZ (200%).

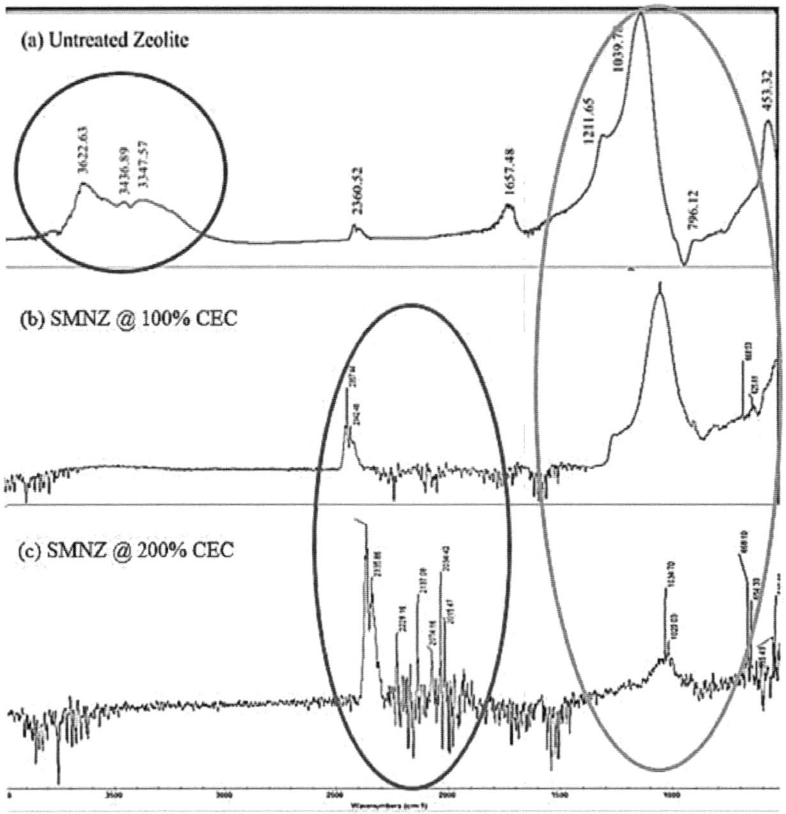

Figure 1. FTIR Curve (a) untreated PNZ, (b) SM-PNZ (100%), and (c) SM-PNZ (200%)

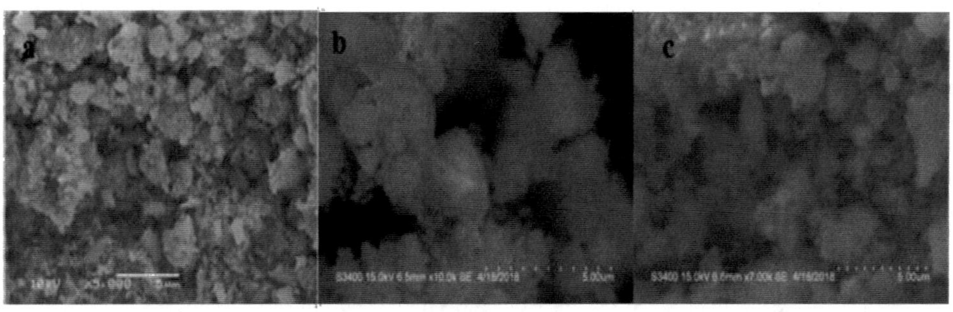

Figure 2. SEM Images of (a) NZ, (b)100%, and (c) 200% ECEC

3.3 Zeta-potential of PNZ and SMNZ

Table 1. Zeta potential of PNZ 100%, and 200% ECEC

Adsorbent	Zeta potential (mV, +/- 2.00 mV)	
	pH 10	pH 11.5
PNZ	-37.50	40.00
SM-PNZ (100%)	-16.20	25.00
SM-PNZ (200%)	-16.20	16.80

The zeta potential values of PNZ, SM-PNZ (100%) and SM-PNZ (200%) are compared. Although the zeta potential values weren't reversed, it can be compared that the zeta potential values are close to the zeta potential of activated carbon (20-40% difference) [13]. A 100% decrease in zeta potential was observed from NZ to SM-PNZ samples which signifies that the PNZ was successfully surface modified using CTAB. This then resulted to a surface charge which is less negative than the unmodified material, PNZ.

3.4 Recovery performance of the adsorbents

The efficacy of the adsorbents, PNZ, SM-PNZ (100) and SM-PNZ (200) for the recovery of gold cyanide solution are compared and shown at Fig 3. At three (3) hours contact time, the SM-PNZ (100) has the highest % recovery. The recovery of gold cyanide almost doubled when PNZ was surface modified with CTAB. This means that surface modification is important for the adsorption of gold cyanide. However, comparing modified SM-PNZ, the recovery of gold cyanide is greater in SM-PNZ (100) than in SM-PNZ (200). This can be attributed to the fact that more aggregates of CTAB were formed on the surface of the zeolite SM-PNZ (200) making less polar than SM-PNZ (100). Thus, the external surface of SM-PNZ (200) is less attracted to anionic forms of the gold cyanide solution.

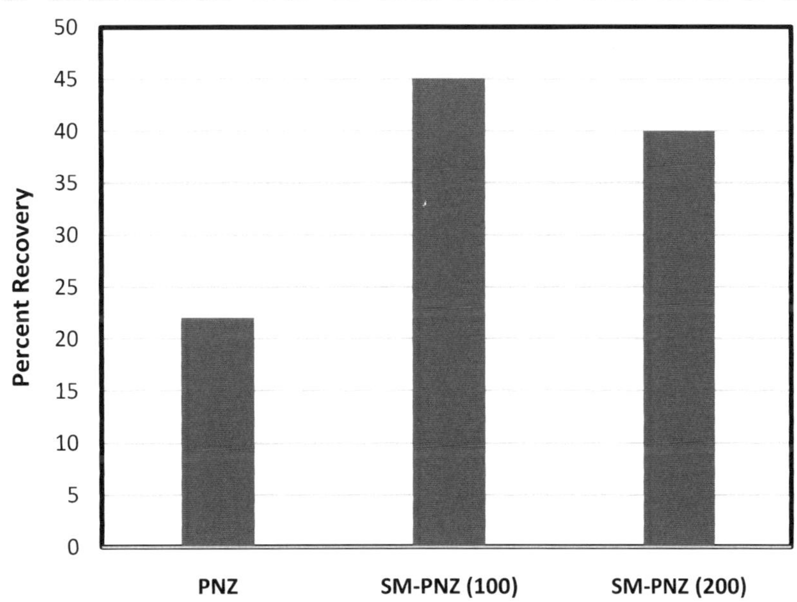

Figure 3. Percent recovery of PNZ, SM-PNZ (100%) and SM-PNZ (200%)

4. Conclusion

Results from the experiment show that zeolite has a potential in adsorbing gold cyanide ions in an anionic solution. Based from the recovery as a function of time, it indicates that in the long run, zeolites either modified or not has the same adsorption %recovery. Although the zeta potential of zeolite was not reversed in modification, the presence of CTAB molecules attached at the zeolite surface and pores may have amplified and assisted in diffusion of gold cyanide into pores and voids present in zeolite.

The study shows that PNZ can be used to treat gold cyanide wastewater from mine operations. Knowing that AC and SMNZ have close % recovery values, much research can be undertaken to manipulate zeolite such as using other modifiers or increasing the dosage of surfactants to increase its adsorption rate and if possible, obtain higher % recovery than that attained by activated carbon. Another recommendation would be studies on how to increase the rate of adsorption, since the kinetics of this process is relatively slow compared to activated carbon.

Acknowledgment
The contribution of SAILE Industries, Inc. for the zeolite materials and technical data provided in this study is thus acknowledged. The faculty and staff of the Pilot Plant of the Department of Mining, Metallurgical and Materials Engineering is also recognised for their assistance and guidance to our study. The authors would like to extend their deepest gratitude to Dr. Ronald T. Dy the financial support.

References

[1] Polatoglu I 2005 Chemical behavior of natural zeolite in aqueous media *Graduate thesis* Izmir Institute of Technology, Izmir, Turkey
[2] Ersoy B and Celik M 2002 *Microporous and Mesoporous Materials* **55** (3) 305-312
[3] Carland R M and Aplan FF 1996 *Conf. on Int Min Proc Cong* (California) **4** (United Stated/OSTI) 210
[4] Choi H J, Yu S W, and Kim J H 2016 *J. Taiwan Inst. of Chem. Eng.* **63** 482-489
[5] Hor K Y, Chee J M C, Chong M N, Jin B, Saint C, Poh P E and Aryal 2016 *J. of Cleaner Prod.* **118** 197-209
[6] Hunter R 1991 *Zeta Potential in Colloidal Science* 10.1016/B978-0-12-361961-7.50007-9.
[7] Barczyk K, Mozgawa W, and Krol M 2014 *Spectrochimica Acta Part A: Molecular and Biomolecular Spectroscopy* **133** 876-882
[8] Olegario-Sanchez E M, Felizco J C, and Mulimbayan F 2017 *AIP Conference Proceedings* **1901** 07000
[9] Olegario-Sanchez E M and Pelicano C M 2017 *Key Engineering Materials* **737** 407
[10] Osonio P and Olegario-Sanchez E M 2017 *AIP Conference Proceedings* **1901** 080003
[11] Ming D W and Dixon J B 1987 *Clays and Clay Minerals* **35** (6) 463-468
[12] Taffarel S R and Rubio J 2010 *J. Mineral Engineering* **23** 771-779
[13] Jones W G, Klauber C and Linge H G 1989 *Fundamental aspect of gold cyanide adsorption on activated carbon* **32** (Bentley, Western Australia 6102) 278-281

Process of andesite stone as material of cement substitution in making paving block

M Amin[1], Y I Supriyatna[1], K Isnugroho[1], Sudibyo[1], R Septiana[2], and Syafriadi[2]

[1]Research Unit for Mineral Technology, Indonesian Institute of Sciences, Lampung, Indonesia
[2]Department of Physics, Faculty of Math and Science, Lampung University, Lampung, Indonesia

Abstract. This study was conducted to determine the effect of andesite stone substitution on compressive strength paving block. The percentage of andesite substitution of cement by 10, 20, 30, 40, and 50% by weight of cement. Andesite rock after calcination was characterized using XRD (X-Ray Diffraction). The results of XRD analysis after calcination at the calcination temperature of 900°C have formed muscovite, hedenbergite, and microcline phases. The highest compressive strength of the paving block is found in the substitution of andesite stone of 10% of age 28 days and the lowest compressive strength is in the substitution of andesite 50% of 7 days. This decrease is due to the excessive SiO_2 content will bind to the free CaO contained in the cement and form $Ca(OH)_2$. $Ca(OH)_2$ causes reduced paving block density due to the formation of air cavities. Besides the high FAS factor and the gradation of sand size also affect the decrease of compressive strength of paving block.

1. Introduction

Paving blocks or concrete bricks for floors are a component of building material made from a mixture of hydraulic cement or the like, aggregates and water with or without other additives which do not reduce the quality of the paving block [1]. The advantages of paving block that has good water absorption, through the installation of paving block can keep the ground water balance [2]. However, the high consumer demand for paving is not matched by the availability of adequate quality in terms of strength, wear life, and durability paving. Many paving found on the surface of the road are cracked, easily broken, overgrown with moss, because paving is brittle. This is due to the quality of materials that are not appropriate, scouring rainwater, material composition that does not comply with the standards, the different levels of compaction (pressing) paving, even shock loads (impact resistance) is very large on the track wheels of the vehicle [3]. With reference to the above explanation, as the growing of the cement industry while the raw materials will also decrease as it continues to be used so that alternative material is needed as substitution of cement [4].

A lot of research has been developed to get new materials that can be used as an alternative or substitution of materials to reduce the use of cement. Triyono [5] conducted research on palm oil shell waste as cement substitution for making paving block with percentage 0%, 5%, 10%, 15%, 20% and 25% of cement weight. Soehardjono et al [5] conducted a study on the effect of using bottom ash as a substitute of cement to the compressive strength and water absorption ability of structure paving, the test object was made 9 variations ie 0%, 25%, 30%, 35%, 40%, 45%, 50 %, 55% and 60% by weight of cement. Nurzal and Mahmud [6] utilize fly ash for mixed paving block, composition fly ash in making paving block is 0%, 5%, 10%, 15% by weight of fly ash plus material paving block (cement and sand). According to Sari [7] the contents of andesite stones contain the chemical composition of $SiO_2$by 62.30%, Al_2O_3 by 14.70%, Fe_2O_3 by 4.04%, MgO by 2.78%, CaO by 4.26%, Na_2O by 2.95%, K_2O of 6.06%,TiO_2, 0.98 P_2O_5 of 0.81%, 0.07% MnO and Cr_2O_3 of 0.014%. Andesite added to the cement material with clinker preparation will have an effect on decreasing the combustion temperature which should not be high at a temperature of 800°C longer time combustion [8]. With the chemical composition of SiO_2 + Al_2O_3 + Fe_2O_3 > 70%, the andesite can be used as pozzolan in cement substitution in accordance with ASTM C 618-92a.

2. Experimental Procedures

2.1. Materials

This study began with andesite stone milling using ball miles for 10 hours until the andesite rocks were completely smooth. After going through the process of ball milling, andesite stone that has passed mesh 200 through the process of calcination using electrical furnace panel until the temperature of 900°C then with a hold time of 2 hours. After cold, the result of andesite stone calcination is characterized by XRD (X-Ray Diffraction) to determine the phase of crystal structure that is formed. Andesite stone before milling and calcined calcite andesite as shown in Figure 1.

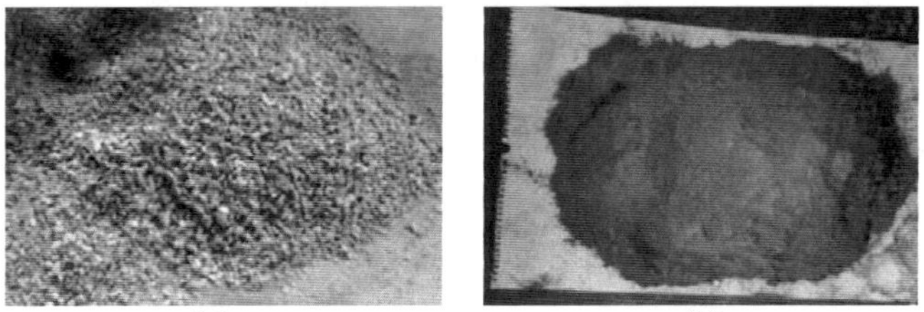

(a) **(b)**

Figure 1. (a) Rocks of despite before milling, (b) After milling and calcination

Subsequently performed gradation test on sand beach by sifting 400 gr with mesh 2 mm, 0.5 mm, 0,42 mm, 0,354 mm, 0,25 mm and 0,149 mm. Gradation of sand obtained by calculating the cumulative percentage of sand grains that pass on each sieve. Sand fineness modulus can be calculated by the following equation.

$$\text{MHB} = \frac{Total\ \ cumulative\ \ weight}{100} \quad (1)$$

In general, the sand can be grouped into three kinds, namely the level of refinement of fine sand with MHB 2.20 to 2.60, the sand being with MHB 2.60 to 2.90 and sand rough with MHB 2.90 - 3.20.

2.2. Manufacture of test objects

The composition of the specimen used in this study is the ratio of cement to sand is 1: 4 with andesite substitution of 10, 20, 30, 40 and 50% of the weight of cement and FAS (Cement Factor) of 0.8. After each ingredient is weighed, then put all the ingredients into the mixer and stir until homogeneous and the dough is formed. Then put the dough into the cube mold (size 5 x 5 x 5 cm). Charging is done layer by layer and each layer is solidified 25 times. After 24 hours, open the mold and code the sample on the specimen. Treatment of the test object (curing) is done by immersing the test object into a bucket that already contains water until the age of 7 days, 14 days and 28 days. After soaking the specimen is ready to be tested for compressive strength. Calculation of compressive strength with equation 2 as follows:

$$f'c = \frac{P}{A} \quad (2)$$

With: A = the cross-sectional area of the specimen (cm^2)
P = load (kg)
f'c = compressive strength (kg/cm^2)

3. Results and discussion

3.1. XRD characterization results (X-Ray diffraction)

The XRD characteristic results on andesite rocks that have been calcined at 900 °C are shown in Figure 2.

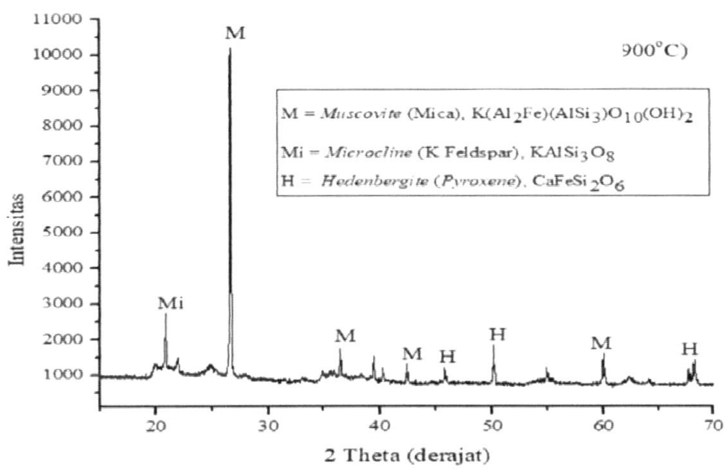

Figure 2. XRD results of andesite stone after calcination

Figure 2 shows the highest peak characteristics of fine andesite rocks after calcination is dominated by the Muscovite phase (Mica) with the chemical formula $K(Al_2Fe)(Si_3Al)O_{10}(OH)_2$, followed by the hedenbergite phase (pyroxene) with $CaFeSi_2O_6$ chemical formula and the Microcline phase (K Feldspar) with chemical formula $KAlSi_3O_8$. According to Stepanus et al (2014), the mineral content present in andesite rocks is potassium feldspar of <10% of total feldspart content, sodium plagioclase, quartz <10%, feldspatoid <10%, bornblende, biotite and pyroxene with mineral composition andesite is required in the formation of cement reactions.

3.2. Sand gradation test result

The result of sand grading test included in fine sand with the fine modulus of grain (MHB) of 2.3. The results of sand grading test are shown in Table 1.

Table 1. Sand Gradation Test Results

No	The Sieve Hole		Weight is Left Behind		Cumulative Weight (%)
	Mesh no.	mm	gr	%	
1	10	2.00	0.00	0.00	0.00
2	35	0.50	43.30	10.99	10.99
3	40	0.42	47.49	12.05	23.04
4	45	0.35	90.14	22.88	45.92
5	60	0.25	57.05	14.48	60.80
6	100	0.15	117.00	29.70	90.50
8	Left		38.93	9.88	
	Cummulative		393.91	99.97	231.25

MHB = $\frac{Total\ cumulative\ weight}{100}$ = 231/ 100 = 2.31 included in the fine sand group = 2.20 – 2.60

The sand gradation or sand beach size distribution used has an MHB (fine modulus of grain) of 2.31, with the MHB, the sand of the beach we use is included into the fine sand which tends to have uniform grains of sand. The grain size distribution influences the compressive strength of the paving block. In accordance with research conducted by Tugino [9], if the grain size of sand has the same size (uniform) then the volume of sand will be large. Conversely, if the size of the grains of sand varies, it occur small pore volume. This is because small grains will fill the pores between large grains so that the pores become slightly or in other words high congestion. In making paving block required sand that has granules with a high compression or which has a small pore volume.

3.3. Results test strong press

Results calculation test strong press on object test ages 7, 14 and 28 days at the show on Figure 3.

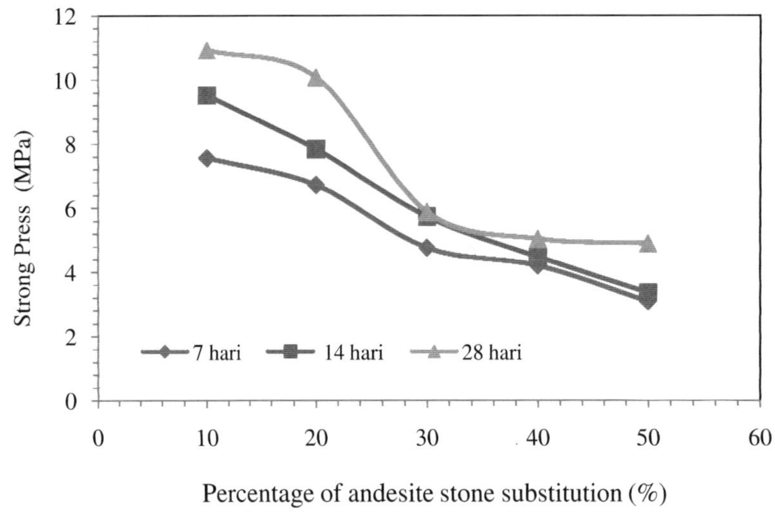

Figure 3. Relation percentage of andesite rock substitution with strong press paving block at age 7, 14 and 28 days

Figure 3 showing substitution relationship between the percentage of andesite with a compressive strength of paving blocks namely that the compressive strength of the paving block decreases with the addition of andesite stone content in the mixture. The highest compressive strength occurs in 10% andesite rock substitution, then the compressive strength will decrease to 50% of andesite substitution. The decrease in compressive strength of paving block in this study was due to the amount of SiO_2 content in the mixture disproportionate/excessive. SiO_2 is mostly contained in sand and andesite stones. In accordance with a study carried out by Hambali [10], the SiO_2 compound reacts with Ca $(OH)_2$ and produces calcium hydrate (CSH) which contributes to hardness in the paving block. The binding reaction of $Ca(OH)_2$ by SiO_2 takes place as follows.

$$x\,Ca(OH)_2 + SiO_2 + nH_2O \rightarrow xCaO.SiO_2.nH_2O \qquad (3)$$

The excessive SiO_2 content will bind to the free CaO contained in the cement and form Ca $(OH)_2$.$Ca(OH)_2$causes reduced *paving block* density due to the formation of air cavities. These air cavities will be filled by water during the sample immersion period. $SiO_2$is the C_3S and C_2S forming oxide which is the main component in the cement. When C_3S and C_2S react with water again form Ca $(OH)_2$ compounds. The ongoing reaction is shown in equation 4 and 5 that is:

$$2(3CaO.SiO_2) + 6H_2O \rightarrow 3Ca(OH)_2 + 3CaO.2SiO_2.3H_2O \qquad (4)$$

$$2(2CaO.SiO_2) + 6H_2O \rightarrow Ca(OH)_2 + 3CaO.2SiO_2.3H_2O \qquad (5)$$

Therefore, the higher percentage of andesite rock substitution causes the compressive strength of the paving block to decrease. In addition, other factors that affect the compressive strength of the paving block are FAS (Cement Water Factor). The water-cement factor (FAS) is a number that shows the ratio between the water weight and the weight of the cement. In high-quality concrete, the definition of FAS can be defined as water to cementitious ratio, ie the ratio of water weight to total weight of cement and cementitious additives commonly added to high quality concrete mixtures [11]. FAS used in this research is 0.8, in accordance with the statement Mulyono [12] generally a minimum FAS value is given about 0.4 and a maximum of 0.65. The low water-cement factor is the most decisive factor in producing high quality paving blocks, with the aim of reducing to a minimum the porosity of the resulting paving block. The larger the water-cement factor (FAS) creates the lower the compressive strength of the concrete [11]. However, too low a water-cement factor (FAS) results in insufficient cement paste to coat aggregate [13].

4. Conclusion

Based on results research could be concluded that s increasing the substitution of andesite rocks makes the compressive strength of the paving block to decrease. This matter was due increasingly substitution stone andesite cause SiO_2 content increases. Gynecology SiO_2 the excess will bind to the free CaO contained in the cement and form $Ca(OH)_2$. $Ca(OH)_2$ causes reduced paving block density due to the formation of air cavities. These air cavities will be filled by water during the sample immersion period. In addition, other factors that affected the compressive strength of paving blocks namely FAS (Factors Water Cement) is high and gradation size sand.

Acknowledge

The author would like to thanks to Researh Unit for Mineral Technology – Indonesian Institute of Sciences which has provided facilities during the research.

References

[1] Adibroto F 2014 *Jurnal Rekayasa Sipil* **10(1)** 1-11
[2] Mulyati dan Maliar S 2015 *Jurnal Momentum* **17(1)**
[3] Riyaldi A 201 *Studi Ekspererimental Paving Block Komposit Concrete Foam Diperkuat Serat TKKS Akibat Beban Tekan Statik*. Medan: Universitas Sumatera Utara
[4] Triyono D D 2010 *Pemanfaatan Limbah Tempurung Kelapa Sawit untuk Pembuatan Paving Block*. Semarang : Universitas Negeri Semarang
[5] Soehardjono A, Prastumi, Hidayat T, dan Prawito G S *Jurnal Rekayasa Sipil* **7(1)** 74-80

[6] Nurzal dan Mahmud, J. 2013 *Jurnal Teknik Mesin* **3(2)** 41-48
[7] Sariisik A, Sariisik G, and Ahmet S 2011 *Construction and Building Materials* **25** 3694-3702
[8] Moussa H, Larbi K, and Pierre C 2018 *MATEC Web of Conferences* **149** 02048
[9] Tugino 2010 *Jurnal Teknik Sipil dan Perencanaan* **12(1)** 1-10
[10] Hambali M, Lesmania I, and Midkasna A 2013 *Jurnal Teknik Kimia* **19(4)** 14-21
[11] Afif M 2013 *Pengaruh Penambahan Silica Fume Dan Superplasticizer Dengan Pemakaian Semen Tipe PPC Dan Tipe PCC Terhadap Peningkatan Mutu Beton* [*Skripsi*]. Semarang: Universitas Negeri Semarang
[12] Mulyono T 2003 *Teknologi Beton* Yogyakarta, Penerbit Andi
[13] Ginting A 2015 *Jurnal Teknik Sipil* **11(2)** 76-168

The influence of number of grid points and radius increments in determining safety factor and estimated sliding volume on three-dimensional slope stability analysis

M A Azizi, I Marwanza, A Anugrahadi, A A Faradiba, and N A Hartanti

Mining Engineering Department – FTKE Universitas Trisakti
Jl. Kyai Tapa No. 1, Tomang, Grogol Petamburan, West Jakarta, DKI Jakarta 11440

Email: masagus.azizi@trisakti.ac.id

Abstract. The existence of geometric complexity in the longitudinal direction makes 3D Limit Equilibrium Method (LEM) able to depict conditions in the field well and obtain Safety Factor (SF) that is more representative than the 2D limit equilibrium method. Safety factor calculations are not only influenced by the Direction of Sliding (DoS), but also by the location of the slip surface. Therefore, determining the location of the slip surface is important in the 3D limit equilibrium method. In this study, the method used for generating slip surfaces is a grid search where variations are made on the number of grid points and the radius increments to determine the effect of these two variables on the determination of safety factor and estimated sliding volume. Based on the results of the study, it can be stated that the greater the number of grid points and the radius increments, the estimated sliding volume will be more precise, while the calculation of safety factor is only affected by the number of grid points where the greater the number of grid points, the search of critical slip surface will be more accurate.

1. Introduction

PT Bukit Asam Tbk (PTBA) applies an open pit mining system. The production and operations of an open pit mine are largely determined by slope stability. Slope stability is influenced by many factors, both internal and external factors. One of the internal factors that influence slope stability is its geometry. Slope geometry must be designed with regard to geotechnical aspects to minimize the occurrence of landslides that could potentially interfere with mining activities or the occurrence of failure. Therefore, slope stability analysis is needed to evaluate the overall slope plan design. Slope stability is represented by the value of safety factor. One of the methods that can be used to calculate safety factor values is limit equilibrium method.

Slope stability analysis with limit equilibrium method is commonly carried out in two-dimensional analysis. But in modeling complex geometries, 2D analysis cannot simulate them properly. The 2D analysis assumes that the slope width is infinitely wide. This assumption is contrary to the original state where slope width is finite and there is a change in the longitudinal direction of the slope. Slope stability analysis with three-dimensional method can depict slope conditions including changes in longitudinal direction and also include spatial which is not calculated by 2D method. Therefore, the 3D analysis is considered to be able to describe the conditions in the field better than the 2D analysis.

Analysis of slope stability with 3D limit equilibrium method starts by assuming the geometry of the sliding mass. The results of the calculation of slope stability using limit equilibrium method can be expressed in safety factor. In this method, safety factor is not only influenced by the direction of sliding, but also by the slip surface [1]. Naderi [2] states that safety factor is sensitive to critical slip surface locations. Therefore the determination of a critical slip surface is very important. Safety factor can be obtained correctly if the determination of critical slip surface is accurate. One of the methods that can be used to determine the critical slip surface is the grid search method. This method is used because based on the results of kinematic analysis conducted by Simatupang [3] and Murtejo [4] in the same research area, the sliding type most likely to occur in this area is circular sliding type. In the grid search method, the first thing to do is to determine the size of the grid box with dimensions x, y, and z. After the grid box is available, the user determines the number of grid points that you want to use in the x, y, and z directions. This point serves as the center of rotation. Each center of rotation can produce a number of circles that are used as slip surfaces. The number of circles produced at each center of rotation from the minimum radius to the maximum radius is called the radius increment. Illustration of the number of grid points and radius increment can be seen in Figure 1 and Figure 2

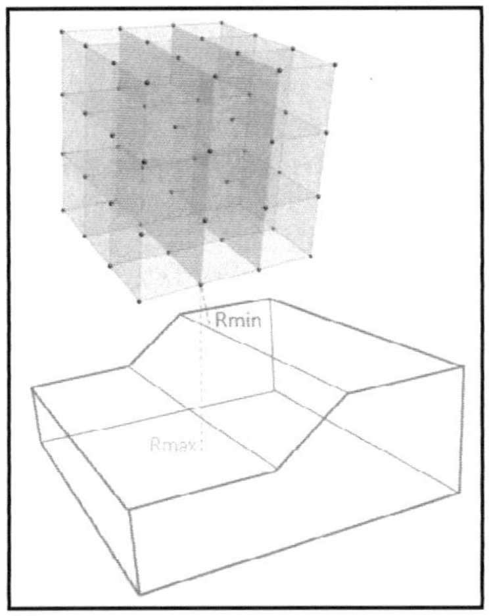

Figure 1. Illustration of grid search method [5]

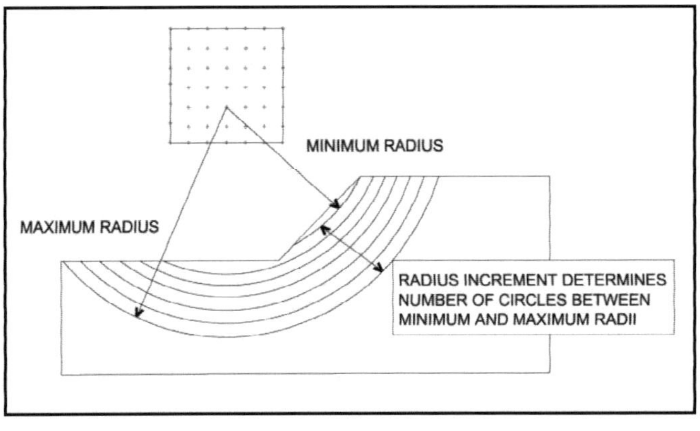

Figure 2. Illustration of the radius increment in the grid search method

After assuming the field geomaterial collapsed, the next stepis mass discretization of the sliding mass into a number of columns [6]. Square nets are applied to the sliding mass so that the sliding mass is divided into columns. There are two kinds of columns; the active column where the column is inside the sliding mass boundary line, and the inactive column where these columns are outside the sliding mass boundary line. In the calculation, the inactive columns are ignored so that the discrete sliding mass is determined only from the sum of the active columns. Figure 3 shows the illustration of the discretization of the sliding mass using a square grid.

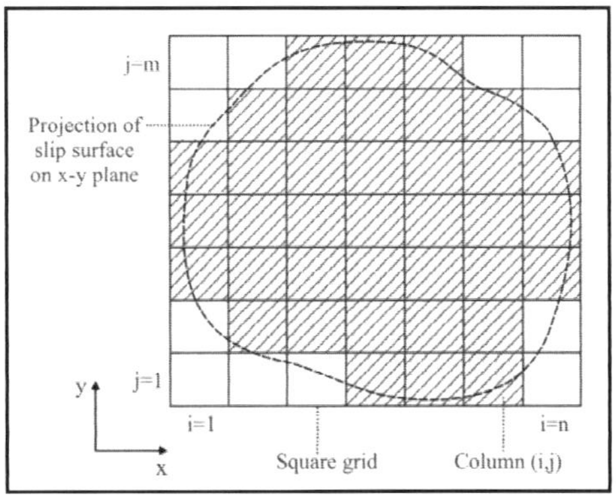

Figure 3. Discretization of the sliding mass using a square grid [1]

After discretizing the sliding mass, internal and external forces in each column can be calculated based on moment equilibrium, force equilibrium, or both depending on what method of calculation is used. Figure 4 shows the internal and external forces acting on the column where Where N_i, U_i = effective normal force and pore pressure force on column base; S_i = mobilized shear force on column base; a_i= space shear angle; a' = projected shear angle; E_i= intercolumn normal force; X_i = intercolumn vertical shear forces; H_i = intercolumn horizontal shear forces; and P_{vi} = vertical external force.

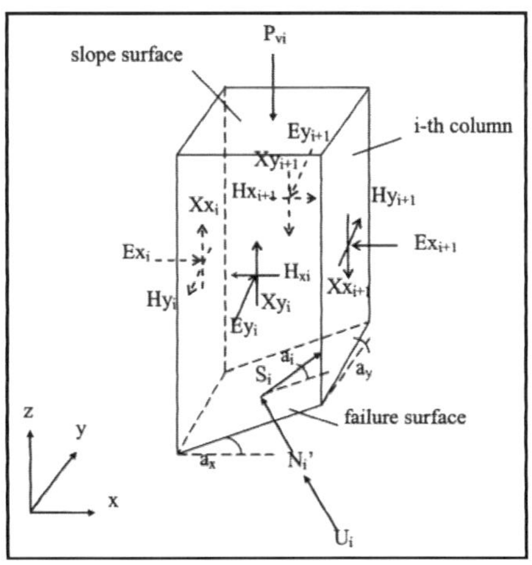

Figure 4. Internal and external forces that work in a column [7]

2. Methodology

The study was conducted in Pit 3 Timur, Banko Barat, PT Bukit Asam Tbk., South Sumatra, Indonesia. The data used in this study are secondary data owned by PT Bukit Asam . The data used in the study are: 2018 pit plan design, lithology maps, rocks unit weights, rock cohesions and friction angles. Then the data is processed using Slide 3 from Rocscience [5] to analyze slope stability with limit equilibrium method. The grid search method is used to find the critical slip surface. The calculation method used in this study is Bishop's simplified where this method assumes that the sliding mass has a circular slip surface. This method only fulfills moment equilibrium [8]. The Bishop's simplified method in 3D analysis is developed from the Bishop's simplified 2D analysis. By considering the overall moment equilibrium in the direction of x and the axis that crosses and parallels the y axis, the following equation is obtained:

$$F_{my} = \frac{\Sigma\{[K_{yi}[f_{1i}RZ_i + f_{3i}RX_i]\}}{\Sigma(W_i + P_{vi})RX_i + \Sigma N_i(g_{1i}RZ_i - g_{3i}RX_i)} \quad (1)$$

$$F_{mx} = \frac{\Sigma\{[K_{xi}[f_{2i}RZ_i + f_{3i}RX_i]\}}{\Sigma(W_i + P_{vi})RY_i + \Sigma N_i(g_{2i}RZ_i - g_{3i}RY_i)} \quad (2)$$

in which:

$$K_{yi} = \frac{\left\{C_i + \left[\frac{(W_i + P_{vi})}{(g_{3i})} - U_i\right]\tan\phi_i\right\}}{1 + \frac{(f_{3i}\tan\phi_i)}{g_{3i}F_{my}}} \quad (3)$$

$$K_{xi} = \frac{\left\{C_i + \left[\frac{(W_i + P_{vi})}{(g_{3i}) - U_i}\right]\tan\phi_i\right\}}{1 + \frac{(f_{3i}\tan\phi_i)}{g_{3i}F_{mx}}} \quad (4)$$

By considering the overall equilibrium moment passing and parallel to the z axis, the following equation is obtained:

$$F_{mz} = \frac{\Sigma\{[K_{zi}[f_{2i}RX_i + f_{3i}RY_i]\}}{\Sigma N(g_{2i}RX_i - g_{1i}RY_i)} \quad (5)$$

in which:

$$K_{zi} = \frac{\left\{C_i + \left[\frac{(W_i + P_{vi})}{(g_{3i}) - U_i}\right]\tan\phi_i\right\}}{1 + \frac{(f_{3i}\tan\phi_i)}{g_{3i}F_{mz}}} \quad (6)$$

Where F_{mx}, F_{my}, and F_{mz} = moment equilibrium in x, y and z directions; R_{xi}, R_{yi}, and R_{zi} = lever arm in the in x, y and z directions; W_i = column weight; $C_i = c'A_i$ and c' and A_i = effective cohessive strenth and the base area of the column, respectively; $g_1, g_2, g_3$and f_1, f_2, f_3= unit vectors adopted from Huang and Tsai [9] and Cheng and Yip[7].

Based on the moment equilibrium in this method, global safety factor (F_m) can be determined as follows:

$$F_m = F_{mx} = F_{my} = F_{mz} \quad (7)$$

Safety factor can be determined by changing the projected sliding direction at a certain angle interval so that F_{mx}, F_{my} dan F_{mz} generate the same value.

3. Result and discussion

A few factors that influence the slope stability are rock properties. Properties of rocks which include physical properties and mechanical properties are input parameters in analyzing slope stability and calculating safety factor in limit equilibrium method. There are four lithologies in the study area, which are claystone, siltstone, sandstone, and coal. The mechanical properties of claystone and coal used in this study are obtained from the line mapping so that the rock mass strength is obtained. The rock mass strength is used because it has taken into account the discontinuities that exist in rock masses, thus representing rock strength in the field better than intact rock. Meanwhile, since there is no line mapping in siltstone and sandstone, the mechanical properties used for the two lithologies are intact rock. The intack rock strengths are obtained from direct shear tests conducted by PT Bukit Asam. Rock mass strength possessed by claystone and coal is considered sufficient to represent rock strength in the study area because Pit 3 Timur consists of 65% claystone and 10% coal [10]. Rock properties used in this study can be seen in Table 1.

3D limit equilibrium analysis is done with Slide3. Since there is no groundwater parameter, Ru coefficient is used as a substitute for it. Ru coefficient is able to model pore pressure as a fraction of vertical soil pressure for each column. The Ru coefficient used in this analysis is 0.7 as an assumption of slope conditions in a natural state. The output of 3D limit equilibrium analysis is safety factor and estimated sliding volume. Figure 6 shows the 3D model used for slope stability analysis. The green color of the model symbolizes claystone lithology, blue is siltstone, yellow is sandstone, while black is coal.

Table 1. Rock properties used in the slope stability analysis

Lithology	Properties	Value	Strength
Claystone	C (KPa)	45.81	Rock mass
	ϕ (°)	34.65	
	Unit weight (kN/m^3)	15.44	
Siltstone	C (KPa)	284.00	Intact rock
	ϕ (°)	32.11	
	Unit weight (kN/m^3)	17.37	
Sandstone	C (KPa)	346.44	Intact rock
	ϕ (°)	44.95	
	Unit weight (kN/m^3)	14.76	
Coal	C (KPa)	161.60	Rock mass
	ϕ (°)	58.73	
	Unit weight (kN/m^3)	10.47	

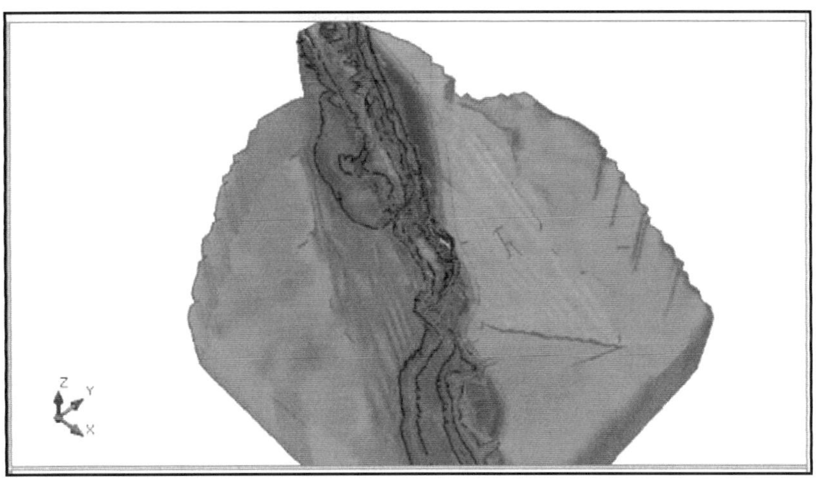

Figure 6. Pit 3 Timur 3D model for slope stability analysis [10]

The grid search method is used to find critical slip surface. The grid search method starts by specifying the grid box dimension. The location and dimensions of the grid box must cover the entire study area so that the search for critical slip surface can be performed optimally. The location and dimensions of the grid box in the direction of x, y, and z can be seen in Table 2.

Table 2. Dimensions and location of the grid box in the analysis of 3D limit equilibrium method

$\mathbf{X_{max}}$=	371808	$\mathbf{Y_{max}}$=	9586510	$\mathbf{Z_{max}}$=	500
$\mathbf{X_{min}}$=	370696	Y_{min} =	9585570	Z_{min} =	50
ΔX =	1112	ΔY =	940	ΔZ =	450

To determine the effect of the number of grid points and the radius increments on the safety factor and the estimated sliding volume produced, an analysis of slope stability is carried out several times by varying the two variables. Variations in the number of grid points used in the direction of x, y, and z are 20x20x10, 30x30x15, and 40x40x20, while the radius increments used are 10, 20, 30, 40, and 50. Table 3 shows the distance between each grid point in the x, y, and z direction.

Table 3. The number of grid points used in slope stability analysis and the distance between each point

Grid 20x20x10	Grid 30x30x15	Grid 40x40x20
X (20) = 59 m	X (30) = 38 m	X (40) = 29 m
Y (20) = 49 m	Y (30) = 32 m	Y (40) = 24 m
Z (10) = 50 m	Z (15) = 32 m	Z (20) = 24 m

Examples of limit equilibrium analysis results with a grid number of 40x40x20 and a radius increment 40 can be seen in Figure 7, and the global minimum location can be seen in Figure 8. Figure 8 shows that the global minimum is located at the northern highwall. As we can see in Figure 7, the highwall is dominated by orange color which represents the range of safety factor values of 0.8 - 1.2. This is because the highwall slope is dominated by claystone which has alow cohesion and internal friction angle compared to the other lithologies. Recapitulation of slope stability analysis results with variations in the number of grid points and radius increments are shown in Table 4.

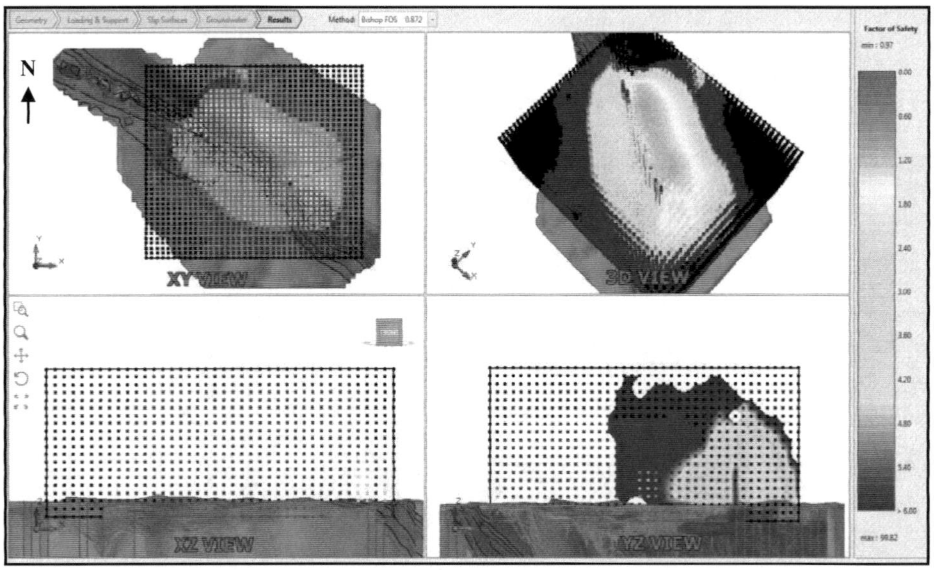

Figure 7. The results of 3D slope stability analysis with 40x40x20 grid points and 40 increment radius [10]

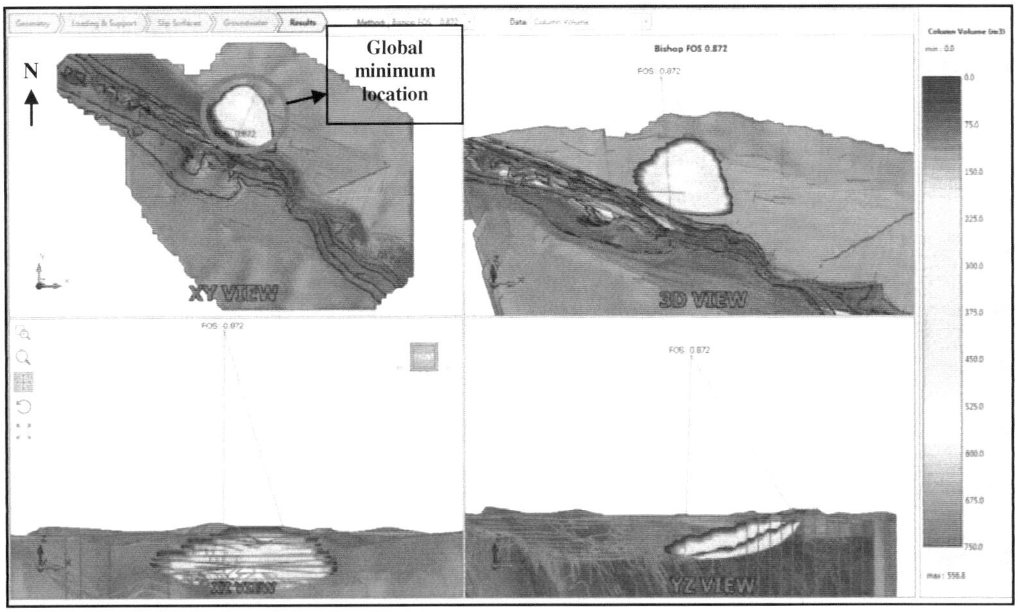

Figure 8. The global minimum with 40x40x20 grid points and 40 increment radius is located in the northern highwall [10]

Table 4. Recapitulation of the results of 3D slope stability analysis

Radius Incremet	Number of Grid Points	Safety Factor	Volume (m^3)	Location	Direction of Sliding (°)	Center of Rotation x	Center of Rotation y	Center of Rotation z	Running time (hour)
10	20x20x10	0.868	2,245,470	North-HW	246.7	371070	9586190	470	0.84
	30x30x15	0.874	1,998,950	North-HW	246.6	371069	9586190	446	2.54
	40x40x20	0.873	2,009,660	North-HW	246.5	371068	9586190	443	6.90
20	20x20x10	0.868	2,245,470	North-HW	246.7	371070	9586190	470	1.52
	30x30x15	0.874	2,028,600	North-HW	246.5	371069	9586200	429	5.30
	40x40x20	0.872	1,934,080	North-HW	246.4	371069	9586200	428	17.23
30	20x20x10	0.868	2,245,470	North-HW	246.7	371070	9586190	470	2.29
	30x30x15	0.906	1,628,780	North-HW	246.2	371092	9586210	352	8.83
	40x40x20	0.867	1,937,390	North-HW	246.1	371076	9586170	459	31.94
40	20x20x10	0.868	2,245,470	North-HW	246.7	371070	9586190	470	2.82
	30x30x15	0.876	1,756,790	North-HW	246.2	371076	9586200	404	13.22
	40x40x20	0.872	1,934,080	North-HW	246.4	371069	9586200	428	49.64
50	20x20x10	0.868	2,245,470	North-HW	246.7	371070	9586190	470	3.53
	30x30x15	0.876	2,068,950	North-HW	246.3	371077	9586180	459	17.77
	40x40x20	0.879	1,968,990	North-HW	246.3	371078	9586190	421	72.72

Slope stability calculations with variations in the number of grid points and radius increments produce various safety factor values and estimated sliding volumes. Based on these results, the relationship between the number of grid points and safety factor for each increment radius can be graphed. The chart can be seen in Figure 9.

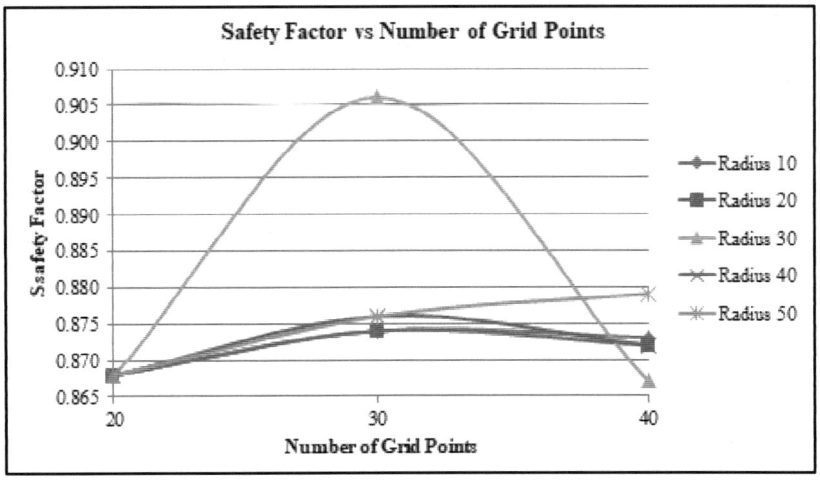

Figure 9. The relationship between safety factor with the number of grid points for each increment radius

It can be observed in Figure 9 that the greater number of grid points used, the safety factor value will be more headed to the convergent point and more representatives. At the number of grid points of 20x20x10 in each radius increment produces the same safety factor because the distance between the grid points is still far apart from a distance of 50 - 59 meters per point. This far distance causes the point taking for the center of rotation to fall at the same point because each point has a wide area of influence. The resemblance of the safety factor on the number of grid points of 20x20x10 is also caused by the direction of sliding that have the same value.

However, there is an anomaly in the number of grid points 30x30x15 and radius increment 30 where there is a significant increase in safety factor compared to the other radius increments on the same number of grid points. The increase in safety factor value is not affected by the radius increment. As described by Naderi [2] that there is no relationship between the length of the sliding arc, which in this study is represented by the number of radius increments, to the value of safety factor. The increase in safety factor on the number of grid points 30x30x15 and radius increment 30 is affected by the location of the center of rotation where the center of rotation is at the lowest altitude (352 meters) compared to the other number of grid points and radius incement resulting a difference distance in the critical slip surface than the other hence the safety factor produced is too high. The relationship between sliding arc length and safety factor according to Naderi [2] can be seen in Figure 10.

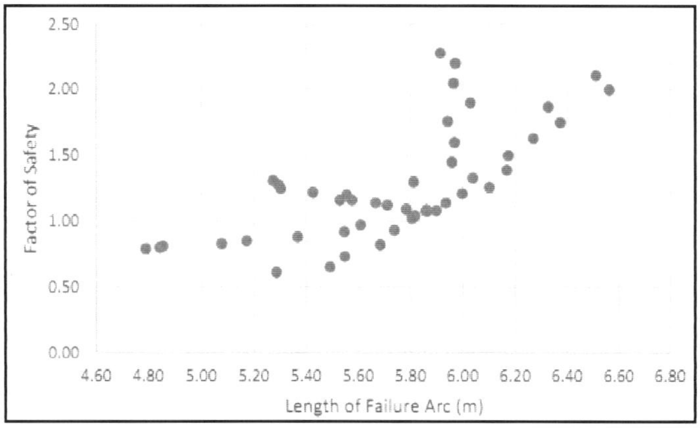

Figure 10. The relationship of sliding arc length and safety factor [2]

In the number of grid points of 40x40x20, safety factor drops but not much different from the safety factor resulting from the number of 30x30x15 grid points. At the number of grid points of 40x40x20, there are two data that produce safety factor values that are not uniform compared to the others, which are safety factor at radius increment 30 and 50. Safety factor produced at radius increment 30 is too low because the center of rotation at the this radius increment lies in the highest elevation (459 meters) compared to other radius increments on the same number of grid points resulting in a low safety factor value. In contrast, the center of rotation at the radius increment 50 has the lowest elevation (421 meters) compared to the others so as to produce a high safety factor. The difference in elevation at the center of rotation at the radius increment 30 and 50 affects the location of the critical slip surface so that the safety factor produced is also different

The output of 3D slope stability analysis is not only in the form of safety factor, but also estimated sliding volume. Just like determining the safety factor, the estimated sliding volume is done by varying the number of grid points and the radius increment to see the effect of the number of grid points and the radius increments on the estimated sliding volume. Figure 11 shows a chart of the relationship of the estimated sliding volume with the number of grid points for each increment radius.

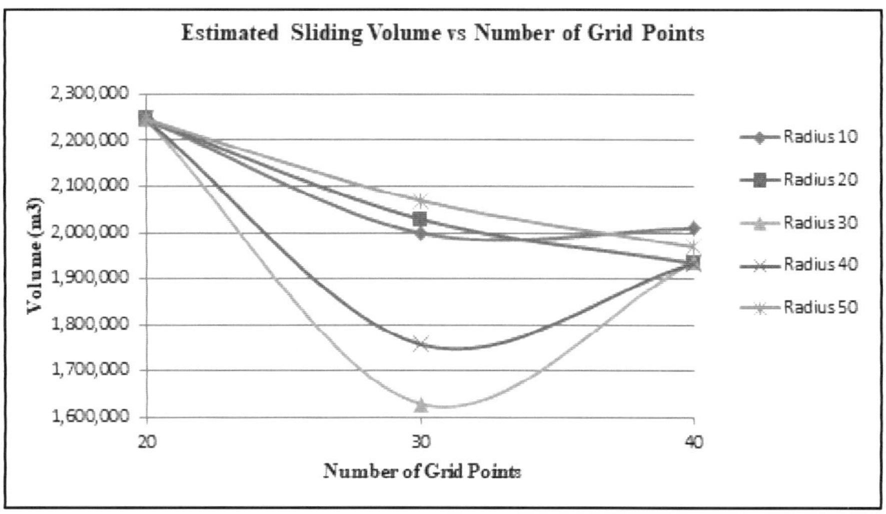

Figure 11. Number of grid points vs. estimated sliding volume of 3D analysis for each radius increment

It can be observed from Figure 11 that at the number of grid points of 20x20x10 the estimated sliding volume produced is too high. This is because the distance between each grid point is still far apart. The distance between points on the x axis is 59 meters, y 50 meters, and z 50 meters which means that the search for the critical slip surface has a less thorough coverage resulting in a less accurate estimation of sliding volumes. From Figure 11 it can also be seen that along with the increase in the number of grid points, it will go towards the convergent point. The distance between the grid points is close enough and can cover the entire area of the study thoroughly so that a more precise estimation of sliding volume is obtained.

Estimated sliding volume is not only affected by the number of grid points, but also influenced by the location of the center of rotation. It can be seen in Table 4 that the center of rotation in the radius increment of 10, 20, 30, 40, and 50 in the number of grid points of 20x20x10 is in the same location so that the same sliding volume estimation is obtained. Whereas in the radius increments 30 and 40 at the number of grid points 30x30x15 estimates a low sliding volume compared to the other radius increments. This is because in the radius increment of 30 and 40 in the 30x30x15 grid points, the rotation center is at a low elevation resulting in a low estimated sliding volume.

4. Conclusions

Based on the results of research and discussions in this study, the influence of the number of grid points and radius increments used in calculating safety factor and estimating sliding volume using Bishop's simplified method on 3D limit equilibrium method can be concluded as follow:

1. Calculation of the safety factor in 3D analysis is only influenced by the number of grid points because it determines the location of the center of rotation. The greater number of grid points, the more precise the search for critical slip surface will be. On the other hand, the safety factor is not affected directly by the length of the sliding arc which represented by the radius increments in this study.
2. The greater number of grid points and the radius increment used in the 3D analysis, the more accurate the estimated sliding volume will be. Just like the safety factor calculation, the location of the center of rotation holds an important role in calculating the estimated sliding volume.

Acknowledgment

The authors would like to thank Trisakti University Mining Engineering Department, PT Bukit Asam, and Rocscience which has provided Slide 3, and everyone who contributed in writing and finishing this research.

References

[1] Kalatehjari R, Rashid A S A, Hajihassani M, Kholghifard M, and Ali N 2014 *Engineering Geology* **182 (A)** 97-108

[2] Naderi M A 2013 *Determination of the Critical Slip Surface in Slope Stability Analysis* Master Thesis Eastern Mediterranean University

[3] Simatupang A R 2018 *Penentuan Kekuatan Massa Batuan Lapisan Batubara Menggunakan Metode Scanline Mapping pada Pit 3 Timur Banko Barat di PT Bukit Asam Tbk.* Skripsi Fakultas Teknologi Kebumian dan Energi, Universitas Trisakti

[4] Murtejo E 2018 *Penentuan Kekuatan Massa Batuan pada Batulempung dengan Menggunakan Metode Scanline di PT Bukit Asam, Tbk., Tanjun Enim – Sumatera Selatan* Skripsi, Fakultas Teknologi Kebumian dan Energi, Universitas Trisakti.

[5] Rocscience 2017 *Define Search Grid* https://www.rocscience.com/help/slide3/#t=slide3Dmodel%2FDefine_Search_Grid.htm. Downloaded on October 2nd 2018

[6] Lu H H, Xu L M, Fredlund M D, and Fredlund D G 2013 *Comparison of 3D Finite Element Method Slope Stability with 3D Limit Equilibirum Method* Proceedings of the 18th International Conference on Soil Mechanics and Geotechnical Engineering
[7] Cheng Y M and Yip C J 2007 *Journal of Geotechnical and Geoenviromental Engineering* **133(12)**
[8] Fredlund D G and Krahn J 1977 *Canadian Geotechnical Journal* **14(3)** 429-439
[9] Huang C C and Tsai C C 2000 *Journal of Geotechnical and Geoenviromental Engineering* **126(10)**
[10] Faradiba A A 2018 *Analisis Kestabilan Lereng Dua Dimensi dan Tiga Dimensi denga Metode Kesetimbangan Batas di PT Bukit Asam Tbk* Skripsi, Fakultas Teknologi Kebumian dan Energi, Universitas Trisakti.

Institute of Physics
Dirac House, Temple Back
Bristol BS1 6BE UK

ISSN: 1757-8981
ISBN 978-1-5108-8215-7

CURRAN ASSOCIATES INC.
proceedings.com